FAO/WHO Food Standards
CODEX alimentarius

国际食品法典标准

——水产品及水产加工品卷

农业部农产品质量安全监管局
农业部科技发展中心　编译
国际食品法典中国联络处

中国农业科学技术出版社

图书在版编目（CIP）数据

国际食品法典标准．水产品及水产加工品卷/农业部农产品质量安全监管局等编译．—北京：中国农业科学技术出版社，2009.4
ISBN 978-7-80233-900-2

Ⅰ.国… Ⅱ.农… Ⅲ.①食品卫生法-法规-汇编-世界②食品卫生-标准-汇编-世界③水产品-加工-食品卫生法-汇编-世界④水产品-加工-食品卫生-标准-汇编-世界 Ⅳ.D912.109 TS207.2

中国版本图书馆 CIP 数据核字(2009)第 086726 号

责任编辑 鲁卫泉
责任校对 贾晓红

出 版 者 中国农业科学技术出版社
北京市中关村南大街 12 号 邮编:100081
电　　话 (010)82106636(编辑室)(010)82109704(发行部)
传　　真 (010)82106636
网　　址 http://www.castp.cn
经 销 者 新华书店北京发行所
印 刷 者 北京华忠兴业印刷有限公司
开　　本 880 mm×1 230 mm 1/16
印　　张 27.625
字　　数 760 千字
版　　次 2009 年 4 月第 1 版 2009 年 4 月第 1 次印刷
定　　价 78.00 元

前言

国际食品法典委员会（CAC）作为联合国粮农组织（FAO）和世界卫生组织（WHO）联合成立的政府间国际组织，负责制定国际协调一致的农产品及食品标准体系，即《国际食品法典》，其宗旨是保护消费者的健康，促进农产品及食品公平贸易。《国际食品法典》汇集了国际食品法典委员会已经批准的标准、规范、准则和其他建议。规范对象包括农产品及食品原料、加工及半加工农产品及食品，具体内容涉及卫生和质量，包括微生物指标、食品添加剂、农药与兽药残留、污染物、标准及产品说明、抽样和分析方法等。

目前，国际食品法典委员会作为世界贸易组织（WTO）确认的三个农产品及食品国际标准化机构之一，其制定的国际食品法典标准已被WTO认可为国际农产品及食品贸易仲裁的重要参考依据，同时也是国际上农产品及食品质量安全问题最重要的参考资料。中国作为WTO的成员，同时也是国际农产品及食品贸易大国，目前正在履行WTO各项协定的相关承诺。在这种形势下，全面系统地将国际食品法典标准介绍给我国的广大利益相关者和用户是非常适时和必要的。于此，我们组织相关专家对目前国际食品法典委员会制定的水产品及水产品加工品方面的标准进行了收集、整理，并编译成《国际食品法典标准——水产品及水产加工品卷》奉献给广大读者。我们真诚希望该书的出版发行能对从事农产品及食品生产、加工、检测、贸易、研究和标准制定的人士提供帮助。

本书英文资料来源于CAC官方网站，资料收集截止时间为2008年7月。由于编译者水平有限，对原资料内容的理解和翻译方面可能存在疏漏和不当之处，欢迎各位读者批评指正！

编译者

2008年12月

目　录

鲑鱼罐头（*CODEX STAN 3－1991，REV. 1－1995*） ……………………………………………… （1）
CODEX STANDARD FOR CANNED SALMON ………………………………………………… （5）

速冻鱼（*CODEX STAN 36－1981，REV. 1－1995*） ……………………………………………… （10）
CODEX STANDARD FOR QUICK FROZEN FINFISH，UNEVISCERATED AND EVISCERATED ………………………………………………………………………… （15）

虾或对虾罐头（*CODEX STAN 37－1991，REV. 1－1995*） ………………………………………… （21）
CODEX STANDARD FOR CANNED SHRIMPS OR PRAWNS ……………………………………… （26）

金枪鱼罐头和鲣鱼罐头（*CODEX STAN 70－1981，REV. 1－1995*） …………………………… （32）
CODEX STANDARD FOR CANNED TUNA AND BONITO ……………………………………… （38）

蟹肉罐头（*CODEX STAN 90－1981，REV. 1－1995*） ……………………………………………… （46）
CODEX STANDARD FOR CANNED CRAB MEAT ……………………………………………… （50）

速冻虾或对虾（*CODEX STAN 92－1981，REV. 1－1995*） ……………………………………… （55）
CODEX STANDARD FOR QUICK FROZEN SHRIMPS OR PRAWNS …………………………… （61）

沙丁鱼和沙丁鱼类制品罐头（*CODEX STAN 94－1981，REV. 1－1995*） …………………… （68）
CODEX STANDARD FOR CANNED SARDINES AND SARDINE－TYPE PRODUCTS …………… （74）

速冻龙虾（*CODEX STAN 95－1981，REV. 1－1995*） ……………………………………………… （80）
CODEX STANDARD FOR QUICK FROZEN LOBSTERS ……………………………………… （86）

鱼罐头（*CODEX STAN 119－1981，REV. 1－1995*） ……………………………………………… （93）
CODEX STANDARD FOR CANNED FINFISH …………………………………………………… （98）

块冻鱼片及碎鱼肉标准（*CODEX STAN 165－1989，REV. 1－1995*） ………………………… （104）
CODEX STANDARD FOR QUICK FROZEN BLOCKS OF FISH FILLET，MINCED FISH FLESH AND MIXTURES OF FILLETS AND MINCED FISH FLESH ……………………………… （112）

冻沾面包屑或挂浆鱼条（鱼棒）、鱼片和鱼块（*CODEX STAN 166－1989, REV. 1－1995*）……（121）
CODEX STANDARD FOR QUICK FROZEN FISH STICKS（FISH FINGERS）, FISH PORTIONS AND FISH FILLETS－BREADED OR IN BATTER ……（128）

盐渍和盐干鳕鱼（*CODEX STAN 167－1989，REV. 2－2005*）……（137）
CODEX STANDARD FOR SALTED FISH AND DRIED SALTED FISH OF THE GADIDAE FAMILY OF FISHES ……（145）

鱼翅（*CODEX STAN 189－1993*）……（154）
CODEX STANDARD FOR DRIED SHARK FINS ……（157）

速冻鱼片（*CODEX STAN 190－1995*）……（161）
CODEX GENERAL STANDARD FOR QUICK FROZEN FISH FILLETS ……（166）

速冻生（原条）鱿鱼 *CODEX STAN 191－1995*）……（173）
CODEX STANDARD FOR QUICK FROZEN RAW SQUID ……（177）

海淡水鱼类、甲壳类以及软体动物类制成的脆片标准（*CODEX STAN 222－2001*）……（182）
STANDARD FOR CRACKERS FROM MARINE AND FRESHWATER FISH, CRUSTACEAN AND MOLLUSCAN SHELLFISH ……（186）

煮盐干鳀鱼（*CODEX STAN 236－2003*）……（190）
CODEX STANDARD FOR BOILED DRIED SALTED ANCHOVIES ……（194）

盐渍大西洋鲱和盐渍黍鲱鱼（*CODEX STAN 244－2004*）……（199）
STANDARD FOR SALTED ATLANTIC HERRING AND SALTED SPRAT ……（204）

水产及水产加工品操作规范（*CAC/RCP 52－2003，REV，2－2005*）……（211）
CODE OF PRACTICE FOR FISH AND FISHERY PRODUCTS ……（288）

鱼贝类实验室感官评价指南（*CAC/GL 31－1999*）……（385）
CODEX GUIDELINES FOR THE SENSORY EVALUATION OF FISH AND SHELLFISH IN LABORATORIES ……（403）

水产及水产加工品产品认证证书模式（*CAC/GL48－2004*）……（424）
MODEL CERTIFICATE FOR FISH AND FISHERY PRODUCTS ……（429）

鲑鱼罐头

CODEX STANDARD FOR CANNED SALMON

(*CODEX STAN 3－1991*, *REV. 1－1995*)

1 范围

本标准适用于鲑鱼罐头。

2 说明

2.1 产品定义

2.1.1 鲑鱼罐头由各属种的去除头、内脏、鳍及尾部的鱼制成，并且在制作中可添加食盐、水、鲑油或其他食用油。包括大西洋鲑（*Salmon salar*）、红大麻哈鱼（*Oncorhynchus nerka*）、银大麻哈鱼（*Oncorhynchus kisutch*）、大鳞大麻哈鱼（*Oncorhynchus tschawytscha*）、红鳞大麻哈鱼（*Oncorhynchus gorbuscha*）、大麻哈鱼（*Oncorhynchus keta*）、马苏大麻哈鱼（*Oncorhynchus masou*）。

2.2 加工过程的定义

将鲑鱼装罐并密封，然后通过一系列加工处理以确保其达到商业无菌状态。

2.3 产品介绍

2.3.1 鲑鱼罐头是由横向切割并垂直装罐的鱼块组成。这些鱼块横断面大致与容器底部平行。

2.3.2 符合下列要求的产品介绍，也是被准许的：

(1) 较2.3.1条所述的形式更具特色；

(2) 达到本标准的所有要求；

(3) 在产品标签中对产品进行详细描述，以免引起混淆或误导消费者。

3 基本成分及质量因素

3.1 鲑鱼

产品必须由2.1所列出的品种的鱼制备，且鱼的品质良好，可作为鲜品供人类消费。

3.2 其他成分

所使用的其他成分应具有食品级的质量，并且符合所有相应法规标准的规定。

3.3 成品

产品应符合本标准要求，根据第9款进行成品批次检查时，其质量应符合第8款的规定。检查方法应符合第7款的规定。

4 食品添加剂

本产品不允许添加任何添加剂。

5 卫生及处理

5.1 成品中不能含有任何危害人体健康的外来杂质

5.2 应用CAC规定的抽样及检测方法进行检验时，产品应具备下列条件：

（1）不含有任何在正常的贮存条件下可能生长的微生物；
（2）其他危害人体健康的有害物质的含量也不能超过CAC标准的规定；
（3）不存在可能危及罐的密封性的缺陷。

5.3 建议本标准中涉及产品预处理的条款应符合推荐性国际操作规范《CAC/RCP1－1969，Rev. 3－1997 食品卫生总则》和以下相关标准：

（1）推荐性国际操作规范《CAC/RCP 10－1976 鱼罐头》；
（2）推荐性国际操作规范《CAC/ RCP 23－1979，Rev. 2－1993 低酸和酸化低酸罐头》；
（3）《水产品国际操作规程草案》中水产养殖品的相关章节（正在完善中）[1]。

6 标签

标签除了应符合《CODEX STAN1－1985，Rev. 1－1991 预包装食品标签通用标准》的要求外，还应遵守以下规定：

6.1 食品名称

产品名称要根据出售该产品的国家的法律、习惯、实际情况及其对原料鱼的种名的称呼而定。

6.2 介质

食品名称中应包含填装介质的名称。

6.3 性状

2.3.2条中所提到的性状应用通俗名称标示出。

---

[1] 操作规程建议草案，在定稿后将替代所有现行的《水产品操作规程》。

7 抽样、检测和分析

7.1 取样

（1）如3.3中指定的一样，产品批次检验用样品的抽样方法应符合FAO/WHO的食品法典《CODEX STAN 233－1969 预包装食品的抽样方案》（AQL－6.5）。

（2）对需检测净重和干重的批次的抽样，抽样计划应以食品法典委员会的相关标准为依据。

7.2 感官与物理检验

产品的感官与物理指标须由经过此类检验培训的人员进行检验，并依据本标准7.3条和附录A以及《CAC/GL 31－1999 实验室中水产品感官评价指南》所述程序进行。

7.3 净重的测定

样品的净重应通过以下程序测定：

（1）称量未开封的罐头；

（2）开罐并除去内容物；

（3）称量空罐重量（包括开启下来的罐底）；

（4）从未开封罐头的重量中减去空罐的重量，得到的就是内容物的净重。

7.4 由食用油（除鲑油外）制成的产品干重的测定

样品单位的内容物沥干后的重量（干重）应通过以下程序测定：

（1）测定前使罐头的温度至少在过去12小时内维持在20～30℃。

（2）开罐并使之倾斜，将内容物倾倒在预先称重的圆形滤网上，该滤网由金属网丝构成，网孔大小：2.8mm×2.8mm。

（3）以17°～20°的角度倾斜滤网以使鱼沥水，从将产品全部倾倒于滤网上开始计时，沥水2min。

（4）称量沥水后的鱼及滤网的总重。

（5）从沥水后的鱼和滤网的总重中减去滤网的重量就可以获得鱼的干重。

8 缺陷的定义

当样品单位呈现下列任何一项特征时，则认定其有“缺陷”。

8.1 外来杂质

样品单位中存在的任何不是来自于鱼体的物质。这些物质虽不会对人体健康造成危害，但用肉眼可直接辨别，或采用某些方法（包括放大）可以确定其存在。出现外来杂质表明不符合良好操作和卫生惯例。

8.2 气味或风味

样品散发的持久、明显、令人厌恶的由腐败、酸败引起的气味或风味。

8.3 质地

（1）含有过多非该产品特征的糊状鱼肉；
（2）含有过多非该产品特征的硬（韧）鱼肉；
（3）含有重量超过内容物干重5%的蜂巢状鱼肉。

8.4 变色

样品受腐败、酸败的影响发生明显的鱼肉变色，或超过内容物干重5%的样品受到硫化物着色。

8.5 异物

样品中发现长度超过5mm的磷酸铵镁结晶（鸟粪石）。

9 批次验收

当满足以下条件时，可以认为此批次产品符合本标准的要求：
（1）根据本标准第8款中规定分类，缺陷总数不超过《CODEX STAN 233 - 1969 预包装食品的抽样方案》（AQL - 6.5）中相应抽样方案规定的可接受数（c）；
（2）不符合本标准2.3条中规定数量的样品总数不超过《CODEX STAN233 - 1969 预包装食品的抽样方案》（AQL - 6.5）中相应抽样方案规定的可接受数（c）；
（3）所有样品单位平均净重不少于标示量，在任何一个包装单位中没有不合理的重量短缺；
（4）符合本标准第4、5、6中对食品添加剂、卫生和处理以及标签的要求。

附录A 感官和物理的检测

1. 对罐外部进行检测以确定其是否存在密封性缺陷，有无罐身外部扭曲。
2. 开罐，并且根据7.3条和7.4条中的程序完成重量测定。
3. 检测产品的变色、外来杂质和异物等方面的情况。出现硬骨表明加工不完全，这时，要求对无菌状态进行评估。
4. 依据《CAC/ GL 31 - 1999 实验室中鱼、贝类感官评价指南》对产品的气味、风味以及质地进行评定。

CODEX STANDARD FOR CANNED SALMON

(*CODEX STAN 3 - 1991, REV. 1 - 1995*)

1 SCOPE

This standard applies to canned salmon.

2 DESCRIPTION

2.1 PRODUCT DEFINITION

2.1.1 Canned Salmon is the product prepared from headed and eviscerated fish of any of the species listed below from which the fins and tails have been removed, and to which salt, water, salmon oil and/or other edible oils may have been added.

- *Salmo salar*;
- *Oncorhynchus nerka*;
- *Oncorhynchus kisutch*;
- *Oncorhynchus tschawytscha*;
- *Oncorhynchus gorbuscha*;
- *Oncorhynchus keta*;
- *Oncorhynchus masou.*

2.2 PROCESS DEFINITION

Canned salmon is packed in hermetically sealed containers and shall have received a processing treatment sufficient to ensure commercial sterility.

2.3 PRESENTATION

2.3.1 Canned salmon shall consist of sections which are cut transversely from the fish and which are filled vertically into the can. The sections shall be packed so that the cut surfaces are approximately parallel with the ends of the container.

2.3.2 Any other presentation shall be permitted provided that it:

(i) is sufficiently distinctive from the form of presentation laid down under 2.3.1;

(ii) meets all other requirements of this standard; and

(iii) is adequately described on the label to avoid confusing or misleading the consumer.

3 ESSENTIAL COMPOSITION AND QUALITY FACTORS

3.1 SALMON

The product shall be prepared from sound fish of the species in Section 2.1 and of a quality fit to be sold fresh for human consumption.

3.2 OTHER INGREDIENTS

All other ingredients used shall be of food grade quality and conform to all applicable Codex standards.

3.3 FINAL PRODUCT

Products shall meet the requirements of this standard when lots examined in accordance with Section 9 comply with the provisions set out in Section 8. Products shall be examined by the methods given in Section 7.

4 FOOD ADDITIVES

No additives are permitted in this product.

5 HYGIENE AND HANDLING

5.1 The final product shall be free from any foreign material that poses a threat to human health

5.2 When tested by appropriate methods of sampling and examination prescribed by the Codex Alimentarius Commission (CAC), the product:

(i) shall be free from microorganisms capable of development under normal conditions of storage; and

(ii) shall not contain any other substance derived from microorganisms in amounts which may represent a hazard to health in accordance with standards established by the CAC; and

(iii) shall be free from container integrity defects which may compromise the hermetic seal.

5.3 It is recommended that the product covered by the provisions of this standard be prepared and handled in accordance with the appropriate sections of the Recommended International Code of Practice – General Principles of Food Hygiene (CAC/RCP 1 – 1969, Rev. 3 – 1997) and the following relevant Codes:

(i) the Recommended International Code of Practice for Canned Fish (CAC/RCP 10 – 1976);

(ii) the Recommended International Code of Hygienic Practice for Low – Acid and Acidified Low – Acid Canned Foods (CAC/RCP 23 – 1979);

(iii) The sections on the Products of Aquaculture in the Proposed Draft International Code of Practice for Fish and Fishery Products (under elaboration)[1].

---

[1] The Proposed Draft Code of Practice, when finalized, will replace all current Codes of Practice for Fish and Fishery Products.

6 LABELLING

In addition to the provisions of the Codex General Standard for Labelling of Prepackaged Foods (CODEX STAN 1 – 1985, Rev. 3 – 1999) the following specific provisions shall apply.

6.1 THE NAME OF THE FOOD

The name of the product shall be the designation appropriate to the species of the fish according to the law, custom or practice in the country in which the product is to be distributed.

6.2 PACKING MEDIUM

The packing medium shall form part of the name of the food.

6.3 PRESENTATION

The presentation provided for in Section 2.3.2 shall be declared in close proximity to the common name.

7 SAMPLING, EXAMINATION AND ANALYSES

7.1 SAMPLING

(i) Sampling of lots for examination of the final product as prescribed in Section 3.3 shall be in accordance with the FAO/WHO Codex Alimentarius Sampling Plans for Prepackaged Foods (1969) (AQL – 6.5) (Ref. CAC/RM 42 – 1977).

(ii) Sampling of lots for examination of net weight shall be carried out in accordance with an appropriate sampling plan meeting the criteria established by the CAC.

7.2 SENSORY EVALUATION AND PHYSICAL EXAMINATION

Samples taken for sensoric and physical examination shall be assessed by persons trained in such examination and in accordance with Section 7.3, Annex A and the *Guidelines for the Sensory Evaluation of Fish and Shellfish in Laboratories* (*CAC/GL* 31 – 1999).

7.3 DETERMINATION OF NET WEIGHT

Net contents of all sample units shall be determined by the following procedure:

(i) Weigh the unopened container.

(ii) Open the container and remove the contents.

(iii) Weigh the empty container, (including the end) after removing excess liquid and adhering meat.

(iv) Subtract the weight of the empty container from the weight of the unopened container. The resultant figure will be the net content.

7.4 DETERMINATION OF DRAINED WEIGHT FOR PRODUCTS PACKED WITH EDIBLE OILS OTHER THAN SALMON OIL

The drained weight of all sample units shall be determined by the following procedure:

(i) Maintain the container at a temperature between 20℃ and 30℃ for a minimum of 12 hours prior to examination.

(ii) Open and tilt the container to distribute the contents on a pre-weighed circular sieve which consists of wire mesh with square openings of 2.8 mm × 2.8 mm.

(iii) Incline the sieve at an angle of approximately 17° ~ 20° and allow the fish to drain for two minutes, measured from the time the product is poured into the sieve.

(iv) Weigh the sieve containing the drained fish.

(v) The weight of drained fish is obtained by subtracting the weight of the sieve from the weight of the sieve and drained product.

8 DEFINITION OF DEFECTIVES

A sample unit will be considered defective when it exhibits any of the properties defined below.

8.1 FOREIGN MATTER

The presence in the sample unit of any matter, which has not been derived from salmon or the packing medium does not pose a threat to human health, and is readily recognized without magnification or is present at a level determined by any method including magnification that indicates non-compliance with good manufacturing and sanitation practices.

8.2 ODOUR/FLAVOUR

A sample unit affected by persistent and distinct objectionable odours or flavours indicative of decomposition or rancidity.

8.3 TEXTURE

(i) Excessive mushy flesh uncharacteristic of the species in the presentation; or

(ii) Excessively tough flesh uncharacteristic of the species in the presentation; or

(iii) Honey combed flesh in excess of 5% of the net contents.

8.4 DISCOLOURATION

A sample unit affected by distinct discolouration indicative of decomposition or rancidity or by sulphide staining of the meat exceeding 5% of the net contents.

8.5 OBJECTIONABLE MATTER

A sample unit affected by struvite crystals - any struvite crystal greater than 5 mm in length.

9 LOT ACCEPTANCE

A lot shall be considered as meeting the requirements of this standard when:

(i) the total number of defectives as classified according to Section 8 does not exceed the acceptance number (c) of the appropriate sampling plan in the Sampling Plans for Prepackaged Foods (AQL – 6.5) (CAC/RM 42 – 1977);

(ii) the total number of sample units not meeting the form of presentation as defined in Section 2.3 does not exceed the acceptance number (c) of the appropriate sampling plan in the Sampling Plans for Prepackaged Foods (AQL – 6.5) (CAC/RM 42 – 1977);

(iii) the average net weight and the average drained weight where appropriate of all sample units examined is not less than the declared weight or drained weight as appropriate, and provided there is no unreasonable shortage in any individual container;

(iv) the Food Additives, Hygiene and Labelling requirements of Sections 4, 5.1, 5.2 and 6 are met.

ANNEX A SENSORY AND PHYSICAL EXAMINATION

1. Complete external can examination for the presence of container integrity defects or can ends which may be distorted outward.

2. Open can and complete weight determination according to defined procedures in Section 7.3 and 7.4.

3. Examine product for discolouration, foreign and objectionable matter. The presence of hard bone is an indicator of underprocessing and will require an evaluation for sterility.

4. Assess odour, flavour and texture in accordance with the *Guidelines for the Sensory Evaluation of Fish and Shellfish in Laboratories* (*CAC/GL* 31 – 1999).

速 冻 鱼

CODEX STANDARD FOR QUICK FROZEN FINFISH, UNEVISCERATED AND EVISCERATED

(CODEX STAN 36 - 1981, REV. 1 - 1995)

1 适用范围

本标准适用于速冻的去内脏和未去内脏的鱼类❶。

2 说明

2.1 产品定义

速冻鱼是指适合于人类消费的冷冻的鱼，可以是有头或无头的，其内脏或其他器官可能已被全部或部分地清除。

2.2 加工定义

产品经过适当预处理后，应在符合以下规定的条件下进行冻结加工：冻结应在合适的设备中进行，并使产品迅速通过最大冰晶生成温度带。速冻加工只有在产品的中心温度达到并稳定在 -18℃（0 ℉）或更低的温度时才算完成。产品在运输、贮存、分销过程中应保持在深度冻结状态，以保证产品质量。

在保证质量的条件下，允许按规定要求对速冻产品再次速冻加工，并按照被认可的操作进行再包装。

在产品的加工和包装过程中应尽量减少脱水和氧化作用的影响。

2.3 产品介绍

产品介绍应符合下列要求：

2.3.1 达到本标准的所有要求

2.3.2 在产品标签中对产品进行详细描述，以免引起混淆或误导消费者

---

❶ 不适用于为进一步深加工而在盐水中冷冻的鱼类。

3 基本成分及质量因素

3.1 鱼

速冻鱼应由品质良好、可作为鲜品供人类消费的个体完整的鱼制备。

3.2 冰衣

用于镀冰衣或制备镀冰衣所用的水应达到饮用水或清洁海水的质量要求。饮用水应符合 WHO 最新版本的《国际饮用水质量规范》要求，清洁海水应是达到饮用水的微生物标准且不含异物的海水。

3.3 其他成分

所使用的其他成分应具有食品级的质量，并且符合所有相应法规标准的规定。

3.4 腐败

被测样品单位的组胺平均含量不应超过 10mg/100g。这仅适用于鲱科（*Clupeidae*）、鲭科（*Scombridae*）、秋刀鱼科（*Scombresocidae*）、鯥科（*Pomatomidae*）以及鳍鳅科（*Coryphaenedae*）等科的鱼类。

3.5 成品

产品应符合本标准要求，根据第 9 款对成品批进行检验时，其质量应符合第 8 款的规定。检验方法应按第 7 款的规定执行。

4 食品添加剂

只允许使用以下添加剂。

添加剂		成品中的最高含量
抗氧化剂	300 抗坏血酸（维生素 C） 301 抗坏血酸钠 303 抗坏血酸钾	GMP

5 卫生及处理

5.1 成品中不能含有任何危害人体健康的外来杂质

5.2 应用 CAC 规定的抽样及检测方法进行检验时，达到以下标准

（1）产品中微生物含量和由微生物产生的危害人体健康的有害物质的含量不能超过 CAC 标准的规定；

（2）组胺含量不应超过 20mg/100g，这仅适用于鲱科（*Clupeidae*）、鲭科（*Scombridae*）、秋刀

鱼科（*Scombresocidae*）、鲑科（*Pomatomidae*）以及鳍鳅科（*Coryphaenedae*）等科的鱼类；

（3）其他危害人体健康的有害物质的含量也不能超过 CAC 标准的规定。

5.3 建议本标准中涉及产品预处理的条款应符合推荐性国际规范《CAC/RCP1 - 1969，Rev. 3 - 1997 食品卫生操作总则》中的相关章节和以下相关标准：

（1）推荐性国际操作规范《CAC/RCP16 - 1978 冻鱼》；

（2）推荐性国际操作规范《CAC/RCP8 - 1976 速冻食品的加工和处理》；

（3）《水产品国际操作规程草案》中水产养殖品的相关章节（正在完善中）[1]。

6 标签

标签除应符合《CODEX STAN1 - 1985，Rev. 3 - 1999 预包装食品标签通用标准》的要求外，还应遵守以下规定：

6.1 食品名称

6.1.1 除该品种的常用名外，对于已去内脏的鱼类，标签应表明该产品是已去除内脏的鱼类，并且注明“有头”或“无头”

6.1.2 如果产品用海水镀冰衣，则应予以说明

6.1.3 在产品标签上应标明“速冻”，除非在有的国家“冻”即代表速冻

6.1.4 标签应注明产品须在运输、贮藏、分销过程中保持的条件，以保证其质量

6.2 净含量（镀冰衣产品）

镀冰衣的产品，其内容物净含量不包括冰衣重。

6.3 贮藏说明

标签应注明产品须在 -18℃或更低的温度下贮藏。

6.4 非零售包装的标签

上述要求应既在包装上又在附文中出现，除了食品名称、批号、制造或分装厂名、地址外，还包括贮藏条件。但批号、制造或分装厂名、地址也可用同一证明标志代替，只要证明标志能在辅助文件中说明清楚。

7 抽样、检验和分析

7.1 取样

（1）批次检验用样品的抽样方法应符合 FAO/WHO 的食品法典《CODEX STAN 233 - 1969 预包装食品的抽样方案》（AQL - 6.5）。样品单位是单条鱼或初级包装。

（2）对需检测净重批次的抽样，抽样计划应以食品法典委员会的相关标准为依据。

---

[1] 操作规程建议草案，在定稿后将替代所有现行的《水产品操作规程》。

7.2　感官与物理检验

产品的感官与物理指标须由经过此类检验培训的人员进行检验，并依据本标准7.3至7.5和附录A以及《CAC/GL 31－1999 实验室中水产品感官评价指南》所述程序进行。

7.3　净重的测定

7.3.1　未镀冰衣产品净重的测定

代表每批次的样品单位的净重（不包括包装材料）应在冷冻状态下测定。

7.3.2　镀冰衣产品净重的测定

（待述）。

7.4　解冻

（待述）。

7.5　凝胶状态的检验

按照AOAC方法883.18“肉及肉制品中水分含量测定时抽样的准备工作”和950.46“肉中的水分含量”检验（方法A）。

7.6　蒸煮方法

蒸煮使产品内部温度达到65～70℃。不能过度蒸煮，蒸煮时间随产品大小和采用的温度而不同。准确的蒸煮时间和条件应依据预先实验来确定。

烘焙：用铝箔包裹产品，并将其均匀放入扁平锅或浅平锅上。

蒸：用铝箔包裹产品，并将其置于带盖容器中沸水之上的金属架上。

袋煮：将产品放入可煮薄膜袋中加以密封，浸入沸水中煮。

微波：将产品放入适于微波加热的容器中，若用塑料袋，应检查确定塑料袋不会发出任何气味。根据设备说明加热。

7.7　组胺检测

按照AOAC 977.13进行检测。

8　缺陷的定义

当样品呈现下列任何一项特征时，则认定其有“缺陷”。

8.1　深度脱水

样品单位中超过表面积10%或大于鱼重10%的部分，出现水分的过度损失，明显表现为鱼体表面呈现异常的白色或黄色，覆盖了肌肉本身的颜色，并已渗透至表层以下，在不过分影响产品外观的情况下，不能轻易地用刀或其他利器刮掉。

8.2　外来杂质

样品单位中存在的任何不是来自于鱼体的物质（包装材料除外），这些物质虽不会对人体健康造成危害，但用肉眼可直接辨别，或采用某些方法（包括放大）可以确定其存在。出现外来杂质表明不符合良好操作和卫生习惯。

8.3 气味或风味

样品散发的持久、明显、令人厌恶的由腐败、酸败或饵料引起的气味或风味。

8.4 组织

8.4.1 鱼肉组织全面下降，组织呈现糊状、膏状或出现鱼肉与鱼骨分离等特征而显示出腐败

8.4.2 鱼肉异常

样品出现过量凝胶状态的鱼肉并伴有任何单条鱼中水分达86%以上，或按重量计算5%以上的样品被寄生虫感染导致质地呈现糊状。

8.5 腹部爆裂

未去除内脏鱼类产品出现腹部破裂的情况，表明发生腐败。

9 批次验收

当满足以下条件时，可以认为此批次产品符合本标准的要求：

（1）根据本标准第8章中规定分类，缺陷总数不超过《CODEX STAN 233－1969 预包装食品的抽样方案》（AQL－6.5）中相应抽样方案规定的可接受数（c）；

（2）所有样品单位平均净重不少于标示量，在任何一个包装单位中没有不合理的重量短缺；

（3）符合本标准第4、5、6款中对食品添加剂、卫生和处理以及标签的要求。

附录A 感官及物理检验

1. 净含量的测定按本标准7.3条规定执行（按要求去冰衣）。

2. 通过测定只能用小刀或其他利器除去的面积，检查冻结的样品中脱水的情况。测量样品单位的总表面积，计算受影响的面积百分比。

3. 解冻并且逐个地检测样品单位中每条鱼有无外来杂质。

4. 使用第8章列出的标准检测每条鱼。在鱼颈部背后撕开或切开裂缝，从而可以对暴露的鱼肉表面进行鱼肉气味的检测和评价。

5. 对在解冻后未蒸煮状态下无法最终判定其气味的样品，则应从样品单位中截取一小块可疑部分（约200g），并立即使用7.6条中限定的某一种蒸煮方法，确定其气味和风味。

6. 对在解冻后未蒸煮状态下无法最终确定其凝胶状态的产品，则应从产品中截取可疑部分，按7.6条规定的蒸煮方法，或用7.5条测定在任何一鱼片中是否含有超过86%的水分，即可确认其凝胶情况。如果无法根据蒸煮后的评估得到结论，则使用7.5条中的程序准确测定水分含量。

CODEX STANDARD FOR QUICK FROZEN FINFISH, UNEVISCERATED AND EVISCERATED

(CODEX STAN 36 – 1981, REV. 1 – 1995)

1 SCOPE

This standard shall apply to frozen finfish uneviscerated and eviscerated❶.

2 DESCRIPTION

2.1 PRODUCT DEFINITION

Frozen finfish suitable for human consumption, with or without the head, from which the viscera or other organs may have been completely or partially removed.

2.2 PROCESS DEFINITION

The product, after any suitable preparation, shall be subjected to a freezing process and shall comply with the conditions laid down hereafter. The freezing process shall be carried out in appropriate equipment in such a way that the range of temperature of maximum crystallization is passed quickly. The quick freezing process shall not be regarded as complete unless and until the product temperature has reached – 18℃ (0 ℉) or colder at the thermal centre after thermal stabilization. The product shall be kept deep frozen so as to maintain the quality during transportation, storage and distribution.

Industrial repacking of quick frozen products under controlled conditions which maintain the quality of the products followed by the reapplication of the quick freezing process is permitted.

Quick frozen finfish, shall be processed and packaged so as to minimize dehydration and oxidation.

2.3 PRESENTATION

Any presentation of the product shall be permitted provided that it:

2.3.1 meets all requirements of this standard; and

2.3.2 is adequately described on the label to avoid confusing or misleading the consumer.

---

❶ It does not apply to fish frozen in brine intended for further processing.

3 ESSENTIAL COMPOSITION AND QUALITY FACTORS

3.1 FISH

Quick frozen finfish shall be prepared from sound fish which are of a quality fit to be sold fresh for human consumption.

3.2 GLAZING

If glazed, the water used for glazing or preparing glazing solutions shall be of potable quality or shall be clean sea - water. Potable water is fresh - water fit for human consumption. Standards of potability shall not be less than those contained in the latest edition of the WHO "International Guidelines for Drinking Water Quality".

Clean sea - water is sea - water which meets the same microbiological standards as potable water and is free from objectionable substances.

3.3 OTHER INGREDIENTS

All other ingredients used shall be of food grade quality and conform to all applicable Codex and WHO standards.

3.4 DECOMPOSITION

The products shall not contain more than 10 mg/100 g of histamine based on the average of the sample unit tested. This shall apply only to species of *Clupeidae*, *Scombridae*, *Scombresocidae*, *Pomatomidae and Coryphaenedae families.*

3.5 FINAL PRODUCT

Products shall meet the requirements of this standard when lots examined in accordance with Section 9 comply with the provisions set out in Section 8. Products shall be examined by the methods given in Section 7.

4 FOOD ADDITIVES

Only the use of the following additives is permitted.

Additive		Maximum Level in the Final Product
Antioxidants	300 Ascorbic acid 301 Sodium ascorbate 303 Potassium ascorbate	GMP

5 HYGIENE AND HANDLING

5.1 The final product shall be free from any foreign material, that poses a threat to human health

5. 2 When tested by appropriate methods of sampling and examination prescribed by the Codex Alimentarius Commission, the product:

(i) shall be free from microorganisms or substances originating from microorganisms in amounts which may present a hazard to health in accordance with standards established by the Codex Alimentarius Commission;

(ii) shall not contain histamine that exceeds 20 mg/100 g. This applies only to species of Clupeidae, Scombridae, Scombresocidae, Pomatomidae and Coryphaenedae families;

(iii) shall not contain any other substance in amounts which may present a hazard to health in accordance with standards established by the Codex Alimentarius Commission.

5. 3 It is recommended that the product covered by the provisions of this standard be prepared and handled in accordance with the appropriate sections of the Recommended International Code of Practice – General Principles of Food Hygiene (CAC/RCP 1 – 1969, Rev. 3 – 1997) and the following relevant Codes:

(i) the Recommended International Code of Practice for Frozen Fish (CAC/RCP 16 – 1978);

(ii) The Recommended International Code of Practice for the Processing and Handling of Quick Frozen Foods (CAC/RCP 8 – 1976);

(iii) The sections on the Products of Aquaculture in the Proposed Draft International Code of Practice for Fish and Fishery Products (under elaboration)[1]

6 LABELLING

In addition to the provisions of the Codex General Standard for the Labelling of Prepackaged Foods (CODEX STAN 1 – 1985, Rev. 3 – 1999) the following specific provisions apply:

6. 1 THE NAME OF THE FOOD

6. 1. 1 In addition to the common or usual name of the species, the label, in the case of eviscerated fish, shall include terms indicating that the fish has been eviscerated and whether presented as "head – on" or "headless"

6. 1. 2 If the product has been glazed with sea – water, a statement to this effect shall be made

6. 1. 3 The term "quick frozen", shall also appear on the label, except that the term "frozen" may be applied in countries where this term is customarily used for describing the product processed in accordance with subsection 2. 2 of this standard

6. 1. 4 The label shall state that the product should be maintained under conditions that will maintain the quality during transportation, storage and distribution

6. 2 NET CONTENTS (GLAZED PRODUCTS)

Where the food has been glazed the declaration of net contents of the food shall be exclusive of the glaze.

6. 3 STORAGE INSTRUCTIONS

The label shall include terms to indicate that the product shall be stored at a temperature of – 18℃ or

[1] The Proposed Draft Code of Practice, when finalized, will replace all current Codes of Practice for Fish and Fishery Products.

colder.

6.4 LABELLING OF NON-RETAIL CONTAINERS

Information specified above shall be given either on the container or in accompanying documents, except that the name of the food, lot identification, and the name and address, as well as storage instructions shall always appear on the container.

However, lot identification, and the name and address may be replaced by an identification mark, provided that such a mark is clearly identifiable with the accompanying documents.

7 SAMPLING, EXAMINATION AND ANALYSES

7.1 SAMPLING

(i) Sampling of lots for examination of the product shall be in accordance with the FAO/WHO Codex Alimentarius Sampling Plans for Prepackaged Foods (AQL-6.5) CAC/RM 42-1977. A sample unit is the individual fish or the primary container.

(ii) Sampling of lots for examination of net weight shall be carried out in accordance with an appropriate sampling plan meeting the criteria established by the CAC.

7.2 SENSORY AND PHYSICAL EXAMINATION

Samples taken for sensory and physical examination shall be assessed by persons trained in such examination and in accordance with procedures elaborated in Sections 7.3, 7.4 and 7.5, Annex A and the Guidelines for the Sensory Evaluation of Fish and Shellfish in Laboratories (CAC/GL 31-1999).

7.3 DETERMINATION OF NET WEIGHT

7.3.1 Determination of Net Weight of Products not Covered by Glaze

The net weight (exclusive of packaging material) of each sample unit representing a lot shall be determined in the frozen state.

7.3.2 Determination of Net Weight of Products Covered by Glaze

(To be elaborated).

7.4 THAWING

(To be elaborated).

7.5 DETERMINATION OF GELATINOUS CONDITIONS

According to the AOAC Methods - "Moisture in Meat and Meat Products, Preparation of Sample Procedure"; 883.18 and "Moisture in Meat" (Method A); 950.46; AOAC 1990.

7.6 COOKING METHODS

The following procedures are based on heating the product to an internal temperature of 65~70℃. The product must not be overcooked. Cooking times vary according to the size of the product and the temperatures used. The exact times and conditions of cooking for the product should be determined by prior experimentation.

Baking Procedure: Wrap the product in aluminum foil and place it evenly on a flat cookie sheet or shallow flat pan.

Steaming Procedure: Wrap the product in aluminum foil and place it on a wire rack suspended over boiling water in a covered container.

Boil – In – Bag Procedure: Place the product into a boilable film – type pouch and seal. Immerse the pouch into boiling water and cook.

Microwave Procedure: Enclose the product in a container suitable for microwave cooking. If plastic bags are used, check to ensure that no odour is imparted from the plastic bags. Cook according to equipment specifications.

7.7 DETERMINATION OF HISTAMINE

AOAC 977.13 (15th Edition, 1990).

8 DEFINITION OF DEFECTIVES

The sample unit shall be considered defective when it exhibits any of the properties defined below:

8.1 DEEP DEHYDRATION

Greater than 10% of the surface area of the block or greater than 10% of the weight of fish in the sample unit exhibits excessive loss of moisture clearly shown as white or yellow abnormality on the surface which masks the colour of the flesh and penetrates below the surface, and cannot be easily removed by scraping with a knife or other sharp instrument without unduly affecting the appearance of the fish.

8.2 FOREIGN MATTER

The presence in the sample unit of any matter which has not been derived from fish (excluding packaging material), does not pose a threat to human health, and is readily recognized without magnification or is present at a level determined by any method including magnification, that indicates non – compliance with good manufacturing and sanitation practices.

8.3 ODOUR AND FLAVOUR

A sample unit affected by persistent and distinct objectionable odours or flavours indicative of decomposition or of feed.

8.4 TEXTURE

8.4.1 Textural breakdown of the flesh, indicative of decomposition characterized by muscle structure which is mushy or paste – like, or by separation of flesh from the bones.

8.4.2 Flesh abnormalities

A sample unit affected by excessive gelatinous condition of the flesh together with greater then 86% moisture found in any individual fish or sample unit with pasty texture resulting from parasitic infestation affecting more than 5% of the sample unit by weight.

8.5 BELLY BURST

The presence of ruptured bellies in uneviscerated fish, indicative of decomposition.

9 LOT ACCEPTANCE

A lot shall be considered as meeting the requirements of this standard when:

(i) the total number of defectives as classified according to Section 8 does not exceed the acceptance number (c) of the appropriate sampling plan in the Sampling Plans for Prepackaged Foods (AQL - 6.5) (CAC/RM 42 - 1969);

(ii) the average net weight of all sample units is not less than the declared weight, provided there is no unreasonable shortage in any container; and

(iii) the Food Additives, Hygiene and Labelling requirements of Sections 3.4, 4, 5.1, 5.2 and 6 are met.

ANNEX A SENSORY AND PHYSICAL EXAMINATION

1. Complete net weight determination, according to defined procedures in Section 7.3 (de - glaze as required).

2. Examine the frozen sample unit for the presence of deep dehydration by measuring those areas or counting instances which can only be removed with a knife or other sharp instrument. Measure the total surface area of the sample unit, and calculate the percentage affected.

3. Thaw and individually examine each fish in the sample unit for the presence of foreign matter.

4. Examine each fish using the criteria outlined in Section 8. Flesh odours are examined by tearing or making a cut across the back of the neck such that the exposed surface of the flesh can be evaluated.

5. In cases where a final decision regarding the odour or texture can not be made in the thawed uncooked state, a small portion of the flesh (approximately 200 g) is sectioned from the product and the odour, flavour or texture confirmed without delay by using one of the cooking methods defined in Section 7.5.

6. In cases where a final decision on gelatinous condition cannot be made in the thawed uncooked state, the disputed material is sectioned from the product and gelatinous condition confirmed by cooking as defined in Section 7.6 or by using the procedure in Section 7.5 to determine if greater than 86% moisture is present in any fish. If a cooking evaluation is inconclusive, then the procedure in 7.5 would be used to make the exact determination of moisture content.

虾或对虾罐头

CODEX STANDARD FOR CANNED SHRIMPS OR PRAWNS

(*CODEX STAN 37－1991, REV. 1－1995*)

1 范围

本标准适用于虾罐头或对虾罐头❶。不适用于罐中虾肉净重少于罐头内容物净重50%（W/W）的罐头。

2 说明

2.1 产品的定义

虾罐头是由已被去除头、壳、触须的对虾科（*Penaeidae*）、长额虾科（*Pandalidae*）、褐虾科（*Crangonidae*）和长臂虾科（*Palaemonidae*）等科中各种虾的虾肉制备的。

2.2 加工过程的定义

将虾肉装罐并密封，然后通过一系列加工处理以确保其达到商业无菌状态。

2.3 性状

产品描述如下：

2.3.1 “去壳虾”

去头、去壳，但未去除肠腺。

2.3.2 “清除”或“去肠腺”

切开去壳虾的背部，去除肠腺（至少从头一直到临近尾巴的最后一段）。“清除”或“去肠腺”的虾的净重应占虾总量的95%。

2.3.3 “碎虾”

有超过虾总量10%的部分是去壳的碎虾。少于四个节已经或尚未去除肠腺的虾即为碎虾。

2.3.4 性状的其他形式

性状的任何其他形式，应符合下列要求：

（1）与本标准中的所述有明显的区别；

（2）达到本标准的所有要求；

（3）在产品标签中对产品进行详细描述，以免引起混淆或误导消费者。

---

❶ 以下以虾类为例，进行说明。

2.3.5 大小分级

虾罐头可按下列要求进行大小分级：

（1）可在标签上注明确切的数量范围；

（2）符合附录B中的各项条款。

3 基本成分及质量因素

3.1 虾

产品必须由2.1条所列出的品种的虾制备，且虾的品质良好，可作为鲜品供人类消费。

3.2 其他成分

所使用的其他成分应具有食品级的质量，并且符合所有相应法规标准的规定。

3.3 成品

产品应符合本标准要求，根据第9章进行成品批次检查时，其质量应符合第8章的规定。检查方法应符合第7章的规定。

4 食品添加剂

只允许使用下列食品添加剂。

添加剂		成品中的最高含量
色素	102 柠檬黄（酒石红） 110 日落黄 123 苋菜红 124 胭脂红	成品中的含量为30mg/kg单用或混用
螯合剂	385 乙二胺四乙酸二钠钙（EDTA二钠钙）	250mg/kg
调酸剂	330 柠檬酸	GMP
	338 正磷酸	850mg/kg

5 卫生及处理

5.1 成品中不能含有任何危害人体健康的外来杂质

5.2 当应用CAC规定的抽样及检测方法进行检验时，产品应符合下列条件

（1）不含有任何在正常的贮存条件下可能生长的微生物；

（2）其他危害人体健康的有害物质的含量也不能超过CAC标准的规定；

（3）不存在可能危及罐的密封性的缺陷。

5.3 建议本标准中涉及产品预处理的条款应符合推荐性国际规范《CAC/RCP1 - 1969, Rev. 3 - 1997 食品卫生操作总则》中的相关章节和以下相关标准

（1）推荐性国际操作规范《CAC/RCP 10 - 1976 鱼罐头》;

（2）推荐性国际操作规范《CAC/ RCP 23 - 1979, REV. 2 - 1993 低酸和酸化低酸罐头食品》;

（3）推荐性国际操作规范《CAC/RCP 17 - 1987 虾类或龙虾类》;

（4）《水产品国际操作规程草案》中水产养殖品的相关章节（正在完善中）[1]。

6 标签

标签除了应符合《CODEX STAN1 - 1985, Rev. 1 - 1991 预包装食品标签通用标准》的要求外，还应遵守以下规定：

6.1 食品名称

6.1.1 标签上标示的食品名称应注明“虾”或“对虾”，产品名称在其常用名的前面或后面标注，符合出售该产品的国家的法律和习惯，同时不能误导消费者；

6.1.2 产品名称应根据 2.3.1 ~2.3.4 条的描述性状的术语来限定；

6.1.3 如果虾罐头根据大小作标签，则其尺寸必须符合 2.3.5 和附录 B 的规定；

6.1.4 2.3.3 中所述的碎虾也应标注。

7 抽样、检测和分析

7.1 取样

（1）产品批次检验用样品的抽样方法应符合 FAO/WHO 的食品法典《CODEX STAN 233 - 1969 预包装食品的抽样方案》（AQL - 6.5）。

（2）对需检测净重和干重的批次的抽样，抽样计划应以食品法典委员会的相关标准为依据。

7.2 感官和物理检验

产品的感官与物理指标须由经过此类检验培训的人员进行检验，并依据附录 A 以及《CAC/GL 31 - 1999 实验室中水产品感官评价指南》所述程序进行。

7.3 净重的测定

所有样品单位内容物的净重应通过以下程序测定：

（1）称量未开封的罐头；

（2）开罐并除去内容物；

（3）称量空罐重量（包括开启下来的罐底）；

（4）从未开封罐头的重量中减去空罐的重量，得到的数字就是内容物的净重。

---

[1] 操作规程建议草案，在定稿后将替代所有现行的《水产品操作规程》。

7.4 干重的测定

所有样品单位内容物沥干后重量（干重）应通过以下程序测定：

（1）测定前使罐头的温度至少在过去12小时内维持在20～30℃。

（2）开罐并使之倾斜，将内容物倾倒在预先称重的圆形滤网上，该滤网由金属网丝构成，网孔大小：2.8mm×2.8mm。

（3）以17°～20°的角度倾斜滤网以使虾沥水，从将产品全部倾倒于滤网上开始计时，沥水2min。

（4）称量沥水后的虾及滤网的总重。

（5）从沥水后的产品和滤网的总重中减去滤网的重量就得到虾的干重。

7.5 大小分级情况的测定

表示为每100g干重的产品中虾的数量，由以下计算式得出：

样品单位中虾的数量/样品单位的准确干重×100＝虾的数量/100g

8 缺陷的定义

当样品单位出现以下特征时，则认定其有“缺陷”。

8.1 外来杂质

样品中存在的任何不是来自于虾体的物质。这些物质虽不会对人体健康造成危害，但用肉眼可直接辨别，或采用某些方法（包括放大）可以确定其存在。出现外来杂质表明不符合良好操作和卫生惯例。

8.2 气味或风味

样品散发的持久、明显、令人厌恶的由腐败、酸败引起的气味或风味。

8.3 质地

（1）含有过多非该产品特征的糊状虾肉；

（2）含有过多非该产品特征的硬（韧）虾肉。

8.4 变色

样品中15%以上的虾的单体出现超过其表面积10%的部分明显变黑。

8.5 异物

样品中发现长度超过5mm的磷酸铵镁结晶（鸟粪石）。

9 批次验收

当满足以下条件时，可以认为此批次产品符合本标准的要求：

（1）根据本标准第8章中规定分类，缺陷总数不超过《CODEX STAN 233－1969 预包装食品的抽样方案》（AQL－6.5）中相应抽样方案规定的可接受数（c）；

（2）不符合本标准 2.3 条中规定数量的样品总数不超过《CODEX STAN233 - 1969 预包装食品的抽样方案》（AQL - 6.5）中相应抽样方案规定的可接受数（c）；

（3）所有样品单位平均净重不少于标示量，在任何一个包装单位中没有不合理的重量短缺；

（4）符合本标准第 4、5、6 款中对食品添加剂、卫生和处理以及标签的要求。

附录 A　感官和物理检验

1. 对罐外部进行检测以确定其是否存在密封性缺陷，有无罐身外部扭曲。
2. 开罐，并且根据 7.3、7.4 条中确定的程序完成重量测定。
3. 小心移出罐中的内容物，并根据 7.5 条中的程序对大小分级情况进行检测。
4. 检测产品的变色、外来杂质和异物等方面的情况。
5. 依据《CAC/ GL 31 - 1999 实验室中水产品感官评价指南》对产品的气味、风味以及质地进行评定。

附录 B　虾罐头中虾的大小分级

根据下表划分的等级使用“特大（extra large）”，“很大（jumbo）”，“大（large）”，“中（medium）”，“小（small）”，“微小（tiny）”等词汇每 100g 干重的产品中虾（包括大于 4 节碎虾）的数量。

大小等级	范围（只）
特大和巨大	13 以下
大	14 ~ 19
中	20 ~ 34
小	35 ~ 65
微小	65 以上

CODEX STANDARD FOR CANNED SHRIMPS OR PRAWNS

(CODEX STAN 37 – 1991, REV. 1 – 1995)

1 SCOPE

This standard applies to canned shrimps or canned prawns[1]. It does not apply to specialty products where shrimp constitutes less than 50% m/m of the contents.

2 DESCRIPTION

2.1 PRODUCT DEFINITION

Canned shrimp is the product prepared from any combination of species of the families *Penaeidae*, *Pandalidae*, *Crangonidae* and *Palaemonidae* from which heads, shell, antennae have been removed.

2.2 PROCESS DEFINITION

Canned shrimp are packed in hermetically sealed containers and shall have received a processing treatment sufficient to ensure commercial sterility.

2.3 PRESENTATION

The product shall be presented as:

2.3.1 Peeled shrimp

shrimp which have been headed and peeled without removal of the dorsal tract.

2.3.2 Cleaned or de – veined

peeled shrimp which have had the back cut open and the dorsal tract removed at least up to the last segment next to the tail. The portion of the cleaned or de – veined shrimp shall make up 95% of the shrimp contents.

2.3.3 Broken shrimp

more than 10% of the shrimp contents consist of pieces of peeled shrimp of less than four segments with or without the vein removed.

2.3.4 Other Forms of Presentation

Any other presentation shall be permitted provided that it:

2.3.4.1 is sufficiently distinctive from other forms of presentation laid down in this standard;

---

[1] Hereafter referred to as "shrimp".

2.3.4.2 meets all other requirements of this standard;

2.3.4.3 is adequately described on the label to avoid confusing or misleading the consumer.

2.3.5 Size

Canned shrimp may be designated as to size in accordance with:

(i) the actual count range may be declared on the label; or

(ii) provisions given in Annex "B".

3 ESSENTIAL COMPOSITION AND QUALITY FACTORS

3.1 SHRIMP

Shrimp shall be prepared from sound shrimp of the species in sub-section 2.1 which are of a quality fit to be sold fresh for human consumption.

3.2 OTHER INGREDIENTS

The packing medium and all other ingredients used shall be of food grade quality and conform to all applicable Codex standards.

3.3 FINAL PRODUCT

Products shall meet the requirements of this Standard when lots examined in accordance with Section 9 comply with the provisions set out in Section 8. Products shall be examined by the methods given in Section 7.

4 FOOD ADDITIVES

Only the use of the following additives is permitted.

	Additive	Maximum Level in the Final Product
Colours	102 Tartrazine 110 Sunset Yellow FCF 123 Amaranth 124 Ponceau 4R	30mg/kg in the final product, singly or in combination
Sequestrant	385 Calcium disodium EDTA	250mg
Acidity Regulator	330 Citric acid	GMP
	338 Orthophosphoric acid	850mg/kg

5 HYGIENE AND HANDLING

5.1 The final product shall be free from any foreign material that poses a threat to human health

5.2 When tested by appropriate methods of sampling and examination by the Codex Alimentarius Commission, the product:

(i) shall be free from micro-organisms capable of development under normal conditions of storage; and

(ii) shall not contain any other substances including substances derived from micro organisms in amounts which may represent a hazard to health in accordance with standards established by the Codex Alimentaius Commission; and

(iii) shall be free from container integrity defects which may compromise the hermetic seal.

5. 3 It is recommended that the products covered by the provisions of this standard be prepared and handled in accordance with the appropriate sections of the Recommended International Code of Practice General Principles of Food Hygiene (CAC/RCP 1 – 1969, Rev. 3 – 1997) and the following relevant Codes:

(i) the Recommended International Code of Practice for Canned Fish (CAC/RCP 10 – 1976);

(ii) the Recommended International Code of Hygienic Practice for Low – Acid and Acidified Low – Acid Canned Foods (CAC/RCP 23 – 1979);

(iii) the Recommended International Code of Practice for Shrimps or Prawns (CAC/RCP 17 – 1978);

(iv) The sections on the Products of Aquaculture in the Proposed Draft International Code of Practice for Fish and Fishery Products (under elaboration)[1]

6 LABELLING

In addition to provisions of the Codex General Standard for the Labelling of Prepackaged Foods (CODEX STAN 1 – 1985, Rev. 3 – 1999) the following specific provisions apply:

6. 1 THE NAME OF THE FOOD

6. 1. 1 The name of the product as declared on the label shall be "shrimp", or "prawns", and may be preceded or followed by the common or usual name of the species in accordance with the law and custom of the country in which the product is sold and in a manner not to mislead the consumer

6. 1. 2 The name of the product shall be qualified by a term descriptive of the presentation in accordance with Sections 2. 3. 1 to 2. 3. 4

6. 1. 3 If the canned shrimp are labelled as to size, the size shall comply with the provisions of Section 2. 3. 5 and Annex "B"

6. 1. 4 Broken shrimp defined in 2. 3. 3 shall be so labelled.

7 SAMPLING, EXAMINATION AND ANALYSES

7. 1 SAMPLING

(i) Sampling of lots for examination of the final product as prescribed in Section 3. 3 shall be in accordance with the FAO/WHO Codex Alimentarius Sampling Plans for Prepackaged Foods (1969) (AQL – 6. 5) (Ref. CAC/RM 42 – 1969).

(ii) Sampling of lots for examination of net weight and drained weight shall be carried out in accordance with an appropriate sampling plan meeting the criteria established by the CAC.

---

[1] The Proposed Draft Code of Practice, when finalized, will replace all current Codes of Practice for Fish and Fishery Products.

7.2 SENSORIC AND PHYSICAL EXAMINATION

Samples taken for sensoric and physical examination shall be assessed by persons trained in such examination in accordance with Annex A and the *Guidelines for the Sensory Evaluation of Fish and Shellfish in Laboratories* (*CAC/GL* 31 – 1999).

7.3 DETERMINATION OF NET WEIGHT

Net contents of all sample units shall be determined by the following procedure:

(i) Weigh the unopened container;

(ii) Open the container and remove the contents;

(iii) Weigh the empty container, (including the end) after removing excess liquid and adhering meat;

(iv) Subtract the weight of the empty container from the weight of the unopened container. The resultant figure will be the net content.

7.4 DETERMINATION OF DRAINED WEIGHT

The drained weight of all sample units shall be determined by the following procedure:

(i) Maintain the container at a temperature between 20℃ and 30℃ for a minimum of 12 hours prior to examination;

(ii) Open and tilt the container to distribute the contents on a pre – weighed circular sieve which consists of wire mesh with square openings of 2.8 mm × 2.8 mm;

(iii) Incline the sieve at an angle of approximately 17° ~ 20° and allow the shrimps to drain for two minutes, measured from the time the product is poured into the sieve;

(iv) Weigh the sieve containing the drained shrimps;

(v) The weight of drained shrimps is obtained by subtracting the weight of the sieve from the weight of the sieve and drained product.

7.5 DETERMINATION OF SIZE DESIGNATION

The size, expressed as the number of shrimp per 100g of drained product, is determined by the following equation:

$$\frac{\text{Number of whole shrimp in unit}}{\text{Actualdrained weight of unit}} \times 100 = \text{Number of shrimp/100g}$$

8 DEFINITION OF DEFECTIVES

A sample unit will be considered defective when it fails to meet any of the following final product requirements referred to in Section 3.3.

8.1 FOREIGN MATTER

The presence in the sample unit of any matter, which has not been derived from shrimp, does not pose a threat to human health, and is readily recognized without magnification or is present at a level determined by any method including magnification that indicates non – compliance with good manufacturing or sanitation practices.

8.2 ODOUR/FLAVOUR

A sample unit affected by persistent and distinct objectionable odours or flavours indicative of decomposition or rancidity.

8.3 TEXTURE

(i) Excessive mushy flesh uncharacteristic of the species in the presentation; or

(ii) Excessively tough flesh uncharacteristic of the species in the presentation.

8.4 DISCOLOURATION

A sample unit affected by distinct blackening of more than 10% of the surface area of individual shrimp which affects more than 15% of the number of shrimp in the sample unit.

8.5 OBJECTIONABLE MATTER

A sample unit affected by:

(i) struvite crystals – any struvite crystal greater than 5 mm in length.

9 LOT ACCEPTANCE

A lot shall be considered as meeting the requirements of this standard when:

(i) the total number of defectives as classified according to Section 8 does not exceed the acceptance number (c) of the appropriate sampling plan in the Sampling Plans for Prepackaged Foods (AQL – 6.5) (CAC/RM 42 – 1977);

(ii) the total number of sample units not meeting presentation requirements in Section 2.3 does not exceed the acceptance number (c) of the appropriate sampling plan in the Sampling Plans for Prepackaged Foods (AQL – 6.5) (CAC/RM 42 – 1977);

(iii) the average net weight and the average drained weight of all sample units examined is not less than the declared weight and provided there is no unreasonable shortage in any individual container;

(iv) the Food Additives, Hygiene and Labelling requirements of Sections 4, 5.1, 5.2, and 6 are met.

ANNEX A SENSORY AND PHYSICAL EXAMINATION

1. Complete external can examination for the presence of container integrity defects or can ends which may be distorted outwards.

2. Open can and complete weight determination according to defined procedures in Sections 7.3 and 7.4.

3. Carefully remove the product and examine for size designation in accordance with the procedure in Section 7.5.

4. Examine product for discolouration, foreign and objectionable matter.

5. Assess odour, flavour and texture in accordance with the Guidelines for the Sensory Evaluation of Fish and Shellfish in Laboratories (CAC/GL 31 – 1999).

ANNEX "B" SIZE DESIGNATION OF CANNED SHRIMPS

The terms "extra large", "jumbo", "large", "medium", "small", "tiny" may be used provided that the range is in accordance with the following table:

Number of whole shrimp (including pieces greater than 4 *segments) per* 100g *of drained product.*

SIZE DESIGNATION	RANGE
Extra Large or Jumbo	13 or less
Large	14 ~ 19
Medium	20 ~ 34
Small	35 ~ 65
Tiny	more than 65

金枪鱼罐头和鲣鱼罐头

CODEX STANDARD FOR CANNED TUNA AND BONITO

(*CODEX STAN 70 - 1981, REV. 1 - 1995*)

1 范围

本标准适用于金枪鱼罐头和鲣鱼罐头。不适用于罐中鱼肉净重少于罐头内容物净重 50% (W/W) 的罐头。

2 说明

2.1 产品的定义

金枪鱼和鲣鱼罐头采用下列属种的鱼为原料，并采用密封容器进行包装。它们是：长鳍金枪鱼 (*Thunnus alalunga*)、黄鳍金枪鱼 (*Thunnus albacares*)、黑鳍金枪鱼 (*Thunnus atlanticus*)、大眼金枪鱼 (*Thunnus obesus*)、蓝鳍金枪鱼 (*Thunnus maccoyii*)、金枪鱼 (*Thunnus thynnus*)、青甘金枪鱼 (*Thunnus tonggol*)、鲔鱼 (*Euthynnus affinis*)、小鲔 (*Euthynnus alletteratus*)、黑鲔 (*Eutnynnus lineatus*)、鲣鱼 [*Katsuwonus pelamis* (*syn. Eutynnus pelamis*)]、智利狐鲣 (*Sarda chiliensis*)、东方狐鲣 (*Sarda orientalis*)、狐鲣 (*Sarda sarda*)。

2.2 加工过程的定义

产品通过一系列加工处理以确保其达到商业无菌状态。

2.3 产品介绍

产品可描述为以下情形。

2.3.1 “整块鱼”(带皮或不带皮)

将鱼切成横段，并垂直装罐，使横断面平行于罐底。鱼碎片及碎块的比例不超过内容物干重的 18%。

2.3.2 “鱼块”

几乎各边长度均不小于 1.2cm 并且保留原有的肌肉组织的大块鱼肉。各边长度小于 1.2cm 的鱼块的比例不超过干重的 30%。

2.3.3 “鱼片”

各边长度低于 1.2cm 但保留了原有肌肉组织的鱼肉碎屑和鱼肉碎片的混合物。各边长度大于 1.2cm 的鱼块的比例不超过干重的 30%。

2.3.4 碎块和碎丝

蒸煮后鱼体上掉下的颗粒的混合物，这些颗粒基本上已减缩到同样尺寸，但各不相连，未形成膏状。

2.3.5 产品介绍的任何其他形式，应符合下列要求

（1）与本标准中的所述有明显的区别；

（2）达到本标准的所有要求；

（3）在产品标签中对产品进行详细描述，以免引起混淆或误导消费者。

3 基本成分及质量因素

3.1 原料

产品必须由2.1条所列出的品种鱼制备，且鱼的品质良好，可作为鲜品供人类消费。

3.2 其他成分

填装介质和所使用的其他成分应具有食品级的质量，并且符合相应法规标准的规定。

3.3 腐败

在对产品检验中组胺平均浓度应少于10mg/100g。

3.4 成品

产品应符合本标准要求，根据第9章进行成品批次检查时，其质量应符合第8款的规定。检查方法应符合第7章的规定。

4 食品添加剂

只允许使用下列食品添加剂。

添加剂		成品中的最高含量
增稠剂（只用于填装介质）	400 褐藻酸 401 褐藻酸钠 402 褐藻酸钾 404 褐藻酸钙 406 琼脂 407 卡拉胶及其Na，K，NH_4 盐（包括红藻胶） 407a 经热加工处理的麒麟菜（PES） 410 角豆胶（刺槐豆胶） 412 瓜尔豆胶 413 黄蓍胶 415 黄原胶 440 果胶 466 羧甲基纤维素钠	GMP

续表

	添加剂	成品中的最高含量
改性淀粉	1401 酸处理淀粉 1402 碱处理淀粉 1404 氧化淀粉 1410 磷酸单淀粉 1412 磷酸二淀粉酯化三偏磷酸钠；磷酸二淀粉酯化氯氧化磷 1414 磷酸乙酰化二淀粉 1413 磷酸化磷酸二淀粉 1420 醋酸淀粉酯化乙酸酐 1421 醋酸淀粉酯化乙烯基乙酸酯 1422 乙酰化二淀粉己二酸酯 1440 羧丙基淀粉 1442 磷酸羧丙基淀粉	GMP
调酸剂	260 乙酸 270（L－、D－和 DL－）乳酸 330 柠檬酸	GMP
	450（i）二磷酸二钠	≤10g/kg，以 P_2O_5 计，（包括天然磷酸盐）
天然香料	香油（香料油） 香精（香料萃取物） 烟香味（自然烟溶液和萃取物）	GMP

5 卫生及处理

5.1 成品中不能含有任何危害人体健康的外来杂质

5.2 应用 CAC 规定的抽样及检测方法进行检验时，产品应符合以下标准

（1）不含有任何在正常的贮存条件下可能生长的微生物；

（2）样品单位中组胺的含量不超过 20mg/100g；

（3）其他危害人体健康的有害物质的含量也不能超过 CAC 标准的规定；

（4）不存在可能危及罐的密封性的缺陷。

5.3 建议本标准中涉及产品预处理的条款应符合推荐性国际规范《CAC/RCP1－1969，Rev. 3－1997 食品卫生操作总则》中的相关章节，和以下相关标准

（1）推荐性国际操作规范《CAC/RCP 10－1976 鱼罐头》；

（2）推荐性国际操作规范《CAC/ RCP 23－1979，REV. 2－1993 低酸和酸化低酸罐头食品》；

（3）《水产品国际操作规程草案》中水产养殖品的相关章节（正在完善中）[1]。

---

[1] 操作规程建议草案，在定稿后将替代所有现行的《水产品操作规程》。

6 标签

标签除应符合《CODEX STAN1 - 1985，Rev. 3 - 1999 预包装食品标签通用标准》的要求外，还应遵守以下规定：

6.1 食品名称

6.1.1 在标签上标示的食品名称应为“金枪鱼”或者“鲣鱼”，并且在其前面或后面标示出鱼种的俗名或常用名

食品名称符合出售该产品的国家的法律和习惯，并且不会误导消费者。

6.1.2 食品名称中应包含产品的颜色

例如，“白色”只应用于长鳍金枪鱼（Thunnus alalunga），“淡”、“深”、“混合”等词语要遵照当地的法规使用。

6.2 性状形式

2.3 条中所提到的性状应使用接近于通俗名称的形式注明。

6.2.1 食品名称中应包含填装介质的名称。

7 抽样、检测和分析

7.1 取样

（1）如 3.3 中指定的一样，产品批次检验用样品的抽样方法应符合 FAO/WHO 的食品法典《CODEX STAN 233 - 1969 预包装食品的抽样方案》（AQL - 6.5）。

（2）对需检测净重和干重的批次的抽样，抽样计划应以食品法典委员会的相关标准为依据。

7.2 感官与物理检验

产品的感官与物理指标须由经过此类检验培训的人员进行检验，并依据本标准 7.3 至 7.5 条和附录 A 以及《CAC/GL 31 - 1999 实验室中水产品感官评价指南》所述程序进行。

7.3 净重的测定

样品内容物的净重应通过以下程序测定：

（1）称量未开封的罐头；

（2）开罐并除去内含物；

（3）称量空罐重量（包括开启下来的罐底）；

（4）从未开封罐头的重量中减去空罐的重量，得到的就是内容物的净重。

7.4 干重的测定

所有样品单位内容物沥干后重量（干重）应通过以下程序测定：

（1）测定前使罐头的温度至少在过去 12 小时内维持在 20 ~ 30℃。

（2）开罐并使之倾斜，将内容物倾倒在预先称重的圆形滤网上，该滤网由金属网丝构成，网孔大小：2.8mm × 2.8mm。

（3）以17°～20°的角度倾斜滤网以使鱼沥水，从将产品全部倾倒于滤网上开始计时，沥水2min。

（4）称量沥水后的鱼及滤网的总重。

（5）从沥水后的产品和滤网的总重中减去滤网的重量就得到鱼的干重。

7.5 冲洗和沥水后产品重量的测定（对以调味汁填装的产品）

（1）测定前使罐头的温度至少在过去12小时内维持在20～30℃；

（2）开罐并使之倾斜，将所有的内容物置于预先称量过皮重的圆形滤网上，使用洗瓶（例如，塑料洗瓶）以热自来水（约40℃）冲洗覆盖的调味汁；

（3）用热水冲洗滤网上的内容物直到其不再含有附着的调味汁；必要时，用钳子分离出可挑选出的配料（香料、蔬菜、水果等）。以17°～20°的角度倾斜滤网以使鱼沥出水分，从冲洗程序结束时开始计时，沥水2min；

（4）用纸巾除去滤网底部附着的水。称量经冲洗和沥水后的鱼及滤网的总重；

（5）用经过冲洗和沥水后的产品及滤网的总重减去滤网的重量就得到冲洗和沥水后的产品重量。

7.6 性状的确定

样品的性状应通过以下程序测定：

（1）根据7.4开罐并排干内容物中的水分；

（2）将内容物放到预称过皮重的滤网（网孔孔径为1.2cm）上。滤网下面放置一个平盘；

（3）用刮刀小心分散鱼肉，尽可能不使鱼肉外形受到破坏。确保小块鱼肉及鱼碎片、鱼肉丝屑可以通过滤网，并被收集于下面放置的盘子中；

（4）将平盘上滤得的内容物分为鱼片、碎块（碎丝）及膏状鱼浆，并分别称重；

（5）如果产品标示的性状为“鱼块”，则称量滤网及上面留存的鱼肉的总重，并记录之。所得数字减去滤网的重量，即得到鱼块与整块鱼的重量；

（6）如果产品标示的性状为“整块鱼”，则先将滤网上留存的鱼块去除，在进行称量并记录之。所得数字减去滤网的重量，即得到整块鱼的重量。

计算方法：

（1）碎片及碎块（包括碎丝和鱼浆）的重量百分比：

鱼碎片含量的百分比（%）＝鱼碎片重量/总的干重×100%

（2）计算整块鱼及大块鱼肉的重量百分比：

整块鱼及大块鱼肉的重量百分比（%）＝整块鱼及大块鱼肉的重量/总的干重×100%

（3）计算滤网上整块鱼的重量百分比，并用a%表示：

整块鱼重量百分比（%）＝整块鱼重/总的干重×100%

7.7 组胺的确定

参见AOAC 977.13。

8 缺陷的定义

当样品单位呈现下列任何一项特征时，则认定其有“缺陷”。

8.1 外来杂质

样品单位中存在的任何不是来自于鱼体的物质。这些物质虽不会对人体健康造成危害，但用肉眼可直接辨别，或采用某些方法（包括放大）可以确定其存在。出现外来杂质表明不符合良好操作和卫生惯例。

8.2 气味或风味

样品散发的持久、明显、令人厌恶的由腐败、酸败引起的气味或风味。

8.3 质地

（1）样品中含有过多非该产品特征的糊状鱼肉；

（2）样品中含有过多非该产品特征的硬（韧）鱼肉；

（3）样品中含有重量超过内容物干重5%的蜂巢状鱼肉。

8.4 变色

样品受腐败、酸败的影响发生明显的鱼肉变色，或超过内容物干重5%的样品受到硫化物着色。

8.5 异物

样品中发现长度超过5mm的磷酸铵镁结晶（鸟粪石）。

9 批次验收

当满足以下条件时，可以认为此批次产品符合本标准的要求：

（1）根据本标准第8章中规定分类，缺陷总数不超过《CODEX STAN 233－1969预包装食品的抽样方案》（AQL－6.5）中相应抽样方案规定的可接受数（c）；

（2）不符合本标准2.3条中规定数量的样品总数不超过《CODEX STAN233－1969预包装食品的抽样方案》（AQL－6.5）中相应抽样方案规定的可接受数（c）；

（3）所有样品单位平均净重不少于标示量，在任何一个包装单位中没有不合理的重量短缺；

（4）符合本标准第4、5、6款中对食品添加剂、卫生和处理以及标签的要求。

附录A 感官和物理的检验

1. 对罐外部的检测以确定其是否存在密封性缺陷，有无罐身外部扭曲。

2. 开罐，并且根据7.3、7.4和7.5条中的程序完成重量测定。

3. 检测产品的变色情况。

4. 根据7.6条中确定的程序，小心移出罐中的内容物，并检测其外观。

5. 检测产品的变色、外来杂质和异物等方面的情况。出现硬骨表明加工不完全，这时，要求对无菌状态进行评估。

6. 依据《CAC/ GL 31－1999 实验室中鱼、肉类感官评价指南》对产品的气味、风味以及质地进行评定。

CODEX STANDARD FOR CANNED TUNA AND BONITO

(*CODEX STAN 70 - 1981, REV. 1 - 1995*)

1 SCOPE

This standard applies to canned tuna and bonito. It does not apply to speciality products where the fish content constitutes less than 50% m/m of the contents.

2 DESCRIPTION

2.1 PRODUCT DEFINITION

Canned Tuna and Bonito are the products consisting of the flesh of any of the appropriate species listed below, packed in hermetically sealed containers.

- *Thunnus alalunga*;
- *Thunnus albacares*;
- *Thunnus atlanticus*;
- *Thunnus obesus*;
- *Thunnus maccoyii*;
- *Thunnus thynnus*;
- *Thunnus tongoe*;
- *Euthynnus affinis*;
- *Euthynnus alleteratus*;
- *Euthynnus lineatus*;
- *Katsuwonus pelamis* (*syn. Euthynnus pelamis*);
- *Sarda chilensis*;
- *Sarda orientalis*;
- *Sarda sarda*.

2.2 PROCESS DEFINITION

The products shall have received a processing treatment sufficient to ensure commercial sterility.

2.3 PRESENTATION

The product shall be presented as:

2. 3. 1 *Solid* (skin – on or skinless)

fish cut into transverse segments which are placed in the can with the planes of their transverse cut ends parallel to the ends of the can. The proportion of free flakes or chunks shall not exceed 18% of the drained weight of the container.

2. 3. 2 *Chunk*

pieces of fish most of which have dimensions of not less than 1. 2cm in each direction and in which the original muscle structure is retained. The proportion of pieces of flesh of which the dimensions are less than 1. 2 cm shall not exceed 30% of the drained weight of the container.

2. 3. 3 *Flake or flakes*

a mixture of particles and pieces of fish most of which have dimensions less than 1. 2 cm in each direction but in which the muscular structure of the flesh is retained. The proportion of pieces of flesh of which the dimensions are less than 1. 2 cm exceed 30% of the drained weight of the container.

2. 3. 4 *Grated or shredded*

a mixture of particles of cooked fish that have been reduced to a uniform size, in which particles are discrete and do not comprise a paste.

2. 3. 5 Any other presentation shall be permitted provided that it:

(i) is sufficiently distinctive from other forms of presentation laid down in this standard;

(ii) meets all other requirements of this standard;

(iii) is adequately described on the label to avoid confusing or misleading the consumer.

3 ESSENTIAL COMPOSITION AND QUALITY FACTORS

3. 1 RAW MATERIAL

The products shall be prepared from sound fish of the species in sub – section 2. 1 and of a quality fit to be sold fresh for human consumption.

3. 2 OTHER INGREDIENTS

The packing medium and all other ingredients used shall be of food grade quality and conform to all applicable Codex standards.

3. 3 DECOMPOSITION

The products shall not contain more than 10 mg/100 g of histamine based on the average of the sample unit tested.

3. 4 FINAL PRODUCT

Products shall meet the requirements of this Standard when lots examined in accordance with Section 9 comply with the provisions set out in Section 8. Products shall be examined by the methodes given in Section 7.

4 FOOD ADDITIVES

Only the use of the following additives is permitted.

	Additive	Maximum level in the Final Product
Thickening or Gelling Agents (for use in packing media only)	400 Alginic acid 401 Sodium alginate 402 Potassium alginate 404 Calcium alginate 406 Agar 407 Carrageenan and its Na, K, and NH_4 salts (including furcelleran 407a Processed *Eucheuma* Seaweed (PES) 410 Carob bean gum 412 Guar gum 413 Tragacanth gum 415 Xanthan gum 440 Pectins 466 Sodium carboxymethylcellulose	GMP
Modified Starches	1401 Acid treated starches (including white and yellow dextrins) 1402 Alkaline treated starches 1404 Oxidized starches 1410 Monostarch phosphate 1412 Distarch phosphate, esterified 1414 Acetylated distarch phosphate 1413 Phosphated distarch phosphate 1420/1421 Starch acetate 1422 Acetylated distarch adipate 1440 Hydroxypropyl starch 1442 Hydroxypropyl starch phosphate	GMP
Acidity Regulators	260 Acetic acid 270 Lactic acid (L-, D-, and DL-) 330 Citric acid	GMP
Natural Flavours	Spice oils Spice extracts Smoke flavours (Natural smoke solutions and extracts)	GMP
	For Canned Tuna and Bonito Only	
Acidity Regulators	450 Disodium diphosphate	10 mg/kg expressed as P_2O_5, (includes natural phosphate)

5 HYGIENE AND HANDLING

5. 1 The final product shall be free from any foreign material that poses a threat to human health
5. 2 When tested by appropriate methods of sampling and examination as prescribed by the Codex Alimentarius Commission, the product:

(i) shall be free from micro – organisms capable of development under normal conditions of storage;

(ii) no sample unit shall contain histamine that exceeds 20 mg per 100 g;

(iii) shall not contain any other substance including substances derived from microorganisms in amounts which may represent a hazard to health in accordance with standards established by the Codex Alimentarius Commission;

(iv) shall be free from container integrity defects which may compromise the hermetic seal.

5. 3 It is recommended that the product covered by the provisions of this standard be prepared and handled in accordance with the appropriate sections of the Recommended International Code of Practice – General Principles of Food Hygiene (CAC/RCP 1 – 1969, Rev. 3 – 1997) and the following relevant Codes:

(i) the Recommended International Code of Practice for Canned Fish (CAC/RCP 10 – 1976);

(ii) the Recommended International Code of Hygienic Practice for Low – Acid and Acidified Low – Acid Canned Foods (CAC/RCP 23 – 1979);

(iii) The sections on the Products of Aquaculture in the Proposed Draft International Code of Practice for Fish and Fishery Products (under elaboration)❶ .

6 LABELLING

In addition to the provisions of the Codex General Standard for the Labelling of Prepackaged Foods (CODEX STAN 1 – 1985, Rev. 3 – 1999) the following specific provisions apply:

6. 1 THE NAME OF THE FOOD

6. 1. 1 The name of the product as declared on the label shall be "tuna" or "bonito", and may be preceded or followed by the common or usual name of the species, both in accordance with the law and custom of the country in which the product is sold, and in a manner not to mislead the consumer.

6. 1. 2 The name of the product may be qualified or accompanied by a term descriptive of the colour of the product, provided that the term "white" shall be used only for Thunnus alalunga and the terms "light" "dark" and "blend" shall be used only in accordance with any rules of the country in which the product is sold.

6. 1. 3 Form of Presentation

The form of presentation provided for in Section 2. 3 shall be declared in close proximity to the common name.

6. 1. 4 The name of the packing medium shall form part of the name of the food

---

❶ The Proposed Draft Code of Practice, when finalized, will replace all current Codes of Practice for Fish and Fishery Products.

7 SAMPLING, EXAMINATION AND ANALYSES

7.1 SAMPLING

(i) Sampling of lots for examination of the final product as prescribed in Section 3.3 shall be in accordance with the FAO/WHO Codex Alimentarius Sampling Plans for Prepackaged Foods (1969) (AQL - 6.5) (Ref. CAC/RM 42 - 1977);

(ii) Sampling of lots for examination of net weight and drained weight where appropriate shall be carried out in accordance with an appropriate sampling plan established by the CAC.

7.2 SENSORY AND PHYSICAL EXAMINATION

Samples taken for sensory and physical examination shall be assessed by persons trained in such examination and in accordance with the procedures set out in Sections 7.3 through 7.5, Annex A and the *Guidelines for the Sensory Evaluation of Fish and Shellfish in Laboratories* (*CAC/GL* 31 - 1999).

7.3 DETERMINATION OF NET WEIGHT

Net contents of all sample units shall be determined by the following procedure:

(i) Weigh the unopened container;

(ii) Open the container and remove the contents;

(iii) Weigh the empty container, (including the end) after removing excess liquid and adhering meat;

(iv) Subtract the weight of the empty container from the weight of the unopened container. The resultant figure will be the net content.

7.4 DETERMINATION OF DRAINED WEIGHT

The drained weight of all sample units shall be determined by the following procedure:

(i) Maintain the container at a temperature between 20℃ and 30℃ for a minimum of 12 hours prior to examination;

(ii) Open and tilt the container to distribute the contents on a pre - weighed circular sieve which consists of wire mesh with square openings of 2.8 mm × 2.8 mm;

(iii) Incline the sieve at an angle of approximately 17° ~ 20° and allow the fish to drain for two minutes, measured from the time the product is poured into the sieve;

(iv) Weigh the sieve containing the drained fish;

(v) The weight of drained fish is obtained by subtracting the weight of the sieve from the weight of the sieve and drained product.

7.5 DETERMINATION OF WASHED DRAINED WEIGHT (FOR PACKS WITH SAUCES)

(i) Maintain the container at a temperature between 20℃ and 30℃ for a minimum of 12 hours prior to examination;

(ii) Open and tilt the container and wash the covering sauce and then the full contents with hot tap water (approx. 40℃), using a wash bottle (e. g. plastic) on the tared circular sieve;

(iii) Wash the contents of the sieve with hot water until free of adhering sauce; where necessary separate optional ingredients (spices, vegetables, fruits) with pincers. Incline the sieve at an angle of approximately 17° ~ 20° and allow the fish to drain two minutes, measured from the time the washing procedure has finished;

(iv) Remove adhering water from the bottom of the sieve by use of paper towel. Weigh the sieve containing the washed drained fish;

(v) The washed drained weight is obtained by subtracting the weight of the sieve from the weight of the sieve and drained product.

7.6 DETERMINATION OF PRESENTATION

The presentation of all sample units shall be determined by the following procedure.

(i) Open the can and drain the contents, following the procedures outlined in 7.4;

(ii) Remove and place the contents onto a tared 1.2 cm mesh screen equipped with a collecting pan;

(iii) Separate the fish with a spatula being careful not to break the configuration of the pieces. Ensure that the smaller pieces of fish are moved to the top of a mesh opening to allow them to fall through the screen onto the collecting pan;

(iv) Segregate the material on the pan according to flaked, grated (shredded) or paste and weigh the individual portions to establish the weight of each component;

(v) If declared as a "chunk" pack weigh the screen with the fish retained and record the weight. Subtract the weight of the sieve from this weight to establish the weight of solid and chunk fish;

(vi) If declared as "solid" pack remove any small pieces (chunks) from the screen and reweigh. Subtract the weight of the sieve from this weight to establish the weight of "solid" fish.

Calculations

(i) Express the weight of flaked, grated (shredded and paste) as a percentage of the total drained weight of fish.

$$\%\ \text{flakes} = \frac{\text{Weight of flakes}}{\text{Total weight of drained fish}} \times 100$$

(ii) Calculate the weight of solid and chunk fish retained on the screen by difference and express as a % of the total drained weight of fish.

$$\%\ \text{solid \& chunk fish} = \frac{\text{Weight of solid \& chunk fish}}{\text{Total weight of drained fish}} \times 100$$

(iii) Calculate the weight of solid fish retained on the screen by difference and express as a% of the total drained weight of the fish.

$$\%\ \text{of solid fish} = \frac{\text{Weight of solid fish}}{\text{Total weight of drained fish}} \times 100$$

7.7 DETERMINATION OF HISTAMINE

According to AOAC 977.13 (15th Edition, 1990).

8 DEFINITION OF DEFECTIVES

A sample unit shall be considered defective when it exhibits any of the properties defined below.

8.1 FOREIGN MATTER

The presence in the sample unit of any matter, which has not been derived from fish, does not pose a threat to human health, and is readily recognized without magnification or is present at a level determined by any method including magnification that indicates non - compliance with good manufacturing practices and sanitation practices.

8.2 ODOUR/FLAVOUR

A sample unit affected by persistent and distinct objectionable odours or flavours indicative of decomposition or rancidity.

8.3 TEXTURE

(i) Excessively mushy flesh uncharacteristic of the species in the presentation; or
(ii) Excessively tough flesh uncharacteristic of the species in the presentation; or
(iii) Honey - combed flesh in excess of 5% of the drained contents.

8.4 DISCOLOURATION

A sample unit affected by distinct discolouration indicative of decomposition or rancidity or by sulphide staining of the meat exceeding 5% of the drained contents.

8.5 OBJECTIONABLE MATTER

A sample unit affected by struvite crystals greater than 5 mm in length.

9 LOT ACCEPTANCE

A lot shall be considered as meeting the requirements of this standard when:

(i) the total number of defectives as classified according to Section 8 does not exceed the acceptance number (c) of the appropriate sampling plan in the Sampling Plans for Prepackaged Foods (AQL - 6.5) (CAC/RM 42 - 1977);

(ii) the total number of sample units not meeting the presentation and colour designation as defined in Section 2.3 does not exceed the acceptance number (c) of the appropriate sampling plan in the Sampling Plans for Prepackaged Foods (AQL - 6.5) (CAC/RM 42 - 1977);

(iii) the average net weight or the average weight of drained meat of all sample units examined is not less than the declared weight, and provided there is no unreasonable shortage in any individual container;

(iv) the Food Additives, Hygiene and Labelling requirements of Sections 4, 5.1, 5.2 and 6 are met.

ANNEX A SENSORY AND PHYSICAL EXAMINATION

1. Complete examination of the can exterior for the presence of container integrity defects or can ends which may be distorted outwards.

2. Open can and complete weight determination according to defined procedures in Sections 7.3 and 7.4.

3. Examine the product for discolouration.

4. Carefully remove the product and determine the presentation according to the defined procedures in Section 7. 5.

5. Examine product for discolouration, foreign matter and struvite crystals. The presence of a hard bone is an indicator of under processing and will require an evaluation for sterility.

6. Assess odour, flavour and texture in accordance with the *Guidelines for the Sensory Evaluation of Fish and Shellfish in Laboratories* (*CAC/GL* 31 – 1999).

蟹肉罐头

CODEX STANDARD FOR CANNED CRAB MEAT

(*CODEX STAN 90-1981, REV. 1-1995*)

1 范围

本标准适用于蟹肉罐头。不适用于蟹肉净重少于罐头内容物总重50%（W/W）的罐头。

2 说明

2.1 产品定义

蟹肉罐头是由十足目（*Decapoda*）、短尾亚目（*Brachyura*）中可食用的种和石蟹科（*lithodidae*）的所有种的蟹，去除外壳后，取出腿、螯、身体和肩部的肉来单独或混合制备而成。

2.2 加工过程的定义

将蟹肉装罐并密封，然后通过一系列加工处理以确保其达到商业无菌状态。

2.3 产品介绍

产品介绍应当满足以下要求：

（1）达到本标准的所有要求；

（2）在产品标签中应对产品进行详细描述，以免引起混淆或误导消费者。

3 基本成分及质量因素

3.1 蟹肉

产品必须由2.1条所列出的品种的蟹制备，且蟹应是品质良好，可作为鲜品供人类消费的活蟹，并且其在加工前仍保持鲜活。

3.2 其他成分

所使用的其他成分应具有食品级的质量，并且符合所有相应法规标准的规定。

3.3 成品

产品应符合本标准要求，根据第9章进行成品批次检查时，其质量应符合第8章的规定。检查方法应符合第7章的规定。

4 食品添加剂

只允许使用下列食品添加剂。

	添加剂	成品中的最高含量
	330 柠檬酸	GMP
调酸剂	338 正磷酸 450（i）二磷氢二钠	≤10g/kg，以 P_2O_5 计， 单用或混用（包括天然磷酸盐）
螯合剂	385 乙二胺四乙酸二钠钙（EDTA 二钠钙）	250mg/kg
增味剂	621 谷氨酸钠	GMP

5 卫生及处理

5.1 成品中不能含有任何危害人体健康的外来杂质

5.2 当应用 CAC 规定的抽样及检测方法进行检验时，产品应符合以下标准

（1）不含有任何在正常的贮存条件下可能生长的微生物；

（2）其他危害人体健康的有害物质的含量也不能超过 CAC 标准的规定；

（3）不存在可能危及罐的密封性的缺陷。

5.3 建议本标准中涉及产品预处理的条款应符合推荐性国际规范《CAC/RCP1 - 1969，Rev. 3 - 1997 食品卫生操作总则》中的相关章节和以下相关标准

（1）推荐性国际操作规范《CAC/RCP 10 - 1976 鱼罐头》；

（2）推荐性国际操作规范《CAC/ RCP 23 - 1979，REV. 2 - 1993 低酸和酸化低酸罐头食品》；

（3）推荐性国际操作规范《CAC/RCP 28 - 1983 蟹类》；

（4）《水产品国际操作规程草案》中水产养殖品的相关章节（正在完善中）[1]。

6 标签

标签除了应符合《CODEX STAN1 - 1985，Rev. 3 - 1999 预包装食品标签通用标准》的要求外，还应遵守以下规定：

6.1 食品名称

6.1.1 标签上标示的食品名称应注明“蟹”或“蟹肉”

6.1.2 另外，标签中应包括其他可以避免混淆或误导消费者的描述性术语

[1] 操作规程建议草案，在定稿后将替代所有现行的《水产品操作规程》。

7 抽样、检测和分析

7.1 取样

（1）产品批次检验用样品的抽样方法应符合 FAO/WHO 的食品法典《CODEX STAN 233－1969 预包装食品的抽样方案》（AQL－6.5）；

（2）对需检测净重和干重的批次的抽样，抽样计划应以食品法典委员会的相关标准为依据。

7.2 感官和物理检验

产品的感官与物理指标须由经过此类检验培训的人员进行检验，并依据附录 A 以及《CAC/GL 31－1999 实验室中水产品感官评价指南》所述程序进行。

7.3 净重的测定

样品内容物的净重应通过以下程序测定：

（1）称量未开封的罐头；

（2）开罐并除去内容物；

（3）称量空罐重量（包括开启下来的罐底）；

（4）从未开封罐头的重量中减去空罐的重量，得到的就是内容物的净重。

7.4 干重的测定

样品内容物沥干后重量（干重）应通过以下程序测定：

（1）测定前使罐头的温度至少在过去 12h 内维持在 20～30℃；

（2）开罐并使之倾斜，将内容物倾倒在预先称重的圆形滤网上，该滤网由金属网丝构成，网孔大小 2.8mm×2.8mm；

（3）以 17°～20°的角度倾斜滤网以使蟹肉沥水，从将产品全部倾倒于滤网上开始计时，沥水 2min；

（4）称量沥水后的蟹肉及滤网的总重；

（5）从沥水后的产品和滤网的总重中减去滤网的重量就可以获得蟹肉的干重。

8 缺陷的定义

当样品单位呈现下列任何一项特征时，则认定其有“缺陷”。

8.1 外来杂质

样品单位中存在的任何不是来自于蟹肉的物质。这些物质虽不会对人体健康造成危害，但用肉眼可直接辨别，或采用某些方法（包括放大）可以确定其存在。存在外来杂质则表明不符合良好操作和卫生惯例。

8.2 气味或口味

样品散发的持久、明显、令人厌恶的由腐败、酸败引起的气味或风味。

8.3 质地

（1）样品中含有过多非该产品特征的糊状蟹肉；

（2）样品中含有过多非该产品特征的硬（韧）蟹肉。

8.4 变色

样品单位中超过干重5%的蟹肉发生明显变色（变黑、变蓝、变棕），或超过内容物干重5%受到硫化物着色。

8.5 异物

样品中发现长度超过5mm的磷酸铵镁结晶（鸟粪石）。

9 批次验收

当满足以下条件时，可以认为此批次产品符合本标准的要求：

（1）根据本标准第8章中规定分类，缺陷总数不超过《CODEX STAN 233－1969 预包装食品的抽样方案》（AQL－6.5）中相应抽样方案规定的可接受数（c）；

（2）不符合本标准2.3条中规定数量的样品总数不超过《CODEX STAN233－1969 预包装食品的抽样方案》（AQL－6.5）中相应抽样方案规定的可接受数（c）；

（3）所有样品单位平均净重不少于标示量，在任何一个包装单位中没有不合理的质量短缺；

（4）符合本标准第4、5、6款中对食品添加剂、卫生和处理以及标签的要求。

附录A 感官和物理检验

1. 对罐外部进行检测以确定其是否存在密封性缺陷，有无罐身外部扭曲。
2. 开罐，并且根据7.3和7.4条中的程序完成重量测定。
3. 检测产品的变色、外来杂质和异物等方面的情况。
4. 依据《CAC/ GL 31－1999 实验室中水产品感官评价指南》对产品的气味、风味以及质地进行评定。

CODEX STANDARD FOR CANNED CRAB MEAT

(CODEX STAN 90 - 1981, REV. 1 - 1995)

1 SCOPE

This standard applies to canned crab meat. It does not apply to specialty products where crab meat constitutes less than 50% m/m of the contents.

2 DESCRIPTION

2.1 PRODUCT DEFINITION

Canned crab meat is prepared singly or in combination from the leg, claw, body and shoulder meat from which the shell has been removed, of any of the edible species of the sub - order Brachyura of the order Decapoda and all species of the family Lithodidae.

2.2 PROCESS DEFINITION

Canned crab meat is packed in hermetically sealed containers and shall have received a processing treatment sufficient to ensure commercial sterility.

2.3 PRESENTATION

Any presentation of the product shall be permitted provided that it:

(i) meets all requirements of this standard; and

(ii) is adequately described on the label to avoid confusing or misleading the consumer.

3 ESSENTIAL COMPOSITION AND QUALITY FACTORS

3.1 CRAB MEAT

Canned crab meat shall be prepared from sound crab of the species designated in 2.1 which are alive immediately prior to the commencement of processing and of a quality suitable for human consumption.

3.2 OTHER INGREDIENTS

The packing medium and all other ingredients used shall be of food grade quality and conform to all applicable Codex standards.

3.3 FINAL PRODUCT

Products shall meet the requirements of this Standard when lots examined in accordance with Section 9 comply with provisions set out in Section 8. Products shall be examined by the methods given in Section 7.

4 FOOD ADDITIVES

Only the use of the following additives is permitted.

	Additive	Maximum Level in the final product
Acidity Regulators	330 Citric acid	GMP
	338 Orthophosphoric acid 450 Disodium diphosphate	10 mg/kg expressed as P_2O_5, singly or in combination (includes natural phosphate)
Sequestrant	385 Calcium disodium EDTA	250 mg/kg
Flavour Enhancer	621 Monosodium glutamate	GMP

5 HYGIENE AND HANDLING

5.1 The final product shall be free from any foreign material that poses a threat to human health.

5.2 When tested by appropriate methods of sampling and examination prescribed by the Codex Alimentarius Commission (CAC), the product:

(i) shall be free from micro – organisms capable of development under normal conditions of storage; and

(ii) shall not contain any other substance including substances derived from microorganisms in amounts which may represent a hazard to health in accordance with standards established by the CAC; and

(iii) shall be free from container integrity defects which may compromise the hermetic seal.

5.3 It is recommended that the product covered by the provisions of this standard be prepared and handled in accordance with the appropriate sections of the Recommended International Code of Practice – General Principles of Food Hygiene (CAC/RCP 1 – 1969, Rev. 3 – 1997) and the following relevant Codes:

(i) the Recommended International Code of Practice for Canned Fish (CAC/RCP 10 – 1976);

(ii) the Recommended International Code of Hygienic Practice for Low – Acid and Acidified Low – Acid Canned Foods (CAC/RCP 23 – 1979);

(iv) the Recommended International Code of Practice for Crabs (CAC/RCP 28 – 1983);

(iii) The sections on the Products of Aquaculture in the Proposed Draft International Code of Practice for Fish and Fishery Products (under elaboration)[1].

6 LABELLING

In addition to provisions of the Codex General Standard for the Labelling of Prepackaged Foods (CO-

[1] The Proposed Draft Code of Practice, when finalized, will replace all current Codes of Practice for Fish and Fishery Products.

DEX STAN 1 – 1985, Rev. 3 – 1999) the following specific provisions apply:

6.1 NAME OF THE FOOD

6.1.1 The name of the product shall be "crab" or "crabmeat"

6.1.2 In addition, the label shall include other descriptive terms that will avoid misleading or confusing the consumer

7 SAMPLING, EXAMINATION AND ANALYSES

7.1 SAMPLING

(i) Sampling of lots for examination of the final product as prescribed in Section 3.3 shall be in accordance with the FAO/WHO Codex Alimentarius Sampling Plans for Prepackaged Foods (1969) (AQL – 6.5) (Ref. CAC/RM 42 – 1969);

(ii) Sampling of lots for examination of net weight and drained weight shall be carried out in accordance with an appropriate sampling plan meeting the criteria established by the CAC.

7.2 SENSORIC AND PHYSICAL EXAMINATION

Samples taken for sensoric and physical examination shall be assessed by persons trained in such examination and in accordance with Annex A and the *Guidelines for the Sensory Evaluation of Fish and Shellfish in Laboratories* (*CAC/GL* 31 – 1999).

7.3 DETERMINATION OF NET WEIGHT

Net weight of all sample units shall be determined by the following procedures:

(i) Weigh the unopened container;

(ii) Open the container and remove the contents;

(iii) Weigh the empty container, including the end and any wrapping material, after removing excess liquid and adhering meat;

(iv) Subtract the weight of the empty container and any wrapping material from the weight of the unopened container. The resultant figure is the net content.

7.4 DETERMINATION OF DRAINED WEIGHT

The drained weight of all sample units shall be determined by the following procedures:

(i) Maintain the container at a temperature of between 20℃ and 30℃ for a minimum of 12 hours prior to examination;

(ii) Open the container and distribute the contents on a pre – weighed circular sieve having a wire mesh with square openings of 2.8 mm × 2.8 mm;

(iii) Remove all wrapping material and incline the sieve at an angle of approximately 17° ~ 20° and allow the meat to drain two minutes, measured from the time the product is poured onto the sieve;

(iv) Weigh the sieve containing the drained crab meat;

(v) Determine the weight of drained crab meat by subtracting the mass of the sieve from the mass of the sieve with drained product.

8 DEFINITION OF DEFECTIVES

A sample unit will be considered defective when it exhibits any of the properties defined below.

8. 1 FOREIGN MATTER

The presence in the sample unit of any matter, which has not been derived from crab meat, does not pose a threat to human health, and is readily recognized without magnification or is present at a level determined by any method including magnification that indicates non – compliance with good manufacturing and sanitation practices.

8. 2 ODOUR/FLAVOUR

A sample unit affected by persistent and distinct objectionable odours or flavours indicative of decomposition or rancidity.

8. 3 TEXTURE

(i) Excessively mushy flesh uncharacteristic of the species in the presentation; or
(ii) Excessively tough flesh uncharacteristic of the species in the presentation.

8. 4 DISCOLOURATION

A sample unit affected by distinct discolourations indicative of decomposition or rancidity or by blue, brown, black discolourations exceeding 5% by weight of the drained contents, or black sulphide staining of the meat exceeding 5% by weight of the drained contents.

8. 5 OBJECTIONABLE MATTER

A sample unit affected by struvite crystals – any struvite crystal greater than 5 mm in length.

9 LOT ACCEPTANCE

A lot shall be considered as meeting the requirements of this standard when:

(i) the total number of defectives as classified according to Section 8 does not exceed the acceptance number (c) of the appropriate sampling plan in the Sampling Plans for Prepackaged Foods (AQL – 6. 5) (CAC/RM 42 – 1977);

(ii) the total number of sample units not meeting the form of presentation defined in Section 2. 3 does not exceed the acceptance number (c) of the appropriate sampling plan in the Sampling Plans for Prepackaged Foods (AQL – 6. 5) (CAC/RM 42 – 1977);

(iii) the average net weight and the average drained weight where appropriate of all sample units examined is not less than the declared weight, and provided there is no unreasonable shortage in any individual container;

(iv) the Food Additives, Hygiene and Labelling requirements of Sections 4, 5. 1, 5. 2 and 6 are met.

ANNEX A SENSORY AND PHYSICAL EXAMINATION

1. Complete external can examination for the presence of container integrity defects or can ends which may be distorted outwards.

2. Open can and complete weight determination according to defined procedures in Sections 7.3 and 7.4.

3. Examine product for discolouration, foreign and objectionable matter.

4. Assess odour, flavour and texture in accordance with the Guidelines for the Sensory Evaluation of Fish and Shellfish in Laboratories (CAC/GL 31 - 1999).

速冻虾或对虾

CODEX STANDARD FOR QUICK FROZEN SHRIMPS OR PRAWNS

(*CODEX STAN 92－1981, REV. 1－1995*)

1 适用范围

本标准适用于生的、完全或不完全煮熟的，去壳或不去壳的速冻虾和对虾❶。

2 说明

2.1 产品定义

2.1.1 速冻虾来自以下科、种

（1）对虾科（*Penaeidae*）；

（2）长额虾科（*Pandalidae*）；

（3）褐虾科（*Crangonidae*）；

（4）长臂虾科（*Palaemonidae*）。

2.1.2 不同属的虾不应混合包装；但同属不同种、外观相近的虾可混合包装

2.2 加工定义

加工用水（蒸煮或冷却用水）应达到饮用水或清洁海水的质量要求。

产品经过适当预处理后，应在符合以下规定的条件进行冻结加工：冻结应在合适的设备中进行，并使产品迅速通过最大冰晶生成温度带。速冻加工只有在产品的中心温度达到并稳定在－18℃（0 ℉）或更低的温度时才算完成。产品在运输、贮存、分销过程中应保持在深度冻结状态，以保证产品质量。

在产品的加工和包装过程中应尽量减少脱水和氧化作用的影响。

2.3 产品介绍

2.3.1 产品介绍应符合下列要求

（1）达到本标准的所有要求；

（2）在产品标签中对产品进行详细描述，以免引起混淆或误导消费者。

2.3.2 虾可按重量单位或包装单位进行包装

---

❶ 以下以虾为例，进行说明。

3 基本成分及质量因素

3.1 虾

速冻虾应由品质良好、可作为鲜品供人类消费的虾制备。

3.2 冰衣

用于镀冰衣或制备镀冰衣所用的水应达到饮用水或清洁海水的质量要求。饮用水应符合 WHO 最新版本的《国际饮用水质量规范》要求，清洁海水应是达到饮用水的微生物标准且不含异物的海水。

3.3 其他成分

所使用的其他成分应具有食品级的质量，并且符合所有相应法规标准的规定。

3.4 成品

产品应符合本标准要求，根据第 9 章进行成品批次检查时，其质量应符合第 8 章的规定。检查方法应符合第 7 章的规定。

4 食品添加剂

只允许使用下列食品添加剂。

添加剂		成品中最高含量
调酸剂	330 柠檬酸	GMP
	450（iii）焦磷酸钠 450（v）焦磷酸钾 451（i）三聚磷酸钠 451（ii）三聚磷酸钾	≤10g/kg，以 P_2O_5 计 单用或混用 （包括天然磷酸盐）
抗氧化剂	300 L－抗坏血酸	GMP
色素	124 胭脂红	30mg/kg，在热处理过的食品中
保鲜剂	221 亚硫酸钠 223 偏亚硫酸氢钠（焦亚硫酸钠） 224 偏亚硫酸氢钾（焦亚硫酸钾） 225 亚硫酸钾	100mg/kg（在生品可食部分）， 30mg/kg（熟产品的可食部分）， 以 SO_2 计，单用或混用

5 卫生及处理

5.1 成品中不能含有任何危害人体健康的外来杂质

5.2 应用 CAC 规定的抽样及检测方法进行检验时，产品应符合以下条件

（1）产品中微生物含量和由微生物产生的危害人体健康的有害物质的含量不能超过 CAC 标准的规定；

（2）其他危害人体健康的有害物质的含量也不能超过CAC标准的规定。

5.3 建议本标准中涉及产品预处理的条款应符合推荐性国际规范《CAC/RCP1－1969，Rev.3－1997 食品卫生操作总则》中的相关章节和以下相关标准

（1）推荐性国际操作规范《CAC/RCP16－1978 冻鱼》；

（2）推荐性国际操作规范《CAC/RCP16－1978 冻虾或对虾》（CAC/RCP17－1978，增补1989年11月）；

（3）推荐性国际操作规范《CAC/RCP8－1976 速冻食品的加工和处理》；

（4）《水产品国际操作规程草案》中水产养殖品的相关章节（正在完善中）[1]。

6 标签

标签除了应符合《CODEX STAN1－1985，Rev.3－1999 预包装食品标签通用标准》的要求外，还应遵守以下规定：

6.1 食品名称

在标签上产品的名称应按出售该食品的国家的法律、习惯、实际情况的名称注明“虾”或“对虾”。

6.1.1 标签上还应在与产品名称紧靠位置出现关于性状形式的参考文字，这些术语应恰当充分地反应产品性状，以免引起混淆或误导消费者

6.1.2 除上述规定的标签名称外，也可加上在出售该产品的国家不会误导消费者的常用名或普遍使用的商品名

6.1.3 标签上应恰当注明产品是“熟的”、“半熟的”或“生的”

6.1.4 如果产品用海水镀冰衣，则应予以说明

6.1.5 在产品标签上应标明“速冻”，除非在有的国家已“冻”即代表“速冻”

6.1.6 标签应注明产品须在运输、贮藏、分销过程中保持的条件，以保证其质量

6.2 净重（镀冰衣产品）

对于镀冰衣的产品，其内容物净含量不包括冰衣重。

6.3 贮藏说明

标签应注明产品须在－18℃或更低的温度下贮藏。

6.4 非零售包装的标签

上述要求应既在包装上又在附文中出现，除了食品名称、批号、制造或分装厂名、地址外，还包括贮藏条件。但批号、制造或分装厂名、地址也可用同一证明标志代替，只要证明标志能在辅助文件中说明清楚。

---

[1] 操作规程建议草案，在定稿后将替代所有现行的《水产品操作规程》。

7 抽样、检验和分析

7.1 取样

（1）产品批次检验用样品的抽样方法应符合 FAO/WHO 的食品法典《CODEX STAN 233 – 1969 预包装食品的抽样方案》（AQL – 6.5）。样品单位是初级包装，对单冻（IQF）产品至少以 1kg 样品为单位。

（2）对需检测净重批次的抽样，抽样计划应以食品法典委员会的相关标准为依据。

7.2 感官与物理检验

产品的感官与物理指标须由经过此类检验培训的人员进行检验，并依据本标准 7.3 至 7.6 条和附录 A 以及《CAC/GL 31 – 1999 实验室中水产品感官评价指南》所述程序进行。

7.3 净重的测定

7.3.1 未镀冰衣产品净重的测定

代表每批次的样品单位的净重（不包括包装材料）应在冷冻状态下测定。

7.3.2 镀冰衣样品净重的测定

（1）从低温贮藏状态下取出速冻虾后立即打开包装。

（a）如果是生虾，把它放入容器中，由容器的底部以 25L/min 的流量通入室温的淡水；

（b）如果是熟虾，把它放入一装有产品标示重量 8 倍的 27℃（80 ℉）新鲜饮用水的容器中。使产品浸在水中，直到冰全部融化。如果是块冻产品，在解冻过程中要多次翻转。可用探针轻探以确定解冻是否彻底完成。

（2）取干燥、清洁的金属滤网一个，选用方孔孔径为 2.8mm（ISO 推荐 R565 号）或 2.38mm（美国 8 号标准滤网）中一种进行称重。

（a）如果包装中的总内容物的重量等于或小于 500g（1.1 磅），则应使用直径为 20cm（8 英寸）的滤网；

（b）如果包装中的总内容物的重量大于 500g（1.1 磅），则应使用直径为 30cm（12 英寸）的滤网。

（3）当可以看到或用手感觉到全部冰衣已被去除，虾与虾之间很容易分开时，清空容器，将内容物置于预先称重的滤网上。使滤网倾斜成 20°角，沥干 2min。

（4）将沥干的产品连滤网一起称重，减去滤网的重量，剩下的数字就是包装中内容物的净重。

7.4 计数测定

当标签上标明时，应通过计数一个初级包装或一个有代表性的样品中所有的虾以测定虾的数量，而后用虾的数量除以其去冰衣的实际重量以得到每单位重量中的虾的数目。

7.5 解冻

解冻时将样品装入薄膜袋中，浸入室温（温度不高于 35℃）水中，不时用手轻捏袋子，至袋中无硬块和冰晶时为止，但不要损坏虾的质地。

7.6 蒸煮方法

蒸煮使产品内部温度达到 65 ~ 70℃。不能过度蒸煮，蒸煮时间随产品大小和采用的温度而不同。准确的蒸煮时间和条件应依据预先实验来确定。

烘焙：用铝箔包裹产品，并将其均匀放入扁平锅或浅平锅上。

蒸：用铝箔包裹产品，并将其置于带盖容器中沸水之上的金属架上。

袋煮：将产品放入可煮薄膜袋中加以密封，浸入沸水中煮。

微波：将产品放入适于微波加热的容器中，若用塑料袋，应检查确定塑料袋不会发出任何气味。根据设备说明加热。

8 缺陷的定义

当样品呈现下列任何一项特征时，则认定其有“缺陷”。

8.1 深度脱水

样品单位中表面积超过10%出现水分的过度损失，明显表现为鱼体表面呈现异常的白色或黄色，覆盖了肌肉本身的颜色，并已渗透至表层以下，在不过分影响产品外观的情况下，不能轻易地用刀或其他利器刮掉。

8.2 外来杂质

样品单位中存在的任何不是来自于鱼体的物质（包装材料除外）。这些物质虽不会对人体健康造成危害，但用肉眼可直接辨别，或采用某些方法（包括放大）可以确定其存在。出现外来杂质表明不符合良好操作和卫生惯例。

8.3 气味或风味

样品散发的持久、明显、令人厌恶的由腐败、酸败或饵料引起的气味或风味。

样品单位受持久、明显、令人厌恶的象征腐败、酸败的或由饵料所致的气味或风味的影响。

8.4 变色

整个样品中25%以上的虾的单体出现超过其表面积10%的部分明显变黑、变绿或变黄（无论是单色变化还是混色变化）。

9 批次验收

当满足以下条件时，可以认为此批次产品符合最后的质量要求：

（1）根据本标准第8章中规定分类，缺陷总数不超过《CODEX STAN 233－1969 预包装食品的抽样方案》（AQL－6.5）中相应抽样方案规定的可接受数（c）；

（2）不符合本标准2.3中规定数量的样品总数不超过《CODEX STAN233－1969 预包装食品的抽样方案》（AQL－6.5）中相应抽样方案规定的可接受数（c）；

（3）所有样品单位平均净重不少于标示量，在任何一个包装单位中没有不合理的重量短缺；

（4）符合本标准第4、5、6款中对食品添加剂、卫生和处理以及标签的要求。

附录 A　感官及物理检验

1. 净含量的测定按本标准 7.3 条的规定执行（按要求去冰衣）。

2. 检查样品单位中冻虾或冻虾块表面的脱水的情况。计算受影响的冻虾或冻虾块面积的百分比。

3. 按本标准 7.5 条的规定解冻，并分别检查样品单位中每一只虾的外来杂质和其他缺陷，测定受缺陷影响的虾的重量。

4. 按本标准 7.4 条的规定检查产品中虾的标示数量。

5. 根据要求评估气味及变色现象。

6. 对在解冻状态下无法最终判定其气味的样品，则应从样品单位中截取一小块可疑部分（100～200g），并立即使用 7.6 条限定的某一种蒸煮方法，确定其气味和风味。

CODEX STANDARD FOR QUICK FROZEN SHRIMPS OR PRAWNS

(CODEX STAN 92 - 1981, REV. 1 - 1995)

1 SCOPE

This standard applies to quick frozen raw or partially or fully cooked shrimps or prawns, peeled or unpeeled. ❶

2 DESCRIPTION

2.1 PRODUCT DEFINITION

2.1.1 Quick frozen shrimp is the product obtained from species of the following families

(*a*) *Penaeidae*;

(*b*) *Pandalidae*;

(*c*) *Crangonidae*;

(*d*) *Palaemonidae.*

2.1.2 The pack shall not contain a mixture of genera but may contain a mixture of species of the same genus which have similar sensory properties.

2.2 PROCESS DEFINITION

The water used for cooking and cooling shall be of potable quality or clean seawater.

The product, after any suitable preparation, shall be subjected to a freezing process and shall comply with the conditions laid down hereafter. The freezing process shall be carried out in appropriate equipment in such a way that the range of temperature of maximum crystallization is passed quickly. The quick freezing process shall not be regarded as complete unless and until the product temperature has reached - 18℃ (0 ℉) or colder at the thermal centre after thermal stabilization. The product shall be kept deep frozen so as to maintain the quality during transportation, storage and distribution.

Quick frozen shrimps shall be processed and packaged so as to minimize dehydration and oxidation.

2.3 PRESENTATION

2.3.1 Any presentation of the product shall be permitted provided that it

2.3.1.1 meets all requirements of this standard; and

---

❶ Hereafter referred to as shrimp.

2.3.1.2 is adequately described on the label to avoid confusing or misleading the consumer.

2.3.2 The shrimp may be packed by count per unit of weight or per package

3 ESSENTIAL COMPOSITION AND QUALITY FACTORS

3.1 SHRIMP

Quick frozen shrimp shall be prepared from sound shrimp which are of a quality fit to be sold fresh for human consumption.

3.2 GLAZING

If glazed, the water used for glazing or preparing glazing solutions shall be of potable quality or shall be clean sea – water. Potable water is fresh – water fit for human consumption. Standards of potability shall not be less than those contained in the latest edition of the WHO "International Guidelines for Drinking Water Quality". Clean sea – water is sea – water which meets the same microbiological standards as potable water and is free from objectionable substances.

3.3 OTHER INGREDIENTS

All other ingredients used shall be of food grade quality and conform to all applicable Codex standards.

3.4 FINAL PRODUCT

Products shall meet the requirements of this standard when lots examined in accordance with Section 9 comply with the provisions set out in Section 8. Products shall be examined by the methods given in Section 7.

4 FOOD ADDITIVES

Only the use of the following additives is permitted.

Additive		Maximum Level in the final product
Acidity Regulators	330 Citric acid	GMP
	450 (iii) Tetrasodium diphosphate 450 (v) Tetrapotassium diphosphate 451 (i) Pentasodium triphosphate 451 (ii) Pentapotassium triphosphate	10 g/kg expressed as P_2O_5, singly or in combination (includes natural phosphate)
Antioxidant	300 Ascorbic acid (L –)	GMP
Colours	124 Ponceau 4R	30 mg/kg in heat – treated products only
Preservatives	221 Sodium sulphite 223 Sodium metabisulphite 224 Potassium metabisulphite 225 Potassium sulphite	100 mg/kg in the edible part of the raw product, or 30 mg/kg in the edible part of the cooked product, singly or in combination, expressed as SO_2

5 HYGIENE AND HANDLING

5.1 The final product shall be free from any foreign material that poses a threat to human health.
5.2 When tested by appropriate methods of sampling and examination prescribed by the Codex Alimentarius Commission , the product:

(i) shall be free from microorganisms or substances originating from microorganisms in amounts which may present a hazard to health in accordance with standards established by the Codex Alimentarius Commission;

(ii) shall not contain any other substance in amounts which may present a hazard to health in accordance with standards established by the Codex Alimentarius Commission.

5.3 It is recommended that the products covered by the provisions of this standard be prepared and handled in accordance with the appropriate sections of the Recommended International Code of Practice – General Principles of Food Hygiene (CAC/RCP 1 – 1969) and the following relevant Codes:

(i) the Recommended International Code of Practice for Frozen Fish (CAC/RCP 16 – 1978);

(ii) the Recommended International Code of Practice for Frozen Shrimps or Prawns (CAC/RCP 17 – 1978 and Supplement November 1989);

(iii) the Recommended International Code of Practice for the Processing and Handling of Quick Frozen Foods (CAC/RCP 8 – 1976);

(iv) The sections on the Products of Aquaculture in the Proposed Draft International Code of Practice for Fish and Fishery Products (under elaboration)[1].

6 LABELLING

In addition to the provisions of the Codex General Standard for the Labelling of Prepackaged Foods (CODEX STAN 1 – 1985) the following specific provisions apply:

6.1 THE NAME OF THE FOOD

The name of the product as declared on the label shall be "shrimps" or "prawns" according to the law, custom or practice in the country in which the product is to be distributed.

6.1.1 There shall appear on the label, reference to the presentation in close proximity to the name of the product in such descriptive terms that will adequately and fully describe the nature of the presentation of the product to avoid misleading or confusing the consumer

6.1.2 In addition to the specified labelling designations above, the usual or common trade names of the variety may be added so long as it is not misleading to the consumer in the country in which the product will be distributed

6.1.3 Products shall be designated as cooked, or partially cooked, or raw as appropriate

6.1.4 If the product has been glazed with sea – water, a statement to this effect shall be made

6.1.5 The term "quick frozen", shall also appear on the label, except that the term "frozen" may be applied in countries where this term is customarily used for describing the product processed in accordance

---

[1] The Proposed Draft Code of Practice, when finalized, will replace all current Codes of Practice for Fish and Fishery Products.

with subsection 2. 2 of this standard

6. 1. 6 The label shall state that the product should be maintained under conditions that will maintain the quality during transportation, storage and distribution

6. 2 NET CONTENTS (GLAZED PRODUCTS)

Where the food has been glazed the declaration of net contents of the food shall be exclusive of the glaze.

6. 3 STORAGE INSTRUCTIONS

The label shall include terms to indicate that the product shall be stored at a temperature of -18℃ or colder.

6. 4 LABELLING OF NON - RETAIL CONTAINERS

Information specified above shall be given either on the container or in accompanying documents, except that the name of the food, lot identification, and the name and address as well as storage instructions shall always appear on the container.

However, lot identification, and the name and address may be replaced by an identification mark, provided that such a mark is clearly identifiable with the accompanying documents.

7 SAMPLING, EXAMINATION AND ANALYSES

7. 1 SAMPLING

(i) Sampling of lots for examination of the product shall be in accordance with the FAO/WHO Codex Alimentarius Sampling Plans for Prepackaged Foods (AQL -6. 5) (CODEX STAN 233 - 1969) . The sample unit is the primary container or for individually quick frozen products is at least a 1 kg portion of the sample unit.

(ii) Sampling of lots for examination of net weight shall be carried out in accordance with an appropriate sampling plan meeting the criteria established by the Codex Alimentarius Commission.

7. 2 SENSORY AND PHYSICAL EXAMINATION

Samples taken for sensory and physical examination shall be assessed by persons trained in such examination and in accordance with procedures elaborated in Sections 7. 3 through 7. 6, Annex A and the Guidelines for the Sensory Evaluation of Fish and Shellfish in Laboratories (CAC/GL 31 - 1999).

7. 3 DETERMINATION OF NET WEIGHT

7. 3. 1 Determination of net weight of Products not Covered by Glaze

The net weight (exclusive of packaging material) of each sample unit representing a lot shall be determined in the frozen state.

7. 3. 2 Determination of Net Weight of Products Covered by Glaze

Procedure

(1) Open the package with quick frozen shrimps or prawns imediately after removal from low tempera-

ture storage.

(i) For the raw product, place the contents in a container into which fresh water at room temperature is introduced from the bottom at a flow of approximately 25 litres per minute;

(ii) For the cooked product place the product in a container containing an amount of fresh potable water of 27℃ (80℉) equal to 8 times the declared weight of the product. Leave the product in the water until all ice is melted. If the product is block frozen, turn block over several times during thawing. The point at which thawing is complete can be determined by gently probing the block apart.

(2) Weigh a dry clean sieve with woven wire cloth with nominal size of the square aperture 2.8 mm (ISO Recommendation R565) or alternatively 2.38 mm (US No. 8 Standard Screen).

(i) If the quantity of the total contents of the package is 500 g (1.1 lbs) or less, use a sieve with a diameter of 20 cm (8 inches);

(ii) If the quantity of the total contents of the package is more than 500 g (1.1 lbs) use a sieve with a diameter of 30 cm (12 inches).

(3) After all glaze that can be seen or felt has been removed and the shrimps or prawns separate easily, empty the contents of the container on the previously weighed sieve. Incline the sieve at an angle of about 20° and drain for two minutes.

(4) Weigh the sieve containing the drained product. Subtract the mass of the sieve; the resultant figure shall be considered to be the net content of the package.

7.4 DETERMINATION OF COUNT

When declared on the label, the count of shrimp shall be determined by counting the numbers of shrimp in the container or a representative sample thereof and dividing the count of shrimp by the actual de-glazed weight to determine the count per unit weight.

7.5 PROCEDURES FOR THAWING

The sample unit is thawed by enclosing it in a film type bag and immersing in water at room temperature (not greater than 35℃). The complete thawing of the product is determined by gently squeezing the bag occasionally so as not to damage the texture of the shrimp, until no hard core or ice crystals are left.

7.6 COOKING METHODS

The following procedures are based on heating the product to an internal temperature of 65 ~ 70℃. The product must not be overcooked. Cooking times vary according to the size of the product and the temperature used. The exact times and conditions of cooking for the product should be determined by prior experimentation.

Baking Procedure: Wrap the product in aluminum foil and place it evenly on a flat cookie sheet or shallow flat pan.

Steaming Procedure: Wrap the product in aluminum foil and place it on a wire rack suspended over boiling water in a covered container.

Boil-in-Bag Procedure: Place the product into a boilable film-type pouch and seal. Immerse the pouch into boiling water and cook.

Microwave Procedure: Enclose the product in a container suitable for microwave cooking. If plastic bags are used, check to ensure that no odour is imparted from the plastic bags. Cook according to equipment instructions.

8 DEFINITION OF DEFECTIVES

The sample unit shall be considered as defective when it exhibits any of the properties defined below.

8.1 DEEP DEHYDRATION

Greater than 10% of the weight of the shrimp in the sample unit or greater than 10% of the surface area of the block exhibits excessive loss of moisture clearly shown as white or yellow abnormality on the surface which masks the colour of the flesh and penetrates below the surface, and cannot be easily removed by scraping with a knife or other sharp instrument without unduly affecting the appearance of the shrimp.

8.2 FOREIGN MATTER

The presence in the sample unit of any matter which has not been derived from shrimp does not pose a threat to human health, and is readily recognized without magnification or is present at a level determined by any method including magnification, that indicates non – compliance with good manufacturing and sanitation practices.

8.3 ODOUR/FLAVOUR

Shrimp affected by persistent and distinct objectionable odours or flavours indicative of decomposition or rancidity or of feed.

8.4 DISCOLOURATION

Distinct blackening or green or yellow discoloration, singly or in combination of more than 10% of the surface area of individual shrimp which affects more than 25% of the sample unit.

9 LOT ACCEPTANCE

A lot shall be considered as meeting the requirements of this standard when:

(i) the total number of defectives as classified according to section 8 does not exceed the acceptance number (c) of the appropriate sampling plan in the Sampling Plans for Prepackaged Foods (AQL – 6.5) (CODEX STAN 233 – 1969);

(ii) the total number of sample units not meeting the count designation as defined in section 2.3 does not exceed the acceptance number (c) of the appropriate sampling plan in the Sampling Plans for Prepackaged Foods (AQL – 6.5) (CODEX STAN 233 – 1969);

(iii) the average net weight of all sample units is not less than the declared weight, provided there is no unreasonable shortage in any individual container;

(iv) the Food Additives, Hygiene and Labelling requirements of Sections 4, 5 and 6 are met.

ANNEX A SENSORY AND PHYSICAL EXAMINATION

1. Complete net weight determination, according to defined procedures in Section 7.3 (de - glaze as required).

2. Examine the frozen shrimp in the sample unit or the surface of the block for the presence of dehydration. Determine the percentage of shrimp or surface area affected.

3. Thaw using the procedure described in Section 7.5 and individually examine each shrimp in the sample unit for the presence of foreign matter and presentation defects. Determine the weight of shrimp affected by presentation defects.

4. Examine product for count declarations in accordance with procedures in Section 7.4.

5. Assess the shrimp for odour and discolouration as required.

6. In cases where a final decision regarding the odour/flavour cannot be made in the thawed state, a small portion of the sample unit (100g to 200 g) is prepared without delay for cooking and the odour/flavour confirmed by using one of the cooking methods defined in Section 7.6.

沙丁鱼和沙丁鱼类制品罐头

CODEX STANDARD FOR CANNED SARDINES AND SARDINE－TYPE PRODUCTS

（*CODEX STAN 94－1981，REV. 1－1995*）

1　范围

本标准适用于以水、油或其他成分作为填装介质的沙丁鱼和沙丁鱼罐头，不适用于罐中鱼肉净重少于罐头内容物净重 50%（W/W）的罐头。

2　说明

2.1　产品的定义

2.1.1　沙丁鱼和沙丁鱼罐头由下列属种的鲜鱼或冻鱼制成

沙丁鱼（*Sardina pilchardus*）：包括远东拟沙丁鱼（*S. melanostictus*），澳大利亚拟沙丁鱼（*S. neopilchardus*），南非拟沙丁鱼（*S. ocellatus*），南美拟沙丁鱼（*S. sagax*），加州拟沙丁鱼（*S. caeruleus*），金色小沙丁鱼（*Sardsinella aurita*），巴西小沙丁鱼（*S. brasiliensis*），短体小沙丁鱼（*S. maderensis*），长头小沙丁鱼（*S. longiceps*），金带小沙丁鱼（*S. gibbosa*）；

太平洋鲱（*Clupea harengus*）；

黍鲱（*Sprattus sprattus*）；

南鲱（*Hyperlophus vitalatus*）；

西澳海（*Nematalosa vlaminghi*）；

蓝背脂眼鲱（*Etrumeus teres*）；

太平洋棱背鲱（*Ethmidim maculatum*）：阿根廷（*Engraulis anchoita*），美洲（*E. mordax*），秘鲁（*E. ringens*）；大西洋后丝鲱（*Opisthonema oglinum*）。

2.1.2　彻底去头、去鳃

可去除鳞和（或）尾。可除内脏。如果除内脏，应去除掉除鱼卵、鱼白和肾脏之外的其他内脏；如果不去除内脏，特别要去除鱼内脏中未被消化或已消化的饵料。

2.2　加工过程的定义

将鱼装罐并密封，然后通过一系列加工处理以确保其达到商业无菌状态。

2.3　产品介绍

产品介绍应符合下列要求：

（1）每罐中至少有两条鱼；
（2）达到本标准的所有要求；
（3）在产品标签中对产品进行详细描述，以免引起混淆或误导消费者；
（4）不同种鱼不能混装。

3 基本成分及质量因素

3.1 原料

产品应必须由 2.1 所列出的品种的鱼制备而成，且鱼的品质良好，可作为鲜品供人类消费。

3.2 其他成分

填装介质和所使用的其他成分应具有食品级的质量，并且符合所有相应法规标准的规定。

3.3 腐败

样品单位中组胺平均含量应少于 10mg/100g。

3.4 成品

产品应符合本标准要求，根据第 9 章进行成品批次检查时，其质量应符合第 8 章的规定。检查方法应符合第 7 章的规定。

4 食品添加剂

只允许使用下列各种添加剂。

	添加剂	成品中的最高含量
增稠剂	400 褐藻酸 401 褐藻酸钠 402 褐藻酸钾 404 褐藻酸钙 406 琼脂 407 卡拉胶及其 Na、K、NH_4 盐（包括红藻胶） 407 经热加工处理的麒麟菜（PES） 410 角豆胶（刺槐豆胶） 412 瓜尔豆胶 413 黄蓍胶 415 黄原胶 440 果胶 466 羧甲基纤维素钠	GMP

续表

添加剂		成品中的最高含量
改性淀粉	1401 酸处理淀粉 1402 碱处理淀粉 1404 氧化淀粉 1410 磷酸单淀粉 1412 磷酸二淀粉酯化三偏磷酸钠；磷酸二淀粉酯化氯氧化磷 1414 磷酸乙酰化二淀粉 1413 磷酸化磷酸二淀粉 1420 醋酸淀粉酯化乙酸酐 1421 醋酸淀粉酯化乙烯基乙酸酯 1422 乙酰化二淀粉己二酸酯 1440 羧丙基淀粉 1442 磷酸羧丙基淀粉	GMP
调酸剂	260 乙酸 270（L－、D－和 DL－）乳酸 330 柠檬酸	GMP
天然香料	香油（香料油） 香精（香料萃取物） 烟香味（自然烟溶液和萃取物）	GMP

5 卫生及处理

5.1 成品中不能含有任何危害人体健康的外来杂质

5.2 应用 CAC 规定的抽样及检测方法进行检验时，产品应符合以下标准

（1）不含有任何在正常的贮存条件下可能生长的微生物；
（2）样品单位中组胺的含量不超过 20mg/100g；
（3）其他危害人体健康的有害物质的含量也不能超过 CAC 标准的规定；
（4）不存在可能危及罐的密封性的缺陷。

5.3 建议本标准中涉及产品预处理的条款应符合推荐性国际规范《CAC/RCP1－1969，Rev. 3－1997 食品卫生操作总则》中的相关章节和以下相关标准

（1）推荐性国际操作规范《CAC/RCP 10－1976 鱼罐头》；
（2）推荐性国际操作规范《CAC/ RCP 23－1979 低酸和酸化低酸罐头食品》。

6 标签

除了应符合《CODEX STAN1－1985，Rev. 3－1998 预包装食品标签通用标准》的要求外，还应遵守以下规定：

6.1 食品名称

产品名称应遵循以下规定：

（1）“沙丁鱼”仅指沙丁鱼（Sardina pilchardus）；

（2）“X 沙丁鱼”，“X”指国家、地区、种名或是符合出售该产品的国家的法律或习惯的该种鱼的通用名，并且不会误导消费者。

6.1.1 食品名称中应包含填装介质的名称

6.1.2 如果鱼曾被熏制或添加熏烟风味，则应在标签上最接近名字的地方注明

6.1.3 另外，标签中应包括其他可以避免混淆或误导消费者的描述性术语

7 抽样、检测和分析

7.1 取样

（1）如3.4中指定的一样，产品批次检验用样品的抽样方法应符合 FAO/WHO 的食品法典《CODEX STAN 233－1969 预包装食品的抽样方案》（AQL－6.5）。

（2）对需检测净重和干重的批次的抽样，抽样计划应以食品法典委员会的相关标准为依据。

7.2 感官与物理检验

产品的感官与物理指标须由经过此类检验培训的人员进行检验，并依据附录 A 以及《CAC/GL 31－1999 实验室中水产品感官评价指南》所述程序进行。

7.3 净重的测定

样品内容物的净重应通过以下程序测定：

（1）称量未开封的罐头；

（2）开罐并除去内容物；

（3）称量空罐重量（包括开启下来的罐底）；

（4）从未开封罐头的重量中减去空罐的重量，得到的就是内容物的净重。

7.4 干重的测定

所有样品单位内容物沥干后重量（干重）应通过以下程序测定：

（1）测定前使罐头的温度至少在过去 12h 内维持在 20～30℃；

（2）开罐并使之倾斜，将内容物倾倒在预先称重的圆形滤网上，该滤网由金属网丝构成，网孔大小：2.8mm×2.8mm；

（3）以 17°～20°的角度倾斜滤网以使鱼沥水，从将产品全部倾倒于滤网上开始计时，沥水 2min；

（4）称量沥水后的鱼及滤网的总重；

（5）从沥水后的产品和滤网的总重中减去滤网的重量就得到鱼的干重。

7.5 冲洗和沥水后产品重量的测定（适用于用调味汁填装的产品）

（1）测定前使罐头的温度至少在过去 12h 内维持在 20～30℃；

（2）开罐并使之倾斜，将所有的内容物置于预先称量过皮重的圆形滤网上，使用洗瓶（例如：塑料洗瓶）以热自来水（约 40℃）冲洗覆盖的调味汁；

（3）用热水冲洗滤网上的内容物直到其不再含有附着的调味汁；必要时，用钳子分离出可挑选出的配料（香料、蔬菜、水果等）。以 17°～20°的角度倾斜滤网以使鱼沥出水分，从冲洗程序结

束时开始计时，沥水 2min；

（4）用纸巾除去滤网底部附着的水。称量经冲洗和沥水后的鱼及滤网的总重；

（5）用经过冲洗和沥水后的产品及滤网的总重减去滤网的重量就得到冲洗和沥水后的产品重量。

7.6 组胺的测定

参见 AOAC 977.13。

8 缺陷的定义

当样品单位呈现下列任何一项特征时，则认定其有“缺陷”。

8.1 外来杂质

样品中存在的任何不是来自于鱼体的物质。这些物质虽不会对人体健康造成危害，但用肉眼可直接辨别，或采用某些方法（包括放大）可以确定其存在。出现外来杂质表明不符合良好操作和卫生惯例。

8.2 气味或风味

样品散发的持久、明显、令人厌恶的由腐败、酸败引起的气味或风味。

8.3 质地

（1）样品中含有过多非该产品特征的糊状鱼肉；

（2）样品中含有过多非该产品特征的硬（韧）鱼肉。

8.4 变色

样品受腐败、酸败的影响发生明显的鱼肉变色，或超过内容物干重 5% 的样品受到硫化物着色。

8.5 异物

样品中发现长度超过 5mm 的磷酸铵镁结晶（鸟粪石）。

9 批次验收

当满足以下条件时，可以认为此批次产品符合本标准的要求：

（1）根据本标准第 8 章中规定分类，缺陷总数不超过《CODEX STAN 233 - 1969 预包装食品的抽样方案》（AQL - 6.5）中相应抽样方案规定的可接受数（c）；

（2）不符合本标准 2.3 条中规定数量的样品总数不超过《CODEX STAN233 - 1969 预包装食品的抽样方案》（AQL - 6.5）中相应抽样方案规定的可接受数（c）；

（3）所有样品单位平均净重不少于标示量，在任何一个包装单位中没有不合理的重量短缺；

（4）符合本标准第 4、5、6 款中对食品添加剂、卫生和处理以及标签的要求。

附录 A　感官和物理检验

1. 对罐外部进行检测以确定其是否存在密封性缺陷，有无罐身外部扭曲。

2. 开罐，并且根据 7.3、7.4 和 7.5 条中的程序完成重量测定。

3. 检测产品的变色、外来杂质和异物等方面的情况。出现硬骨表明加工不完全，这时，要求对无菌状态进行评估。

4. 依据《CAC/ GL 31 – 1999 实验室中水产品感官评价指南》对产品的气味、风味以及质地进行评定。

CODEX STANDARD FOR CANNED SARDINES AND SARDINE - TYPE PRODUCTS

(*CODEX STAN 94 - 1981, REV. 1 -1995*)

1 SCOPE

This standard applies to canned sardines and sardine - type products packed in water or oil or other suitable packing medium. It does not apply to speciality products where fish content constitute less than 50% m/m of the net contents of the can.

2 DESCRIPTION

2.1 PRODUCT DEFINITION

2.1.1 Canned sardines or sardine type products are prepared from fresh or frozen fish of the following species:

- *Sardina pilchardus*:
  Sardinops melanostictus, *S. neopilchardus*, *S. ocellatus*, *S. sagax S. caeruleus*, *Sardinella aurita*, *S. brasiliensis*, *S. maderensis*, *S. longiceps*, *S. gibbosa*;
- *Clupea harengus*;
- *Sprattus sprattus*;
- *Hyperlophus vittatus*;
- *Nematalosa vlaminghi*;
- *Etrumeus teres*;
- *Ethmidium maculatum*: *Engraulis anchoita*, *E. mordax*, *E. ringens*;
- *Opisthonema oglinum.*

2.1.2 Head and gills shall be completely removed; scales or tail may be removed. The fish may be eviscerated. If eviscerated, it shall be practically free from visceral parts other than roe, milt or kidney. If ungutted, it shall be practically free from undigested feed or used feed.

2.2 PROCESS DEFINITION

The products are packed in hermetically sealed containers and shall have received a processing treatment sufficient to ensure commercial sterility.

2.3 PRESENTATION

Any presentation of the product shall be permitted provided that it:

(i) contains at least two fish in each can; and

(ii) meets all requirements of this standard; and

(iii) is adequately described on the label to avoid confusing or misleading the consumer;

(iv) contain only one fish species.

3 ESSENTIAL COMPOSITION AND QUALITY FACTORS

3.1 RAW MATERIAL

The products shall be prepared from sound fish of the species listed under sub – section 2.1 which are of a quality fit to be sold fresh for human consumption.

3.2 OTHER INGREDIENTS

The packing medium and all other ingredients used shall be of food grade quality and conform to all applicable Codex standards.

3.3 DECOMPOSITION

The products shall not contain more than 10 mg/100 g of histamine based on the average of the sample unit tested.

3.4 FINAL PRODUCT

Products shall meet the requirements of this Standard when lots examined in accordance with Section 9 comply with provisions set out in Section 8. Product shall be examined by the methods given in Section 7.

4 FOOD ADDITIVES

Only the use of the following additives is permitted.

	Additive	Maximum Level in the Final Product
Thickening or Gelling Agents (for use in packing media only)	400 Alginic acid 401 Sodium alginate 402 Potassium alginate 404 Calcium alginate 406 Agar 407 Carrageenan and its Na, K, and NH_4 salts (including furcelleran 407 Processed Eucheuma Seaweed (PES) 410 Carob bean gum 412 Guar gum 413 Tragacanth gum 415 Xanthan gum 440 Pectins 466 Sodium carboxymethylcellulose	CMP

Next table

	Additive	Maximum Level in the Final Product
Modified Starches	1401 Acid treated starches (including white and yellow dextrins) 1402 Alkaline treated starches 1404 Oxidized starches 1410 Monostarch phosphate 1412 Distarch phosphate, esterified 1414 Acetylated distarch phosphate 1413 Phosphated distarch phosphate 1420/1421 Starch acetate 1422 Acetylated distarch adipate 1440 Hydroxypropyl starch 1442 Hydroxypropyl starch phosphate	GMP
Acidity Regulators	260 Acetic acid 270 Lactic acid (L - , D - , and DL -) 330 Citric acid	GMP
Natural Flavours	Spice oils Spice extracts Smoke flavours (Natural smoke solutions and extracts)	GMP

5 HYGIENE AND HANDLING

5.1 The final product shall be free from any foreign material that poses a threat to human health

5.2 When tested by appropriate methods of sampling and examination as prescribed by the Codex Alimentarius Commission, the product:

(i) shall be free from micro - organisms capable of development under normal conditions of storage;

(ii) no sample unit shall contain histamine that exceeds 20mg per 100 g;

(iii) shall not contain any other substance including substances derived from microorganisms in amounts which may represent a hazard to health in accordance with standards established by the Codex Alimentarius Commission;

(iv) shall be free from container integrity defects which may compromise the hermetic seal.

5.3 It is recommended that the product covered by the provisions of this standard be prepared and handled in accordance with the appropriate sections of the Recommended International Code of Practice - General Principles of Food Hygiene (CAC/RCP 1 - 1969, Rev. 3 - 1997) and the following relevant Codes:

(i) the Recommended International Code of Practice for Canned Fish (CAC/RCP 10 - 1976);

(ii) the Recommended International Code of Hygienic Practice for Low - Acid and Acidified Low - Acid Canned Foods (CAC/RCP 23 - 1979).

6 LABELLING

In addition to the provisions of the Codex General Standard for the Labelling of Prepackaged Foods (CODEX STAN 1 - 1985, Rev. 3 - 1999) the following specific provisions apply:

6.1 NAME OF THE FOOD

The name of the product shall be:

(i) "Sardines" (to be reserved exclusively for Sardina pilchardus (Walbaum)); or

(ii) "X sardines" of a country, a geographic area, the species, or the common name of the species in accordance with the law and custom of the country in which the product is sold, and in a manner not to mislead the consumer.

6.1.1 The name of the packing medium shall form part of the name of the food

6.1.2 If the fish has been smoked or smoke flavoured, this information shall appear on the label in close proximity to the name

6.1.3 In addition, the label shall include other descriptive terms that will avoid misleading or confusing the consumer

7 SAMPLING, EXAMINATION AND ANALYSES

7.1 SAMPLING

(i) Sampling of lots for examination of the final product as prescribed in Section 3.3 shall be in accordance with the FAO/WHO Codex Alimentarius Sampling Plans for Prepackaged Foods (AQL - 6.5) (Ref. CAC/RM 42 - 1977);

(ii) Sampling of lots for examination of net weight and drained weight where appropriate shall be carried out in accordance with an appropriate sampling plan meeting the criteria established by the CAC.

7.2 SENSORIC AND PHYSICAL EXAMINATION

Samples taken for sensoric and physical examination shall be assessed by persons trained in such examination and in accordance with Annex A and the *Guidelines for the Sensory Evaluation of Fish and Shellfish in Laboratories* (CAC/GL 31 - 1999).

7.3 DETERMINATION OF NET WEIGHT

Net contents of all sample units shall be determined by the following procedure:

(i) Weigh the unopened container;

(ii) Open the container and remove the contents;

(iii) Weigh the empty container, (including the end) after removing excess liquid and adhering meat;

(iv) Subtract the weight of the empty container from the weight of the unopened container. The resultant figure will be the net content.

7.4 DETERMINATION OF DRAINED WEIGHT

The drained weight of all sample units shall be determined by the following procedure:

(i) Maintain the container at a temperature between 20℃ and 30℃ for a minimum of 12 hours prior to examination;

(ii) Open and tilt the container to distribute the contents on a pre - weighed circular sieve which consists of wire mesh with square openings of 2. 8 mm × 2. 8 mm;

(iii) Incline the sieve at an angle of approximately 17° ~ 20° and allow the fish to drain for two minutes, measured from the time the product is poured into the sieve;

(iv) Weigh the sieve containing the drained fish;

(v) The weight of drained fish is obtained by subtracting the weight of the sieve from the weight of the sieve and drained product.

7. 5 PROCEDURE FOR PACKS IN SAUCES (WASHED DRAINED WEIGHT)

(i) Maintain the container at a temperature between 20℃ and 30℃ for a minimum of 12 hours prior to examination;

(ii) Open and tilt the container and wash the covering sauce and then the full contents with hot tap water (approx. 40℃), using a wash bottle (e. g. plastic) on the tared circular sieve;

(iii) Wash the contents of the sieve with hot water until free of adhering sauce; where necessary separate optional ingredients (spices, vegetables, fruits) with pincers. Incline the sieve at an angle of approximately 17° ~ 20° and allow the fish to drain two minutes, measured from the time the washing procedure has finished;

(iv) Remove adhering water from the bottom of the sieve by use of paper towel. Weigh the sieve containing the washed drained fish;

(v) The washed drained weight is obtained by subtracting the weight of the sieve from the weight of the sieve and drained product.

7. 6 DETERMINATION OF HISTAMINE

AOAC 977. 13 (15th Edition, 1990).

8 DEFINITION OF DEFECTIVES

A sample unit will be considered defective when it exhibits any of the properties defined below.

8. 1 FOREIGN MATTER

The presence in the sample unit of any matter, which has not been derived from the fish or the packing media, does not pose a threat to human health, and is readily recognized without magnification or is present at a level determined by any method including magnification that indicates non - compliance with good manufacturing and sanitation practices.

8. 2 ODOUR/FLAVOUR

A sample unit affected by persistent and distinct objectionable odours or flavours indicative of decomposition or rancidity.

8. 3 TEXTURE

(i) Excessively mushy flesh uncharacteristic of the species in the presentation;

(ii) Excessively tough or fibrous flesh uncharacteristic of the species in the presentation.

8.4 DISCOLOURATION

A sample unit affected by distinct discolouration indicative of decomposition or rancidity or by sulphide staining of more than 5% of the fish by weight in the sample unit.

8.5 OBJECTIONABLE MATTER

A sample unit affected by Struvite crystals – any struvite crystal greater than 5 mm in length.

9 LOT ACCEPTANCE

A lot will be considered as meeting the requirements of this standard when:

(i) the total number of defectives as classified according to section 8 does not exceed the acceptance number (c) of the appropriate sampling plan in the Sampling Plans for Prepackaged Foods (AQL – 6.5) (CAC/RM 42 – 1977);

(ii) the total number of sample units not meeting the presentation defined in 2.3 does not exceed the acceptance number (c) of the appropriate sampling plan in the Sampling Plans for Prepackaged Foods (AQL – 6.5) (CAC/RM 42 – 1977);

(iii) the average net weight or the average drained weight where appropriate of all sample units examined is not less than the declared weight, and provided there is no unreasonable shortage in any individual container;

(iv) the Food Additives, Hygiene and Labelling requirements of Sections 3.3, 4, 5.1, 5.2 and 6 are met.

ANNEX A SENSORY AND PHYSICAL EXAMINATION

1. Complete external can examination for the presence of container integrity defects or can ends which may be distorted outwards.

2. Open can and complete weight determination according to defined procedures in Sections 7.3, 7.4 and 7.5.

3. Carefully remove product and examine for discolouration, foreign matter and struvite crystals. The presence of a hard bone is an indicator of underprocessing and will require an evaluation for sterility.

4. Assess odour, flavour and texture in accordance with the Guidelines for the Sensory Evaluation of Fish and Shellfish in Laboratories (CAC/GL 31 – 1999).

速冻龙虾

CODEX STANDARD FOR QUICK FROZEN LOBSTERS

(*CODEX STAN 95－1981，REV. 2－2004*)

1 适用范围

此标准适用于生的或熟的速冻龙虾、岩龙虾、海湾龙虾和挪威龙虾❶。

2 说明

2.1 产品的定义

2.1.1 产品是由海螯虾科（*Nephropsidae*）、巨螯龙虾属（*Homarus*）、龙虾科（*Palinuridae*）或蝉虾科（*Scyllaridae*）的龙虾制备的，也可由挪威海螯虾（*Nephrops norvigcus*），俗称挪威龙虾制成

2.1.2 不同品种的龙虾不应混合包装

2.2 加工定义

加工用水（蒸煮用水）应达到饮用水或清洁海水的质量要求。

产品经过适当预处理后，应在符合以下规定的条件进行冻结加工：冻结应在适当的设备中进行，并使产品迅速通过最大冰晶生成温度带。速冻加工只有在产品的中心温度达到并稳定在－18℃（0 ℉）或更低的温度时才算完成。产品在运输、贮存、分销过程中应保持在深度冻结状态，以保证产品质量。

在产品的加工和包装过程中应尽量减少脱水和氧化作用的影响。

2.3 产品介绍

2.3.1 产品介绍应符合下列要求

（1）达到本标准的所有要求；

（2）在产品标签中对产品进行详细描述，以免引起混淆或误导消费者。

2.3.2 龙虾产品可按单位重量、单位包装重量或规定重量等级进行包装

---

❶ 以下以龙虾为例，进行说明。

3 基本成分及质量因素

3.1 龙虾

速冻虾应由品质良好、可作为鲜品供人类消费的龙虾制备。

3.2 冰衣

用于镀冰衣或制备镀冰衣所用溶液的水应达到饮用水或清洁海水的质量要求。饮用水应符合WHO最新版本的《国际饮用水质量规范》要求，清洁海水应是达到饮用水的微生物标准且不含异物的海水。

3.3 其他成分

所使用的其他成分应具有食品级的质量，并且符合所有相应法规标准的规定。

3.4 成品

产品应符合本标准要求，根据第9章进行成品批次检查时，其质量应符合第8章的规定。检查方法应符合第7章的规定。

4 食品添加剂

只允许应用下列食品添加剂。

添加剂		成品中的最高含量
保水剂	451（i）三聚磷酸钠 451（ii）三聚磷酸钾 452（i）多聚磷酸钠 452（iv）多聚磷酸钙	≤10g/kg，以 P_2O_5 计， 单用或混用 （包括天然磷酸盐）
保鲜剂	221 亚硫酸钠 223 偏亚硫酸氢钠（焦亚硫酸钠） 224 偏亚硫酸氢钾（焦亚硫酸钾） 225 亚硫酸钾 228 亚硫酸氢钾（只用于原料）	100mg/kg（在生品可食部分）， 或 30mg/kg（熟产品的可食部分）， 以 SO_2 计，单用或混用
抗氧化剂	300 抗坏血酸（维生素C） 301 抗坏血酸钠 303 抗坏血酸钾	GMP

5 卫生和处理

5.1 成品中不能含有任何危害人体健康的外来杂质

5.2 应用CAC规定的抽样及检测方法进行检验时，应符合以下标准

（1）产品中微生物含量和由微生物产生的危害人体健康的有害物质的含量不能超过CAC标准

的规定；

（2）其他危害人体健康的有害物质的含量也不能超过 CAC 标准的规定。

5.3 建议本标准中涉及产品预处理的条款应符合推荐性国际规范《CAC/RCP1 - 1969，Rev. 3 - 1997 食品卫生操作总则》中的相关章节和以下相关标准

（1）推荐性国际操作规范《CAC/RCP24 - 1978 龙虾》；

（2）推荐性国际操作规范《CAC/RCP8 - 1976 速冻食品的加工和处理》；

（3）《水产品国际操作规程草案》中水产养殖品的相关章节（正在完善中）❶。

6 标签

标签除应符合《CODEX STAN1 - 1985，Rev. 1 - 1991 预包装食品标签通用标准》的要求外，还应遵守以下规定：

6.1 食品名称

食品名称应指明产品类型：

（1）龙虾，如果产品原料取自巨螯龙虾属（Homarus）；

（2）岩龙虾、真龙虾或螯虾，如果产品原料取自龙虾科（Palinuridae）中的一些属、种；

（3）“海湾龙虾”或“沙底龙虾”，如果产品原料取自蝉虾科（Scyllaridae）；

（4）挪威龙虾，如果产品原料取自挪威海螯虾（Nephrops norvigcus）。

6.1.1 标签上还应在与产品名称紧靠位置出现关于性状形式的参考文字，这些术语应恰当充分地反应产品性状，以免引起混淆或误导消费者

6.1.2 除上述规定的标签名称外，也可加上在出售该产品的国家不会误导消费者的常用名或普遍使用的商品名

6.1.3 标签上应恰当注明产品是“熟的”还是“生的”

6.1.4 如果产品用海水镀冰衣，则应予以说明

6.1.5 在产品标签上应标明“速冻”，除非在有的国家“冻”即代表“速冻”

6.1.6 标签应注明产品须在运输、贮藏、分销过程中保持的条件，以保证其质量

6.2 净含量（镀冰衣产品）

对于镀冰衣的产品，其内容物净含量不包括冰衣重。

6.3 贮藏说明

标签应注明产品须在 -18℃或更低的温度下贮藏。

6.4 非零售包装的标签

上述要求应既在包装上又在附文中出现，除了食品名称、批号、制造或分装厂名、地址外，还包括贮藏条件。但批号、制造或分装厂名、地址也可用同一证明标志代替，只要证明标志能在辅助文件中说明清楚。

---

❶ 操作规程建议草案，在定稿后将替代所有现行的《水产品操作规程》。

7 抽样、检验和分析

7.1 取样

（1）产品批次检验用样品的抽样方法应符合 FAO/WHO 的食品法典《CODEX STAN 233－1969 预包装食品的抽样方案》（AQL－6.5）。样品单位是初级包装，对单冻（IQF）产品至少以 1kg 样品为单位；

（2）对需检测净重批次的抽样，抽样计划应以食品法典委员会的相关标准为依据。

7.2 感官与物理检验

产品的感官与物理指标须由经过此类检验培训的人员进行检验，并依据本标准 7.3 至 7.6 条和附录 A 以及《CAC/GL 31－1999 实验室中水产品感官评价指南》所述程序进行。

7.3 净重的测定

7.3.1 未镀冰衣产品净重的测定

代表每批次的样品单位的净重（不包括包装材料）应在冷冻状态下测定。

7.3.2 镀冰衣产品净重的测定（以下方法可依据具体情况选用）

7.3.2.1 从低温贮藏状态下取出样品后，立即打开包装，将内容物放在冷水喷头下，用温和的水流喷洒内含物，并小心摇动样品，不要弄破样品；直到可以看到或用手感觉到全部冰衣已被去除时，停止喷洒；用纸巾除去黏附的水并放在皮重已知的盘中称重；

7.3.2.2 用手将预称重的镀冰衣样品浸入水浴中，直到全部冰衣都被去除（用手指可以很好地感觉到）。表面一经变粗糙，立即将仍然保持冰冻状态的样品从水浴中移出；用纸巾拭干样品，而后再次称重以计算产品净含量。通过此程序，可以避免解冻汁液的流失和（或）附着的水分再结冰；

（1）从低温贮藏状态下取出样品后，立即打开包装，将产品放入装有 8 倍于产品标示重量的 27℃（80 ℉）新鲜饮用水的容器中。使产品浸在水中，直到冰全部融化。如果产品是块冻产品，则需在解冻过程中多次翻转。可用探针轻探以确定解冻是否彻底完成。

（2）取干燥、清洁的金属滤网一个，选用方孔孔径为 2.8mm（ISO 推荐 R565 号）或 2.38mm（美国 8 号标准滤网）中一种进行称重。

（a）如果包装中的总内容物的重量等于或小于 500g（1.1 磅），则应使用直径为 20cm（8 英寸）的滤网；

（b）如果包装中的总内容物的重量大于 500g（1.1 磅），则应使用直径为 30cm（12 英寸）的滤网。

（3）当可以看到或用手感觉到全部冰衣已被去除，龙虾与龙虾之间很容易分开时，清空容器，将内容物置于预先称重的滤网上。使滤网倾斜成 20°角，沥干 2min。

（4）将沥干的产品连滤网一起称重，减去滤网的重量，剩下的数字就是包装中内容物的净重。

7.4 计数测定

当标签上标明时，应通过计数一个初级包装所有的龙虾或虾尾以测定龙虾的数量，而后用龙虾的数量除以其去冰衣的实际平均重量以得到每单位重量中的龙虾的数目。

7.5 解冻

解冻时将样品装入薄膜袋中，浸入室温（温度不高于 35℃）流动水的水池中，不时用手轻捏袋子，至袋中无硬块和冰晶时为止，但不要损坏虾的质地。

7.6 蒸煮方法

蒸煮使产品内部温度达到65~70℃。不能过度蒸煮，蒸煮时间随产品大小和采用的温度而不同。准确的蒸煮时间和条件应依据预先实验来确定。

烘焙：用铝箔包裹产品，并将其均匀放入扁平锅或浅平锅上。

蒸：用铝箔包裹产品，并将其置于带盖容器中沸水之上的金属架上。

袋煮：将产品放入可煮薄膜袋中加以密封，浸入沸水中煮。

微波：将产品放入适于微波加热的容器中，若用塑料袋，应检查确定塑料袋不会发出任何气味。根据设备说明加热。

8 缺陷的定义

当样品呈现下列任何一项特征时，则认定其有“缺陷”。

8.1 深度脱水

样品单位中超过表面积10%或大于龙虾重量10%的部分，出现水分的过度损失，明显表现为龙虾体表面呈现异常的白色或黄色，覆盖了肌肉本身的颜色，并已渗透至表层以下，在不过分影响产品外观的情况下，不能轻易地用刀或其他利器刮掉。

8.2 外来杂质

样品单位中存在的任何不是来自于龙虾体的物质（包装材料除外）。这些物质虽不会对人体健康造成危害，但用肉眼可直接辨别，或采用某些方法（包括放大）可以确定其存在。出现外来杂质表明不符合良好操作和卫生惯例。

8.3 气味或风味

样品散发的持久、明显、令人厌恶的由腐败、酸败或饵料引起的气味或风味。

8.4 变色

单个整龙虾或半个龙虾外壳上，超过其表面积10%的部分明显变黑；或对于龙虾尾部的肉，超过标示重量10%明显变黑、变褐、变绿或变黄（无论是单色变化还是混色变化）。

9 批次验收

当满足以下条件时，可以认为此批次产品符合最后的质量要求：

（1）根据本标准第8章中规定分类，缺陷总数不超过《CODEX STAN 233－1969 预包装食品的抽样方案》（AQL－6.5）中相应抽样方案规定的可接受数（c）；

（2）不符合本标准2.3中规定数量的样品总数不超过《CODEX STAN233－1969 预包装食品的抽样方案》（AQL－6.5）中相应抽样方案规定的可接受数（c）；

（3）所有样品单位平均净重不少于标示量，在任何一个包装单位中没有不合理的重量短缺；

（4）符合本标准第4、5、6款中对食品添加剂、卫生和处理以及标签的要求。

附录A 感官及物理检验

1. 净含量的测定按本标准7.3条规定执行（按要求去冰衣）。

2. 检测冻龙虾深度脱水的情况。测定受影响龙虾的百分比。

3. 按本标准7.5条的规定解冻，并分别检查样品单位中每一只虾的外来杂质和其他缺陷，测定受缺陷影响的龙虾的重量。

4. 按本标准7.4条的规定检查产品中龙虾的标示数量和重量。

5. 根据要求评估气味及变色现象。

6. 对在解冻状态下无法最终判定其气味或风味的样品，则应从样品单位中截取一小块可疑部分（100g～200g），并立即使用7.6条中限定的某一种蒸煮方法，确定其气味和风味。

CODEX STANDARD FOR QUICK FROZEN LOBSTERS

(CODEX STAN 95 – 1981, REV. 2 – 2004)

1 SCOPE

This standard applies to quick frozen raw or cooked lobsters, rock lobsters, spiny lobsters and slipper lobsters[1]. Furthermore it applies to quick frozen raw or cooked squat lobsters (red and yellow).

2 DESCRIPTION

2.1 PRODUCT DEFINITION

2.1.1 The product is prepared from lobsters from the genus Homarus of the family Nephropidae and from the families Palinuridae and Scyllaridae. It may also be prepared from Nephrops norvegicus provided it is presented as Norway lobster. For squat lobsters the product is prepared from species of Cervimunida johnii, Pleuroncodes monodon and Pleuroncodes planipes of the family Galatheidae

2.1.2 The pack shall not contain a mixture of species

2.2 PROCESS DEFINITION

The water used for cooking shall be of potable quality or clean seawater.

The product, after any suitable preparation, shall be subjected to a freezing process and shall comply with the conditions laid down hereafter. The freezing process shall be carried out in appropriate equipment in such a way that the range of temperature of maximum crystallization is passed quickly. The quick freezing process shall not be regarded as complete unless and until the product temperature has reached – 18℃ (0 ℉) or colder at the thermal centre after thermal stabilization. The product shall be kept deep frozen so as to maintain the quality during transportation, storage and distribution.

Quick frozen lobsters shall be processed and packaged so as to minimize dehydration and oxidation.

2.3 PRESENTATION

2.3.1 Any presentation of the product shall be permitted provided that it:

2.3.1.1 meets all requirements of this standard;

2.3.1.2 is adequately described on the label to avoid confusing or misleading the consumer.

2.3.2 The lobster may be packed by count per unit of weight or per package or within a stated weight

---

[1] Hereafter referred to as lobster.

range.

3 ESSENTIAL COMPOSITION AND QUALITY FACTORS

3.1 LOBSTERS

The product shall be prepared from sound lobsters which are of a quality fit to be sold fresh for human consumption.

3.2 GLAZING

If glazed, the water used for glazing or preparing glazing solutions shall be of potable quality or shall be clean sea – water. Potable water is fresh – water fit for human consumption. Standards of potability shall not be less than those contained in the latest edition of the WHO "International Guidelines for Drinking Water Quality". Clean sea – water is sea – water which meets the same microbiological standards as potable water and is free from objectionable substances.

3.3 OTHER INGREDIENTS

All other ingredients used shall be of food grade quality and conform to all applicable Codex standards.

3.4 FINAL PRODUCT

Products shall meet the requirements of this standard when lots examined in accordance with Section 9 comply with the provisions set out in Section 8. Products shall be examined by the methods given in Section 7.

4 FOOD ADDITIVES

Only the use of the following additives is permitted.

	ADDITIVE	MAXIMUM LEVEL IN THE FINAL PRODUCT
Moisture/Water Retention Agents	451 (i) Pentasodium triphosphate 451 (ii) Pentapotassium triphosphate 452 (i) Sodium polyphosphate 452 (iv) Calcium polyphosphates	10 g/kg expressed as P_2O_5, singly or in combination (includes natural phosphate)
Preservatives	221 Sodium sulphite 223 Sodium metabisulphite 224 Potassium metabisulphite 225 Potassium sulphite 228 Potassium bisulphite (for use in the raw product only)	100 mg/kg in the edible part of the raw product, or 30 mg/kg in the edible part of the cooked product, singly or in combination, expressed as SO_2
Antioxidants	300 Ascorbic acid 301 Sodium ascorbate 303 Potassium ascorbate	GMP

5 HYGIENE AND HANDLING

5.1 The final product shall be free from any foreign material that poses a threat to human health

5.2 When tested by appropriate methods of sampling and examination prescribed by the Codex Alimentarius Commission , the product:

(i) shall be free from microorganisms or substances originating from microorganisms in amounts which may present a hazard to health in accordance with standards established by the Codex Alimentarius Commission;

(ii) shall not contain any other substance in amounts which may present a hazard to health in accordance with standards established by the Codex Alimentarius Commission.

5.3 It is recommended that the products covered by the provisions of this standard be prepared and handled in accordance with the appropriate sections of the Recommended International Code of Practice – General Principles of Food Hygiene (CAC/RCP 1 – 1969, Rev. 3 – 1997) and the following relevant Codes:

(i) The Recommended International Code of Practice for Lobsters (CAC/RCP 24 – 1978);

(ii) The Recommended International Code of Practice for the Processing and Handling of Quick Frozen Foods (CAC/RCP 8 – 1976);

(iii) The sections on the Products of Aquaculture in the Proposed Draft International Code of Practice for Fish and Fishery Products (under elaboration)[1] .

6 LABELLING

In addition to the provisions of the General Standard for the Labelling of Prepackaged Foods (CODEX STAN 1 – 1985, Rev. 1 – 1991) the following specific provisions apply:

6.1 THE NAME OF THE FOOD

The product shall be designated:

(i) Lobster if derived from the genus Homarus;

(ii) Rock Lobster, Spiny Lobster or Crawfish if derived from species of the family Palinuridae;

(iii) Slipper Lobster, Bay Lobster or Sand Lobster if derived from species of the family Scyllaridae;

(iv) Norway Lobster if derived from the species Nephrops norvegicus;

(v) Squat Lobster if derived from the species Cervimunida johnii, Pleuroncodes monodon and Pleuroncodes planipes.

6.1.1 There shall appear on the label, reference to the form of presentation in close proximity to the name of the product in such descriptive terms that will adequately and fully describe the nature of the presentation of the product to avoid misleading or confusing the consumer

6.1.2 In addition to the specified labelling designations above, the usual or common trade names of the variety may be added so long as it is not misleading to the consumer in the country in which the product will be distributed

6.1.3 Products shall be designated as cooked or raw as appropriate

---

[1] The Proposed Draft Code of Practice, when finalized, will replace all current Codes of Practice for Fish and Fishery Products.

6.1.4 If the product has been glazed with sea – water, a statement to this effect shall be made

6.1.5 The term "quick frozen", shall also appear on the label, except that the term "frozen" may be applied in countries where this term is customarily used for describing the product processed in accordance with subsection 2.2 of this standard

6.1.6 The label shall state that the product should be maintained under conditions that will maintain the quality during transportation, storage and distribution

6.2 NET CONTENTS (GLAZED PRODUCTS)

Where the food has been glazed the declaration of net contents of the food shall be exclusive of the glaze.

6.3 STORAGE INSTRUCTIONS

The label shall include terms to indicate that the product shall be stored at a temperature of –18℃ or colder.

6.4 LABELLING OF NON – RETAIL CONTAINERS

Information specified above shall be given either on the container or in accompanying documents, except that the name of the food, lot identification, and the name and address of the manufacturer or packer as well as storage instructions shall always appear on the container.

However, lot identification, and the name and address may be replaced by an identification mark, provided that such a mark is clearly identifiable with the accompanying documents.

7 SAMPLING, EXAMINATION AND ANALYSES

7.1 SAMPLING

(i) Sampling of lots for examination of the product shall be in accordance with the FAO/WHO Codex Alimentarius Sampling Plans for Prepackaged Foods (AQL – 6.5) (CODEX STAN 233 – 1969). In the case of shell on lobster the sample unit is an individual lobster. In the case of shell – off lobster the sample unit shall be at least a 1 kg portion of lobster from the primary container. In the case of squat lobster the sampling unit shall be at least 1 kg portion;

(ii) Sampling of lots for examination of net weight shall be carried out in accordance with an appropriate sampling plan meeting the criteria established by the Codex Alimentarius Commission.

7.2 SENSORY AND PHYSICAL EXAMINATION

Samples taken for sensory and physical examination shall be assessed by persons trained in such examination and using procedures elaborated in Sections 7.3 through 7.6, Annex A and the *Guidelines for the Sensory Evaluation of Fish and Shellfish in Laboratories* (CAC/GL 31 – 1999).

7.3 DETERMINATION OF NET WEIGHT

7.3.1 Determination of net weight of Products not Covered by Glaze

The net weight (exclusive of packaging material) of each sample unit representing a lot shall be determined in the frozen state.

7. 3. 2 Determination of Net Weight of Products Covered by Glaze (Alternate Methods)

7. 3. 2. 1 As soon as the package is removed from frozen temperature storage, open immediately and place the contents under a gentle spray of cold water until all ice glaze that can be seen or felt is removed. Remove adhering water by the use of paper towel and weigh the product.

7. 3. 2. 2 The pre – weighed glazed sample is immersed into a water bath by hand, until all glaze is removed, which preferably can be felt by the fingers. As soon as the surface becomes rough, the still frozen sample is removed from the water bath and dried by use of a paper towel before estimating the net product content by second weighing. By this procedure thaw drip losses and/or re – freezing of adhering moisture can be avoided.

(i) As soon as the package is removed from frozen temperature storage, place the product in a container containing an amount of fresh potable water of 27℃ (80°F) equal to 8 times the declared weight of the product. Leave the product in the water until all ice is melted. If the product is block frozen, turn block over several time during thawing. The point at which thawing is complete can be determined by gently probing the block.

(ii) Weigh a dry clean sieve with woven wire cloth with nominal size of the square aperture 2. 8 mm (ISO Recommendation R565) or alternatively 2. 38 mm (U. S. No. 8 Standard Screen) .

(a) If the quantity of the total contents of the package is 500 g (1. 1 lbs) or less, use a sieve with a diameter of 20 cm (8 inches) .

(b) If the quantity of the total contents of the package is more than 500 g (1. 1 lbs) use a sieve with a diameter of 30 cm (12 inches).

(iii) After all glaze that can be seen or felt has been removed and the lobsters separate easily, empty the contents of the container on the previously weighed sieve. Incline the sieve at an angle of about 20° and drain for two minutes.

(iv) Weigh the sieve containing the drained product. Subtract the mass of the sieve; the resultant figure shall be considered to be part of the net content of the package.

7. 4 DETERMINATION OF COUNT

When declared on the label, the count shall be determined by counting all lobsters or tails in the primary container and dividing the count of lobster by the average deglazed weight to determine the count per unit weight.

7. 5 PROCEDURE FOR THAWING

The sample unit is thawed by enclosing it in a film type bag and immersing in water at room temperature (not greater than 35℃) . The complete thawing of the product is determined by gently squeezing the bag occasionally so as not to damage the texture of the lobster, until no hard core or ice crystals are left.

7. 6 COOKING METHODS

The following procedures are based on heating the product to an internal temperature of 65 ~ 70℃. The product must not be overcooked. Cooking times vary according to the size of the product and the temperature used. The exact times and conditions of cooking for the product should be determined by prior experimentation.

Baking Procedure: Wrap the product in aluminum foil and place it evenly on a flat cookie sheet or shallow flat pan.

Steaming Procedure: Wrap the product in aluminum foil and place it on a wire rack suspended over boiling water in a covered container.

Boil – in – Bag Procedure: Place the product into a boilable film – type pouch and seal. Immerse the pouch into boiling water and cook.

Microwave Procedure: Enclose the product in a container suitable for microwave cooking. If plastic bags are used check to ensure that no odour is imparted from the plastic bags. Cook according to equipment specifications.

8 DEFINITION OF DEFECTIVES

The sample unit shall be considered as defective when it exhibits any of the properties defined below.

8.1 DEEP DEHYDRATION

Greater than 10% of the weight of the lobster in the sample unit or greater than 10% of the surface area of the block exhibits excessive loss of moisture clearly shown as white or yellow abnormality on the surface which masks the colour of the flesh and penetrates below the surface, and cannot be easily removed by scraping with a knife or other sharp instrument without unduly affecting the appearance of the lobster .

8.2 FOREIGN MATTER

The presence in the sample unit of any matter which has not been derived from lobster, does not pose a threat to human health, and is readily recognized without magnification or is present at a level determined by any method including magnification that indicates non – compliance with good manufacturing and sanitation practices.

8.3 ODOUR/FLAVOUR

Lobster affected by persistent and distinct objectionable odours or flavours indicative of decomposition or rancidity, or feed.

8.4 DISCOLOURATION

Distinct blackening of more than 10% of the surface area of the shell of individual whole or half lobster, or in the case of tail meat and meat presentations distinct black, brown, green or yellow discolourations singly or in combination, of the meat affecting more than 10% of the declared weight.

9 LOT ACCEPTANCE

A lot shall be considered as meeting the requirements of this standard when:

(i) the total number of defectives as classified according to section 8 does not exceed the acceptance number (c) of the appropriate sampling plan in the Sampling Plans for Prepackaged Foods (AQL – 6.5) (CODEX STAN 233 – 1969);

(ii) the total number of sample units not meeting the count or weight range designation as defined in Section 2. 3 does not exceed the acceptance number (c) of the appropriate sampling plan in the Sampling Plans for Prepackaged Foods (AQL – 6. 5) (CODEX STAN 233 – 1969);

(iii) the average net weight of all sample units is not less than the declared weight, provided there is no unreasonable shortage in any individual container;

(iv) the Food Additives, Hygiene and Labelling requirements of Sections 4, 5 and 6 are met.

ANNEX A SENSORY AND PHYSICAL EXAMINATION

1. Complete net weight determination, according to defined procedures in Section 7. 3 (de – glaze as required).

2. Examine the frozen lobster for the presence of deep dehydration. Determine the percentage of lobster affected.

3. Thaw using the procedure described in Section 7. 5 and individually examine each sample unit for the presence of foreign and objectionable matter.

4. Examine product count and weight declarations in accordance with procedures in Section 7. 4.

5. Assess the lobster for odour and discolouration as required.

6. In cases where a final decision regarding the odour/flavour cannot be made in the thawed state, a small portion of the sample unit (100 to 200 g) is prepared without delay for cooking and the odour/flavour confirmed by using one of the cooking methods defined in Section 7. 6.

鱼 罐 头

CODEX STANDARD FOR CANNED FINFISH

(*CODEX STAN 119 - 1981, REV. 1 - 1995*)

1 适用范围

本标准适用于以水、油或其他成分作为填装介质的鱼罐头，不适用于罐中鱼肉净重少于罐头内容物净重50%（W/W）的罐头，这类罐头有其他专门适用标准的鱼罐头。

2 说明

2.1 产品的定义

鱼罐头是由任何品种（有专门适用标准的罐头除外）有鳍鱼类的鲜鱼肉制成的，适于人类消费的产品，可以由具有相似感官特性的同属而不同种的鱼混合制成。

2.2 加工过程的定义

将鱼装罐并密封，然后通过一系列加工处理以确保其达到商业无菌状态。

2.3 产品介绍

产品介绍应符合下列要求：

(1) 达到本标准的所有要求；

(2) 在标签中对产品进行详细描述，以免引起混淆或误导消费者。

3 基本成分及质量因素

3.1 鱼

产品制备应用已经去除头、尾及内脏的有鳍鱼，且鱼的品质良好，可作为鲜品供人类消费。

3.2 其他成分

所使用的其他成分应具有食品级的质量，并且符合所有相应法规标准的规定。

3.3 腐败

被测样品单位的组胺平均含量不应超过10mg/100g。这仅适用于鲱科（*Clupeidae*）、鲭科（*Scombridae*）、秋刀鱼科（*Scombresocidae*）、鲑科（*Pomatomidae*）以及鲯鳅科（*Coryphaenedae*）等

科的鱼类。

3.4 成品

产品应符合本标准要求，根据第9章进行成品批次检查时，其质量应符合第8章的规定。检查方法应符合第7章的规定。

4 食品添加剂

只允许使用下列添加剂。

	添加剂	成品中的最高含量
增稠剂（只适用于装罐介质）	400 褐藻酸 401 褐藻酸钠 402 褐藻酸钾 404 褐藻酸钙 406 琼脂 407 卡拉胶及其 Na，K，NH_4 盐（包括红藻胶） 407a 经热加工处理的麒麟菜（PES） 410 角豆胶（刺槐豆胶） 412 瓜尔豆胶 413 黄蓍胶 415 黄原胶 440 果胶 466 羧甲基纤维素钠	GMP
改性淀粉	1401 酸处理淀粉 1402 碱处理淀粉 1404 氧化淀粉 1410 磷酸单淀粉 1412 磷酸二淀粉酯化三偏磷酸钠；磷酸二淀粉酯化氯氧化磷 1414 磷酸乙酰化二淀粉 1413 磷酸化磷酸二淀粉 1420 醋酸淀粉酯化乙酸酐 1421 醋酸淀粉酯化乙烯基乙酸酯 1422 乙酰化二淀粉己二酸酯 1440 羧丙基淀粉 1442 磷酸羧丙基淀粉 1442 羧丙基淀粉磷酸盐	GMP
调酸剂	260 醋酸 270（L－，D－及 DL－）乳酸 330 柠檬酸	GMP
天然香料	香油（香料油） 香精（香料萃取物） 烟香味（自然烟溶液和萃取物）	GMP

5 卫生及处理

5.1 成品中不能含有任何危害人体健康的外来杂质

5.2 应用 CAC 规定的抽样及检测方法进行检验时，产品应具备下列条件

（1）不含有任何在正常的贮存条件下可能生长的微生物；

（2）组胺含量不应超过20mg/100g，这仅适用于鲱科（*Clupeidae*）、鲭科（*Scombridae*）、秋刀鱼科（*Scombresocidae*）、鲑科（*Pomatomidae*）以及鳍鳅科（*Coryphaenedae*）等科的鱼类；

（3）其他危害人体健康的有害物质的含量也不能超过 CAC 标准的规定；

（4）不存在可能危及罐的密封性的缺陷。

5.3 建议本标准中涉及产品预处理的条款应符合推荐性国际规范《CAC/RCP1 -1969，Rev. 3 - 1997 食品卫生操作总则》中的相关章节和以下相关标准

（1）推荐性国际操作规范《CAC/RCP 10 - 1976 鱼罐头》；

（2）推荐性国际操作规范《CAC/ RCP 23 - 1979 低酸和酸化低酸罐头食品》；

（3）《水产品国际操作规程草案》中水产养殖品的相关章节（正在完善中）[1]。

6 标签

除了应符合《CODEX STAN1 - 1985，Rev. 3 - 1999 预包装食品标签通用标准》的要求外，还应遵守以下规定。

6.1 食品名称

6.1.1 在标签上标示的食品名称应是依据出售该产品的国家的法律或习惯而使用的俗名或常用名，并且不会误导消费者；

6.1.2 产品名称应通过叙述性的术语来限定；

6.1.3 食品名称中应包含填装介质的名称；

6.1.4 使用同属中不同种鱼混合作为原料时，应在标签中注明；

6.1.5 标签上还应出现关于外观的参考文字，这些术语应恰如其分地反应产品外观，并应确保它们不会引起混淆或误导消费者。

7 抽样、检验和分析

7.1 取样

（1）如 3.4 中指定的一样，产品批次检验用样品的抽样方法应符合 FAO/WHO 的食品法典《CODEX STAN 233 - 1969 预包装食品的抽样方案》（AQL - 6.5）。

（2）对需检测净重和干重的批次的抽样，抽样计划应以食品法典委员会的相关标准为依据。

---

[1] 操作规程建议草案，在定稿后将替代所有现行的《水产品操作规程》。

7.2　感官与物理检验

产品的感官与物理指标须由经过此类检验培训的人员进行检验，并依据本标准7.3至7.5条和附录A以及《CAC/GL 31－1999 实验室中水产品感官评价指南》所述程序进行。

7.3　净重的测定

样品内容物的净重应通过以下程序测定：

（1）称量未开封的罐头；

（2）开罐并除去内容物；

（3）称量空罐重量（包括开启下来的罐底）；

（4）从未开封罐头的重量中减去空罐的重量，得到的就是内容物的净重。

7.4　干重的测定

样品内容物沥干后重量（干重）应通过以下程序测定：

（1）测定前使罐头的温度至少在过去12小时内维持在20～30℃。

（2）开罐并使之倾斜，将内容物倾倒在预先称重的圆形滤网上，该滤网由金属网丝构成，网孔大小：2.8mm×2.8mm。

（3）以17°～20°的角度倾斜滤网以使鱼沥水，从将产品全部倾倒于滤网上开始计时，沥水2min。

（4）称量沥水后的鱼及滤网的总重。

（5）从沥水后的产品和滤网的总重中减去滤网的重量就得到鱼的干重。

7.5　冲洗和沥水后产品重量的测定（适用于用调味汁填装的产品）

（1）测定前使罐头的温度在至少12小时内维持在20～30℃。

（2）开罐并使之倾斜，将所有的内容物置于预先称量过皮重的圆形滤网上，使用洗瓶（例如：塑料洗瓶）以热自来水（约40℃）冲洗覆盖的调味汁。

（3）用热水冲洗滤网上的内容物直到其不再含有附着的调味汁；必要时，用钳子分离出可挑选出的配料（香料、蔬菜、水果等）。以17°～20°的角度倾斜滤网以使鱼沥出水分，从冲洗程序结束时开始计时，沥水2min。

（4）用纸巾除去滤网底部附着的水。称量含有经冲洗和沥水后的鱼及滤网的总重。

（5）用经过冲洗和沥水后的鱼及滤网的总重减去滤网的重量就得到冲洗和沥水后的产品重量。

7.6　组胺的测定

AOAC 977.13。

8　缺陷的定义

当样品单位呈现下列任何一项特征时，则认定其有“缺陷”。

8.1　外来杂质

样品中存在的任何不是来自于鱼体的物质。这些物质虽不会对人体健康造成危害，但用肉眼可直接辨别，或采用某些方法（包括放大）可以确定其存在。出现外来杂质则表明不符合良好操作和卫生惯例。

8.2 气味/风味

样品散发的持久、明显、令人厌恶的由腐败、酸败引起的气味或风味。

8.3 质地

（1）含有过多非该产品特征的糊状鱼肉；
（2）含有过多非该产品特征的硬（韧）鱼肉；
（3）含有重量超过内容物干重5%的蜂巢状鱼肉。

8.4 变色

样品受腐败、酸败的影响发生明显的鱼肉变色，或超过内容物干重5%的样品受到硫化物着色。

8.5 异物

样品中发现长度超过5mm的磷酸铵镁结晶（鸟粪石）。

9 批次验收

当满足以下条件时，可以认为此批次产品符合本标准的要求：

（1）根据本标准第8章中规定分类，缺陷总数不超过《CODEX STAN 233－1969 预包装食品的抽样方案》（AQL－6.5）中相应抽样方案规定的可接受数（c）；

（2）不符合本标准2.3条中规定数量的样品总数不超过《CODEX STAN233－1969 预包装食品的抽样方案》（AQL－6.5）中相应抽样方案规定的可接受数（c）；

（3）所有样品单位平均净重不少于标示量，在任何一个包装单位中没有不合理的重量短缺；

（4）符合本标准第4、5、6章中对食品添加剂、卫生和处理以及标签的要求。

附录A 感官和物理检验

1. 对罐外部进行检测以确定其是否存在密封性缺陷，有无罐身外部扭曲。
2. 开罐，并且根据7.3、7.4和7.5条中确定的程序完成重量测定。
3. 检测产品的外观。
4. 检测产品的变色、外来杂质和异物等方面的情况。出现硬骨表明加工不完全，这时，要求对无菌状态进行评估。
5. 依据《CAC/ GL 31－1999 实验室中水产品感官评价指南》对产品的气味、风味以及质地进行评定。

CODEX STANDARD FOR CANNED FINFISH

(*CODEX STAN 119 - 1981, REV. 1 - 1995*)

1 SCOPE

This standard applies to canned finfish packed in water, oil or other suitable packing medium. It does not apply to speciality products where the canned finfish constitutes less than 50% m/m of the net contents of the can or to canned finfish covered by other Codex product standards.

2 DESCRIPTION

2.1 PRODUCT DEFINITION

Canned finfish is the product produced from the flesh of any species of finfish (other than canned finfish covered by other Codex product standards) which is suitable for human consumption and may contain a mixture of species, with similar sensoric properties, from within the same genus.

2.2 PROCESS DEFINITION

Canned finfish are packed in hermetically sealed containers and shall have received a processing treatment sufficient to ensure commercial sterility.

2.3 PRESENTATION

Any presentation of the product shall be permitted provided that it

(i) meets all requirements of this standard; and

(ii) is adequately described on the label to avoid confusing or misleading the consumer.

3 ESSENTIAL COMPOSITION AND QUALITY FACTORS

3.1 FISH

The product shall be prepared from sound finfish from which the heads, tails and viscera have been removed. The raw material shall be of a quality fit to be sold fresh for human consumption.

3.2 OTHER INGREDIENTS

The packing medium and all other ingredients used shall be of food grade quality and conform to all applicable Codex standards.

3.3 DECOMPOSITION

Canned finfish of the families Scombridae, Scombresocidae, Clupeidae, Coryphaenidae and Pomatomidae shall not contain more than 10 mg/100 g of histamine based on the average of the sample units tested.

3.4 FINAL PRODUCT

Products shall meet the requirements of this Standard when lots examined in accordance with Section 9 comply with the provisions set out in Section 8. Products shall be examined by the methods given in Section 7.

4 FOOD ADDITIVES

Additive	Maximum Level in the Final Product
Thickening or Gelling Agents (for use in packing media only)	
400 Alginic acid 401 Sodium alginate 402 Potassium alginate 404 Calcium alginate 406 Agar 407 Carrageenan and its Na, K, and NH_4 salts (including furcelleran) 407a Processed *Eucheuma* Seaweed (PES) 410 Carob bean gum 412 Guar gum 413 Tragacanth gum 415 Xanthan gum 440 Pectins 466 Sodium carboxymethylcellulose	GMP
Modified Starches	
1401 Acid treated starches (including white and yellow dextrins) 1402 Alkaline treated starches 1404 Oxidized starches 1410 Monostarch phosphate 1412 Distarch phosphate, esterified 1414 Acetylated distarch phosphate 1413 Phosphated distarch phosphate 1420/1421 Starch acetate 1422 Acetylated distarch adipate 1440 Hydroxypropyl starch 1442 Hydroxypropyl starch phosphate	GMP
Acidity Regulators	
260 Acetic acid 270 Lactic acid (L-, D-, and DL-) 330 Citric acid	

continued

Additive	Maximum Level in the Final Product
Natural Flavours	
Spice oils Spice extracts Smoke flavours (Natural smoke solutions and extracts)	GMP

5 HYGIENE AND HANDLING

5.1 The final product shall be free from any foreign material that poses a threat to human health.

5.2 When tested by appropriate methods of sampling and examination prescribed by the Codex Alimentarius Commission, the product:

(i) shall be free from micro - organisms capable of development under normal conditions of storage; and

(ii) no sample unit shall contain histamine that exceeds 20 mg per 100 g. This applies only to species of the families Scombridae, Clupeidae, Coryphaenidae, Scombresocidae and Pomatomidae.

(iii) shall not contain any other substance including substances derived from microorganisms in amounts which may represent a hazard to health in accordance with standards established by the Codex Alimentarius Commission; and

(iv) shall be free from container integrity defects which may compromise the hermetic seal.

5.3 It is recommended that the product covered by the provisions of this standard be prepared and handled in accordance with the appropriate sections of the Recommended International Code of Practice - General Principles of Food Hygiene (CAC/RCP 1 - 1969, Rev. 3 - 1997) and the following relevant Codes:

(i) the Recommended International Code of Practice for Canned Fish (CAC/RCP 10 - 1976);

(ii) the Recommended International Code of Hygienic Practice for Low - Acid and Acidified Low - Acid Canned Foods (CAC/RCP 23 - 1979);

(iii) The sections on the Products of Aquaculture in the Proposed Draft International Code of Practice for Fish and Fishery Products (under elaboration)[1].

6 LABELLING

In addition to the provisions of the Codex General Standard for the Labelling of Prepackaged Foods (CODEX STAN 1 - 1985, Rev. 3 - 1999) the following specific provisions apply.

6.1 NAME OF THE FOOD

6.1.1 The name of the product declared on the label shall be the common or usual name applied to the species in accordance with the law and custom of the country in which the product is sold, and in a manner not to mislead the consumer.

6.1.2 The name of the product shall be qualified by a term descriptive of the presentation.

---

[1] The Proposed Draft Code of Practice, when finalized, will replace all current Codes of Practice for Fish and Fishery Products.

6. 1. 3 The name of the packing medium shall form part of the name of the food

6. 1. 4 Where a mixture of species of the same genus are used, they shall be indicated on the label

6. 1. 5 In addition, the label shall include other descriptive terms that will avoid misleading or confusing the consumer

7 SAMPLING, EXAMINATION AND ANALYSES

7. 1 SAMPLING

(i) Sampling of lots for examination of the final product as prescribed in Section 3. 3 shall be in accordance with the FAO/WHO Codex Alimentarius Sampling Plans for Prepackaged Foods (1969) (AQL - 6. 5) (Ref. CAC/RM 42 - 1977);

(ii) Sampling of lots for examination of net weight and drained weight, where appropriate, shall be carried out in accordance with an appropriate sampling plan meeting the criteria established by the CAC.

7. 2 SENSORIC AND PHYSICAL EXAMINATION

Samples taken for sensoric and physical examination shall be assessed by persons trained in such examination and in accordance with Sections 7. 3 through 7. 5, Annex A and the *Guidelines for the Sensory Evaluation of Fish and Shellfish in Laboratories* (CAC/GL 31 - 1999).

7. 3 DETERMINATION OF NET WEIGHT

The net weight of all sample units shall be determined by the following procedure:

(i) Weigh the unopened container;

(ii) Open the container and remove the contents;

(iii) Weigh the empty container, (including the end) after removing excess liquid and adhering meat;

(iv) Subtract the weight of the empty container from the weight of the unopened container. The resultant figure will be the net content.

7. 4 DETERMINATION OF DRAINED WEIGHT

The drained weight of all sample units shall be determined by the following procedure:

(i) Maintain the container at a temperature between 20℃ and 30℃ for a minimum of 12 hours prior to examination.

(ii) Open and tilt the container to distribute the contents on a pre - weighed circular sieve which consists of wire mesh with square openings of 2. 8 mm × 2. 8 mm.

(iii) Incline the sieve at an angle of approximately 17° ~ 20° and allow the fish to drain for two minutes, measured from the time the product is poured into the sieve.

(iv) Weigh the sieve containing the drained fish.

(v) The weight of drained fish is obtained by subtracting the weight of the sieve from the weight of the sieve and drained product.

7. 5 DETERMINATION OF WASHED DRAINED WEIGHT (FOR PACKS WITH SAUCES)

(i) Maintain the container at a temperature between 20℃ and 30℃ for a minimum of 12 hours prior to examination.

(ii) Open and tilt the container and wash the covering sauce and then the full contents with hot tap water (approx. 40℃), using a wash bottle (e. g. plastic) on the tared circular sieve.

(iii) Wash the contents of the sieve with hot water until free of adhering sauce; where necessary separate optional ingredients (spices, vegetables, fruits) with pincers. Incline the sieve at an angle of approximately 17° ~ 20° and allow the fish to drain two minutes, measured from the time the washing procedure has finished.

(iv) Remove adhering water from the bottom of the sieve by use of paper towel. Weigh the sieve containing the washed drained fish.

(v) The washed drained weight is obtained by subtracting the weight of the sieve from the weight of the sieve and drained product.

7.6 DETERMINATION OF HISTAMINE

AOAC 977.13 (15th Edition, 1990).

8 DEFINITION OF DEFECTIVES

A sample unit will be considered defective when it exhibits any of the properties defined below.

8.1 FOREIGN MATTER

The presence in the sample unit of any matter, which has not been derived from fish or the packing medium, does not pose a threat to human health, and is readily recognized without magnification or is present at a level determined by any method including magnification that indicates non - compliance with good manufacturing and sanitation practices.

8.2 ODOUR/FLAVOUR

A sample unit affected by persistent and distinct objectionable odours or flavours indicative of decomposition or rancidity.

8.3 TEXTURE

(i) Excessive mushy flesh uncharacteristic of the species in the presentation; or

(ii) Excessively tough flesh uncharacteristic of the species in the presentation; or

(iii) Honey combed flesh in excess of 5% of the drained contents.

8.4 DISCOLOURATION

A sample unit affected by distinct discolouration of the flesh indicative of decomposition or rancidity or by sulphide staining of more than 5% of the drained contents.

8.5 OBJECTIONABLE MATTER

A sample unit affected by Struvite crystals - any struvite crystal greater than 5 mm in length.

9 LOT ACCEPTANCE

A lot shall be considered as meeting the requirements of this standard when:

(i) the total number of defectives as classified according to Section 8 does not exceed the acceptance number (c) of the appropriate sampling plan in the Sampling Plans for Prepackaged Foods (AQL - 6.5) (CAC/RM 42 - 1977);

(ii) the total number of sample units not meeting the presentation defined in 2.3 does not exceed the acceptance number (c) of the appropriate sampling plan in the Sampling Plans for Prepackaged Foods (AQL - 6.5) (CAC/RM 42 - 1977);

(iii) the average net weight and the average drained weight where appropriate of all sample units examined is not less than the declared weight, and provided there is no unreasonable shortage in any individual container.

(iv) the Food Additives, Hygiene and Handling and Labelling requirements of Sections 3.3, 5, and 6 are met.

ANNEX A SENSORY AND PHYSICAL EXAMINATION

1. Complete external can examination for the presence of container integrity defects or can ends which may be distorted outwards.

2. Open can and complete weight determination according to defined procedures in Sections 7.3, 7.4 and 7.5.

3. Examine the product for the form of presentation.

4. Examine product for discolouration, foreign and objectionable matter. The presence of a hard bone is an indicator of underprocessing and will require an evaluation for sterility.

5. Assess odour, flavour and texture in accordance with the *Guidelines for the Sensory Evaluation of Fish and Shellfish in Laboratories* (CAC/GL 31 - 1999).

块冻鱼片及碎鱼肉标准

CODEX STANDARD FOR QUICK FROZEN BLOCKS OF FISH FILLET, MINCED FISH FLESH AND MIXTURES OF FILLETS AND MINCED FISH FLESH

(*CODEX STAN 165 - 1989, REV. 1 - 1995*)

1　适用范围

本标准适用于粘结鱼肉的速冻块。这种速冻块是由鱼片❶、碎鱼肉或鱼片、碎鱼肉的混合物制备而成的，并用于进一步的深加工。

2　说明

2.1　产品定义

速冻块是将鱼片、碎鱼肉或二者的混合物冻结在一起制成的，呈矩形或其他均一的外形，适于人类消费，包括：

（1）单一品种；

（2）具有相似感官特征的鱼的混合物。

2.1.1　鱼片是具有不同尺寸和形状的鱼的切片。它是经过对鱼体平行于脊椎骨进行切割并对这些鱼片进行接合、修补而制成的。这些鱼片可以是有皮或无皮的。

2.1.2　碎鱼肉的大块产品是由不含鱼骨、内脏、皮、鳞的骨骼肌制成的。

2.2　加工定义

产品经过适当预处理后，应在以下规定的条件下进行冻结加工：冻结应在合适的设备中进行，并使产品迅速通过最大冰晶生成温度带；速冻加工只有在产品的中心温度达到并稳定在 -18℃（0℉）或更低的温度时才算完成；产品在运输、贮存、分销过程中应保持在深度冻结状态，以保证产品质量。

在保证质量的受控条件下，允许按规定要求对速冻产品再次进行速冻加工，并按照被认可的操作进行再包装。

在产品的加工和包装过程中应尽可能减少脱水和氧化作用的影响。

2.3　产品介绍

产品介绍应符合下列要求：

---

❶　可包括多个鱼片。

2.3.1 达到本标准的所有要求
2.3.2 在产品标签中对产品进行详细描述，以免引起混淆或误导消费者
2.3.3 如果已经剔除了包括细刺在内的鱼骨，则可作为无骨碎鱼肉块

3 基本成分及质量因素

3.1 鱼

速冻碎鱼肉块应由品质良好、可作为鲜品供人类消费的个体完整的鱼制备。

3.2 冰衣

用于镀冰衣或制备镀冰衣所用溶液的水应达到饮用水或清洁海水的质量要求。饮用水应符合WHO最新版本的《国际饮用水质量规范》要求，清洁海水应是达到饮用水的微生物标准且不含异物的海水。

3.3 其他成分

所使用的其他成分应具有食品级的质量，并且符合所有相应法规标准的规定。

3.4 腐败

被测样品单位的组胺平均含量不应超过10mg/100g。这仅适用于鲱科（*Clupeidae*）、鲭科（*Scombridae*）、秋刀鱼科（*Scombresocidae*）、鲑科（*Pomatomidae*）以及鳍鳅科（*Coryphaenedae*）等科的鱼类。

3.5 成品

产品应符合本标准要求，根据第9章进行成品批次检查时，其质量应符合第8章的规定。检查方法应符合第7章的规定。

4 食品添加剂

只允许使用以下添加剂。

添加剂		成品中的最高含量
保水剂	339（i）磷酸二氢钠	≤10g/kg，以 P_2O_5 计，单用或混用（包括天然磷酸盐）
	340（i）磷酸二氢钾	
	450（iii）焦磷酸钠	
	450（v）焦磷酸钾	
	451（i）三聚磷酸钠	
	451（ii）三聚磷酸钾	
	452（i）多聚磷酸钠	
	452（iv）多聚磷酸钙	
	401 褐藻酸钠	GMP

续表

	添加剂	成品中的最高含量
抗氧化剂	300 抗坏血酸（维生素 C） 301 抗坏血酸钠 303 抗坏血酸钾	GMP
	304 抗坏血酸棕榈酸酯	1g/kg
仅用于碎鱼肉		
调酸剂	330 柠檬酸 331 柠檬酸钠 332 柠檬酸钾	GMP
增稠剂	412 瓜尔豆胶 410 角豆胶（刺槐豆胶） 440 果胶 466 羧甲基纤维素钠 415 黄原胶 407 卡拉胶及其 Na，K，NH_4 盐（包括红藻胶） 407a 经热加工处理的麒麟菜（PES ） 461 甲基纤维素	GMP

5 卫生及处理

5.1 成品中不能含有任何危害人体健康的外来杂质

5.2 应用 CAC 规定的抽样及检测方法进行检验时：

（1）产品中微生物含量和由微生物产生的危害人体健康的有害物质的含量不能超过 CAC 标准的规定；

（2）组胺含量不应超过 20mg/100g，这仅适用于鲱科（*Clupeidae*）、鲭科（*Scombridae*）、秋刀鱼科（*Scombresocidae*）、鲑科（*Pomatomidae*）以及鲳鳅科（*Coryphaenedae*）等科的鱼类；

（3）其他危害人体健康的有害物质的含量也不能超过 CAC 标准的规定。

5.3 建议本标准中涉及产品预处理的条款应符合推荐性国际规范《CAC/RCP1 - 1969 食品卫生操作总则》中的相关章节和以下相关标准：

（1）推荐性国际操作规范《CAC/RCP 16 - 1978 冻鱼》；

（2）推荐性国际操作规范《CAC/ RCP 35 - 1985 冻挂浆或和（或）沾面包屑的水产品》；

（3）推荐性国际操作规范《CAC/ RCP 27 - 1983 机械分离法制备碎鱼肉》；

（4）推荐性国际操作规范《CAC/RCP 8 - 1976 速冻食品的加工和处理》；

（5）《水产品国际操作规程草案》中水产养殖品的相关章节（正在完善中）❶。

---

❶ 操作规程建议草案，在定稿后将替代所有现行的《水产品操作规程》。

6 标签

除应符合《CODEX STAN1 - 1985，Rev. 3 - 1999 预包装食品标签通用标准》的要求外，还应遵守以下规定：

6.1 食品名称

6.1.1 食品名称应根据出售该食品的国家的法律、习惯、实际情况，表示为“X Y 块”，其中 X 为该包装品种的常用名，Y 为鱼肉块形状（见 2.3）。
6.1.2 如果产品用海水镀冰衣，则应予以说明。
6.1.3 在产品标签上应标明“速冻”，除非在有的国家“冻”即代表速冻。
6.1.4 碎块的比例如果超过鱼肉净含量的 10%，则应注明其百分比等级：10 ~ 25、大于 25 ~ 35 等。碎块大于 90% 的鱼块，被认为是碎鱼肉块。
6.1.5 标签应注明产品须在运输、贮藏、分销过程中保持的条件，以保证其质量。

6.2 净含量（镀冰衣产品）

对于镀冰衣的产品，其内容物净含量不包括冰衣重。

6.3 贮藏说明

标签应注明产品须在 -18℃或更低的温度下贮藏。

6.4 非零售包装的标签

上述要求应既在包装上又在附文中出现，除了食品名称、批号、制造或分装厂名、地址外，还包括贮藏条件。但批号、制造或分装厂名、地址也可用同一证明标志代替，只要证明标志能在辅助文件中说明清楚。

7 抽样、检验和分析

7.1 取样

（1）产品批次检验用样品的抽样方法应依据下述的抽样计划进行。抽样单位为整块速冻鱼肉块。

批量（速冻鱼肉块的数目）	样品量（测试用速冻鱼肉块的数目，n）	接受数（c）
<15	2	0
16 ~ 50	3	0
51 ~ 150	5	1
151 ~ 500	8	1
501 ~ 3 200	13	2
3 201 ~ 35 000	20	3
>35 000	32	5

如果抽样样品中有缺陷的速冻鱼块的数目少于或等于 c，则接受此批产品；反之，则拒绝此批产品。

（2）对需检测净重批次的抽样，抽样计划应以食品法典委员会的相关标准为依据。

7.2　感官与物理检验

产品的感官与物理指标须由经过此类检验培训的人员进行检验，并依据本标准 7.3 至 7.7 条和附录 A 以及《CAC/GL 31 - 1999 实验室中水产品感官评价指南》所述程序进行。

7.3　净重的测定

7.3.1　未镀冰衣产品净重的测定

代表每批次的样品单位的净重（不包括包装材料）应在冷冻状态下测定。

7.3.2　镀冰衣产品的净重的测定

从低温贮藏状态下取出样品后，立即打开包装，将内容物放在冷水喷头下，用缓慢的水流喷洒内含物，并小心摇动样品，不要弄破样品。直到可以看到或用手感觉到全部冰衣已被去除时，停止喷洒。用纸巾除去黏附的水并放在已知皮重的盘中称重。

在附录 B 中，介绍了另一种方法。

7.4　无皮鱼片冻块中寄生虫的检验程序（方法 1）

对整个抽样单位的检测可以通过以下方法非破坏性地进行：将适当大小的解冻样品置于一 5mm 厚、透光率为 45% 的丙烯酸塑料薄片上，并在其上 30cm 处应用发出 1 500lux（勒克司）的光源进行灯检。

7.5　鱼片和碎鱼肉混合物的速冻鱼块中鱼片和碎鱼肉的比例的测定[1][2]

依据 AOAC 988. 09 的方法进行测定。

7.6　凝胶情况的测定

按照 AOAC 方法 983. 18“肉及肉制品中水分含量测定时抽样的准备工作” 和 950. 46“肉中的水分含量”（方法 A）进行测定。

7.7　蒸煮方法

蒸煮使产品内部温度达到 65 ~ 70℃。不能过度蒸煮，蒸煮时间随产品大小和采用的温度而不同。准确的蒸煮时间和条件应依据预先实验来确定。

烘焙：用铝箔包裹产品，并将其均匀放入扁平锅或浅平锅上。

蒸：用铝箔包裹产品，并将其置于带盖容器中沸水之上的金属架上。

袋煮：将产品放入可煮薄膜袋中加以密封，浸入沸水中煮。

微波：将产品放入适于微波加热的容器中，若用塑料袋，应检查确定塑料袋不会发出任何气味。根据设备说明加热。

7.8　速冻鱼块的解冻程序

空气解冻法：

---

[1] 此方法只对鳕做过评价，但原则上，应适用于其他品种或混合品种的鱼片产品。

[2] 此方法只对鳕做过评价，但原则上，应适用于其他品种或混合品种的鱼片产品。

去除速冻鱼块的包装。将速冻鱼块分别放入合适的暖和的非渗透性塑料袋中，或将其分别置于相对湿度至少在80%的湿度受控的环境下。尽可能去除袋中的空气，并密封。将被密封于塑料袋中的速冻鱼块分别置于各自盘中，在25℃（77 ℉）或更低的空气温度下进行解冻。当无须撕扯而可以将产品轻松地分离时，解冻完成。鱼块内的温度不应超过7℃（44.6 ℉）。

水浸法：

去除速冻鱼块的包装。将速冻鱼块密封于塑料袋中。尽可能去除袋中的空气，并密封。将速冻鱼块置于循环水浴中，将温度维持在21℃+1.5℃（70 ℉+3 ℉）。当无须撕扯而可以将产品轻松地分离时，解冻完成。鱼块内的温度不应超过7℃（44.6 ℉）。

7.9 组胺检测

参见 AOAC 977.13。

8 缺陷的定义

当样品单位呈现下列任何一项特征时，则认定其是“有缺陷的”：

8.1 深度脱水

样品单位中表面积超过10%出现水分的过度损失，明显表现为鱼体表面呈现异常的白色或黄色，覆盖了肌肉本身的颜色，并已渗透至表层以下，在不过分影响产品外观的情况下，不能轻易地用刀或其他利器刮掉。

8.2 外来杂质

样品单位中存在的任何不是来自于鱼体的物质（包装材料除外）。这些物质虽不会对人体健康造成危害，但用肉眼可直接辨别，或采用某些方法（包括放大）可以确定其存在。出现外来杂质表明不符合良好操作和卫生习惯。

8.3 寄生虫

采用7.4方法，每千克样品个体检测到2个或2个以上直径大于3mm的胶囊状寄生虫，或非胶囊状但长度大于10mm的寄生虫。

8.4 骨刺（在指明无刺的包装中）

每千克产品中检测出多于一个长度为10mm或者直径大于或等于1mm的骨刺；骨刺的长度等于或小于5mm，如果其直径不大于2mm则该产品不被认定为缺陷。如果骨刺根部（与脊椎骨的连接处）的宽度小于或等于2mm，或其可用手指甲轻松地剥除，则其可被忽略。

8.5 气味/风味

样品散发的持久、明显、令人厌恶的由腐败、酸败或饵料引起的气味或风味。

8.6 鱼肉异常

样品单位出现过量凝胶状态的鱼肉并伴有任何单片中水分达86%以上，或按重量计算5%以上的样品被寄生虫感染导致质地呈现糊状。

9 批次验收

当满足以下条件时，可以认为本批次产品符合本标准的要求：

（1）根据本标准第8章中规定分类，缺陷总数不超过《CODEX STAN 233 - 1969 预包装食品的抽样方案》（AQL - 6.5）中相应抽样方案规定的可接受数（c）；

（2）所有样品单位平均净重不少于标示量，在任何一个包装单位中没有不合理的重量短缺；

（3）符合本标准第4、5、6款中对食品添加剂、卫生和处理以及标签的要求。

附录A 感官及物理检验

1. 净含量的测定按本标准7.3条规定执行（按要求去冰衣）。

2. 通过测定只能用小刀或其他利器除去的面积，检查冻鱼片中脱水的情况。测量样品单位的总表面积，计算受影响的面积百分比。

3. 解冻并且逐个地检测样品单位中每个冻鱼块有无外来杂质、寄生虫、骨刺（需要时）、臭味及鱼肉质地异常等缺陷。

4. 对在解冻后未蒸煮状态下无法最终判定其气味的样品，则应从样品单位中截取一小块可疑部分（约200g），并立即使用7.8条中规定的蒸煮方法，确定其气味和风味。

5. 对在解冻后未蒸煮状态下无法最终确定其凝胶状态的产品，则应从产品中截取可疑部分，按7.7条规定的蒸煮方法，或用7.6测定在任何一鱼块中是否含有超过86%的水分，即可确认其凝胶情况。如果无法根据蒸煮后的评估得到结论，则使用7.6条中的程序准确测定水分含量。

附录B 镀冰衣速冻鱼块净含量的测定方法

在白色肉鱼类速冻鱼块的生产中不镀冰衣。只有鲱、鲭及其他有褐色（脂肪）鱼肉鱼类的速冻鱼块的生产中要镀冰衣，它们通常用于深加工（罐装、熏制等）。对于这些速冻鱼块，可采用以下的程序（使用速冻虾块测试方法）：

1. 原则

用手将预称重镀冰衣样品浸入水浴中，直到全部冰衣都被除去（用手感觉）。表面变粗糙后，立刻将仍处于冻结状态的样品从水浴中移出，用纸巾将其拭干后，再次称重并计算产品净含量。通过此程序，可避免解冻汁液的流失和（或）附着的水分的重结冰。

2. 设备

· 天平，感度为1g

· 水浴，温度可调式

· 圆形滤网（ISO R 565），直径20cm、滤网孔径1～3mm

· 纸巾或棉布，表面光滑

· 冷冻箱，在工作地点使到

3. 样品和水浴的准备

· 应将产品温度调整到 - 18℃/ - 20℃以达到标准去冰衣条件（如果确定了规定外形产品的标准去冰衣周期，则尤其必要）。

· 在低温贮存条件下抽样后，如发现冻结产品包装上的外部有冰结晶或冰霜，清除之。

· 水浴中应含有约10倍于产品标示重量的饮用水；水温应调到15～35℃。

4. 毛重“A”的测定

去除包装后，确定镀冰衣产品的重量：对于单个鱼片，记录单体重量（A_1～A_n）。已称重的样品应立即放入冷冻箱中。

5. 去除冰衣：

将预称重样品/（从中抽取的）子样品，移入水浴，浸入水下。小心摇动产品，直到用指尖，感觉到产品表面从光滑变得粗糙，没有冰衣。根据产品所含冰衣的尺寸/外形，所用时间为 10～60s（如果冰衣含量较高或一起冻结，则可更长）。

对于零售包装的块冻产品（也对于贮存期间一起冻结的单个镀冰衣产品）可采用下述（初步的）程序：将预称重鱼块或其中的一部分置于尺寸合适的滤网上，并浸入水浴中。用手指轻压，冰衣部分被去除。如果仍然有冰衣残渣，则重复进行快速的浸入。

6. 净重“B”的测定

在用纸巾（不能按压）清除已去除冰衣的样品/（从中抽取的）子样品表面上附着的水后，立即称重。得到单个子样品的净重：$B_1 \sim B_n$。

（1）冰衣重量“C”的确定

毛重“A”－净重“B”＝冰衣重量“C”

（2）百分比含量的计算

产品中净重的百分含量“F”%＝净重“B”/毛重“A”×100

冰衣的百分含量（对产品的毛重）“G”%＝冰衣重量“C”/毛重“A”×100

冰衣的百分含量（对产品的净重）“H”%＝冰衣重量“C”/净重“B”×100

CODEX STANDARD FOR QUICK FROZEN BLOCKS OF FISH FILLET, MINCED FISH FLESH AND MIXTURES OF FILLETS AND MINCED FISH FLESH

(CODEX STAN 165 – 1989, REV. 1 – 1995)

1 SCOPE

This standard applies to quick frozen blocks of cohering fish flesh, prepared from fillets[1] or minced fish flesh or a mixture of fillets and minced fish flesh, which are intended for further processing.

2 DESCRIPTION

2.1 PRODUCT DEFINITION

Quick frozen blocks are rectangular or other uniformly shaped masses of cohering fish fillets, minced fish or a mixture thereof, which are suitable for human consumption, comprising:

(i) a single species; or

(ii) a mixture of species with similar sensory characteristics.

2.1.1 Fillets are slices of fish of irregular size and shape which are removed from the carcass by cuts made parallel to the back bone and pieces of such fillets, with or without the skin.

2.1.2 Minced fish flesh used in the manufacture of blocks are particles of skeletal muscle which have been separated from and are essentially free from bones, viscera and skin.

2.2 PROCESS DEFINITION

The product after any suitable preparation shall be subjected to a freezing process and shall comply with the conditions laid down hereafter. The freezing process shall be carried out in appropriate equipment in such a way that the range of temperature of maximum crystallization is passed quickly. The quick freezing process shall not be regarded as complete unless and until the product temperature has reached –18℃ or colder at the thermal centre after thermal stabilization. The product shall be kept deep frozen so as to maintain the quality during transportation, storage and distribution.

Industrial repacking or further processing of intermediate quick frozen material under controlled conditions which maintain the quality of the product followed by the reapplication of the quick freezing process is permitted.

---

[1] Including pieces of fillets.

These products shall be processed and packaged so as to minimize dehydration and oxidation.

2. 3 PRESENTATION

Any presentation of the product shall be permitted provided that it:

2. 3. 1 meets all requirements of this standard, and

2. 3. 2 is adequately described on the label to avoid confusing or misleading the consumer.

2. 3. 3 Blocks may be presented as boneless, provided that boning has been completed including the removal of pin – bones.

3 ESSENTIAL COMPOSITION AND QUALITY FACTORS

3. 1 FISH

Quick frozen blocks shall be prepared from fillets or minced flesh of sound fish which are of a quality fit to be sold fresh for human consumption.

3. 2 GLAZING

If glazed, the water used for glazing or preparing glazing solutions shall be of potable quality or shall be clean sea – water. Potable water is fresh – water fit for human consumption. Standards of potability shall not be less than those contained in the latest edition of the WHO "International Guidelines for Drinking Water Quality" .

Clean sea – water is sea – water which meets the same microbiological standards as potable water and is free from objectionable substances.

3. 3 OTHER INGREDIENTS

All other ingredients used shall be of food grade quality and conform to all applicable Codex standards.

3. 4 DECOMPOSITION

The products shall not contain more than 10 mg/100 g of histamine based on the average of the sample unit tested. This shall apply only to species of *Clupeidae*, *Scombridae*, *Scombresocidae*, *Pomatomidae* and *Coryphaenedae families.*

3. 5 FINAL PRODUCT

Products shall meet the reguirements of this standard when lots examined in accordance with Section 9 comply with the provisions set out in Section 8. Products shall be examined by the methods given in Section 7.

4 FOOD ADDITIVES

Only the use of the following additives is permitted.

Additive	Maximum Level in the Final Product
Moisture/Water Retention Agents	
339 (i) Monosodium orthophosphate	10 g/kg expressed as P_2O_5, singly or in combination (includes natural phosphate)
340 (i) Monopotassium orthophosphate	
450 (iii) Tetrasodium diphosphate	
450 (v) Tetrapotassium diphosphate	
451 (i) Pentasodium triphosphate	
451 (ii) Pentapotassium triphosphate	
452 (i) Sodium polyphosphate	
452 (v) Calcium, polyphosphates	
401 Sodium alginate	GMP
Antioxidants	
300 Ascorbic acid	GMP
301 Sodium ascorbate	
303 Potassium ascorbate	
304 Ascorbyl palmitate	1 g/kg
In Minced Fish Flesh Only Acidity Regulator	
330 Citric acid	GMP
331 Sodium citrate	
332 Potassium citrate	
Thickeners	
412 Guar gum	GMP
410 Carob bean (Locust bean) gum	
440 Pectins	
466 Sodium carboxymethyl cellulose	
415 Xanthan gum	
407 Carrageenan and its Na, K, NH_4 salts (including Furcelleran)	
407a Processed *Eucheuma* Seaweed (PES)	
461 Methyl cellulose	

5 HYGIENE AND HANDLING

5.1 The final product shall be free from any foreign material that poses a threat to human health.

5.2 When tested by appropriate methods of sampling and examination prescribed by the Codex Alimentarius Commission, the product:

(i) shall be free from microorganisms or substances originating from microorganisms in amounts which may represent a hazard to health in accordance with standards established by the Codex Alimentarius Commission;

(ii) shall not contain histamine that exceeds 20 mg/100 g in any sample unit. This applies only to species of Clupeidae, Scombridae, Scombresocidae, Pomatomidae and Coryphaenedae families;

(iii) shall not contain any other substances in amounts which may represent a hazard to health in accordance with standards established by the Codex Alimentarius Commission.

5.3 It is recommended that the product covered by the provisions of this standard be prepared and handled in accordance with the appropriate sections of the Recommended International Code of Practice – General Principles of Food Hygiene (CAC/RCP 1 – 1969) and the following relevant Codes:

(i) The Recommended International Code of Practice for Frozen Fish (CAC/RCP 16 – 1978).

(ii) The Recommended International Code of Practice for Frozen Battered and/or Breaded Fishery Products (CAC/RCP 35 – 1985).

(iii) The Recommended International Code of Practice for Minced Fish Prepared by Mechanical Separation (CAC/RCP 27 – 1983).

(iv) The Recommended International Code of Practice for the Processing and Handling of Quick Frozen Foods (CAC/RCP 8 – 1976).

(v) The sections on the Products of Aquaculture in the Proposed Draft International Code of Practice for Fish and Fishery Products (under elaboration)[1].

6 LABELLING

In addition to the provisions of the Codex General Standard for the Labelling of Prepackaged Foods (CODEX STAN 1 – 1985) the following specific provisions apply.

6.1 THE NAME OF THE FOOD

6.1.1 The name of the food shall be declared as "x y blocks" in accordance with the law, custom or practice of the country in which the product is distributed, where "x" shall represent the common name (s) of the species packed and "y" shall represent the form of presentation of the block (see Section 2.3).

6.1.2 If the product has been glazed with sea – water, at statement to this effect shall be made.

6.1.3 The name "quick frozen", shall also appear on the label, except that the term "frozen" may be applied in countries where this term is customarily used for describing the product processed in accordance with subsection 2.2 of this standard.

6.1.4 The proportion of mince in excess of 10% of net fish content shall be declared stating the percentage ranges: 10 ~ 25, > 25 ~ 35, etc. Blocks with more than 90% mince are regarded as mince blocks.

6.1.5 The label shall state that the product should be maintained under conditions that will maintain the quality during transportation, storage and distribution.

6.2 NET CONTENTS (GLAZED BLOCKS)

Where the food has been glazed, the declaration of net contents of the food shall be exclusive of the glaze.

6.3 STORAGE INSTRUCTIONS

The label shall include terms to indicate that the product shall be stored at a temperature of – 18℃ or colder.

6.4 LABELLING OF NON – RETAIL CONTAINERS

Information specified above shall be given either on the container or in accompanying documents, ex-

---

[1] The Proposed Draft Code of Practice, when finalized, will replace all current Codes of Practice for Fish and Fishery Products.

cept that the name of the product, lot identification, and the name and address of the manufacturer or packer as well as storage instructions, shall appear on the container.

However, lot identification, and the name and address of the manufacturer or packer may be replaced by an identification mark provided that such mark is clearly identifiable with the accompanying documents.

7 SAMPLING, EXAMINATION AND ANALYSES

7.1 SAMPLING PLAN FOR FISH BLOCKS

(i) Sampling of lots for examination of the product shall be in accordance with the sampling plan defined below.

The sample unit is the entire block.

Lot Size (Number of blocks)	Sample Size (Number of blocks to be tested, n)	Acceptance number (c)
<15	20	
16 ~ 50	3	0
51 ~ 150	5	1
151 ~ 500	8	1
501 ~ 3 200	13	2
3 201 ~ 35 000	20	3
>35 000	32	5

If the number of defective blocks in the sample is less than or equal to c, accept the lot; otherwise, reject the lot.

(ii) Sampling of lots for examination of net weight shall be carried out in accordance with an appropriate sampling plan meeting the established criteria established by the CAC.

7.2 SENSORY AND PHYSICAL EXAMINATION

Samples taken for sensory and physical examination shall be assessed by persons trained in such examination and in accordance with procedures elaborated in Sections 7.3 through 7.7 and Annex A and in accordance with the Code of Practice for the Sensory Evaluation of Fish and Shellfish (under development) .

7.3 DETERMINATION OF NET WEIGHT

7.3.1 Determination of Net Weight of Product Not Covered by Glaze

The net weight (exclusive of packaging material) of each sample unit representing a lot shall be determined in the frozen state.

7.3.2 Determination of Net Weight of Products Covered by Glaze As soon as the package is removed from frozen temperature storage, open immediately and place the contents under a gentle spray of cold water until all ice glaze that can be seen or felt is removed. Remove adhering water by the use of paper towel and weigh the product.

An alternate method is outlined in Annex B.

7.4 PROCEDURE FOR THE DETECTION OF PARASITES FOR SKINLESS BLOCKS OF FISH FILLETS (TYPE I METHOD)

The entire sample unit is examined non – destructively by placing appropriate portions of the thawed sample unit on a 5 mm thick acryl sheet with 45% translucency and candled with a light source giving 1500 lux 30 cm above the sheet.

7.5 DETERMINATION OF PROPORTIONS OF FILLET AND MINCED FISH IN QUICK FROZENBLOCKS PREPARED FROM MIXTURES OF FILLETS AND MINCED FISH❶❷

According to the AOAC Method – "Physical Separation of Fillets and Minced Fish", AOAC 1988, 71, 206 (Type Ⅱ).

7.6 DETERMINATION OF GELATINOUS CONDITION

According to the AOAC Methods – "Moisture in Meat and Meat Products, Preparation of Sample Procedure"; AOAC 1990, 983.18 and "Moisture in Meat" Method A, 950.46; AOAC 1990.

7.7 COOKING METHODS

The following procedures are based on heating the product to an internal temperature of 65 ~ 70℃. The product must not be overcooked. Cooking times vary according to the size of the product and the temperatures used. The exact times and conditions of cooking for the products should be determined by prior experimentation.

Baking Procedure: Wrap the product in aluminum foil and place it evenly on a flat cookie sheet or shallow flat pan.

Steaming Procedure: Wrap the product in aluminum foil and place it on a wire rack suspended over boiling water in a covered container.

Boil – In – Bag Procedure: Place the product into a boilable film – type pouch and seal. Immerse the pouch into boiling water and cook.

Microwave Procedure: Enclose the product in a container suitable for microwave cooking. If plastic bags are used, check to ensure that no odour is imparted from the plastic bags. Cook according to equipment instructions.

7.8 THAWING PROCEDURE FOR QUICK FROZEN BLOCKS

Air Thaw Method:

Frozen fish blocks are removed from the packaging. The frozen fish blocks are individually placed into snug fitting impermeable plastic bags or a humidity controlled environment with a relative humidity of at least 80%. Remove as much air as possible from the bags and seal. The frozen fish blocks sealed in plastic bags are placed on individual trays and thawed at air temperature of 25℃ (77 ℉) or lower. Thawing is completed when the product can be readily separated without tearing. Internal block temperature should not

---

❶ This method has been evaluated for cod only but, in principle, should be appropriate to other fish species or mixed species.

❷ This method is accurate for levels of mince greater than 10%.

exceed 7℃ (44.6 ℉).

Water Immersion Method:

Frozen fish blocks are removed from the packaging. The frozen fish blocks are sealed in plastic bags. Remove as much air as possible from the bags and seal. The frozen fish blocks are placed into a circulating water bath with temperatures maintained at 21℃ + 1.5℃ (70 ℉ + 3 ℉). Thawing is completed when the product can be easily separated without tearing. Internal block temperature should not exceed 7℃ (44.6 ℉).

7.9 DETERMINATION OF HISTAMINE

According to AOAC 977.13 (15th Edition, 1990).

8 DEFINITION OF DEFECTIVES

The sample unit shall be considered defective when it exhibit any of the properties defined below.

8.1 DEEP DEHYDRATION

Greater than 10% of the surface area of the sample unit exhibits excessive loss of moisture clearly shown as white or yellow abnormality on the surface which masks the colour of the flesh and penetrates below the surface, and cannot be easily removed by scraping with a knife or other sharp instrument without unduly affecting the appearance of the block.

8.2 FOREIGN MATTER

The presence in the sample unit of any matter which has not been derived from fish (excluding packing material), does not pose a threat to human health, and is readily recognized without magnification or is present at a level determined by any method including magnification that indicates non - compliance with good manufacturing and sanitation practices.

8.3 PARASITES

The presence of two or more parasites per kg of the sample unit detected by a method described in 7.4 with a capsular diameter greater than 3 mm or a parasite not encapsulated and greater than 10 mm in length.

8.4 BONES (IN PACKS DESIGNATED BONELESS)

More than one bone per kg of product greater or equal to 10mm in length, or greater or equal to 1 mm in diameter; a bone less than or equal to 5 mm in length, is not considered a defect if its diameter is not more than 2 mm. The foot of a bone (where it has been attached to the vertebra) shall be disregarded if its width is less than or equal to 2 mm, or if it can easily be stripped off with a fingernail.

8.5 ODOUR AND FLAVOUR

A sample unit affected by persistent and distinct objectionable odours or flavours indicative of decomposition or rancidity or of feed.

8.6 FLESH ABNORMALITIES

A sample unit affected by excessive gelatinous condition of the flesh together with greater than 86%

moisture found in any individual fillet, or a sample unit with pasty texture resulting from parasitic infestation affecting more than 5% of the sample unit by weight.

9 LOT ACCEPTANCE

A lot shall be considered as meeting the requirements of this standard when:

(i) the total number of defective sample units as classified according to Section 8 does not exceed the acceptance number (c) of the sampling plan in Section 7; and

(ii) the average net weight of all sample units is not less than the declared weight, provided there is no unreasonable shortage in any container; and

(iii) the Food Additives, Hygiene and Labelling requirements of Sections 3.4, 4, 5.1, 5.2 and 6 are met.

ANNEX A SENSORY AND PHYSICAL EXAMINATION

1. Complete net weight determination, according to defined procedures in Section 7.3 (de-glaze as required).

2. Examine the frozen block for the presence of dehydration by measuring those areas which can only be removed with a knife or other sharp instrument. Measure the total surface area of the sample unit, and calculate the percentage affected.

3. Thaw and individually examine each block in the sample unit for the presence of foreign matter, bone where applicable, odour, and textural defects.

4. In cases where a final decision on odour can not be made in the thawed uncooked sate, a small portion of the disputed material (approximately 200 g) is sectioned from the block and the odour and flavour confirmed without delay by using one of the cooking methods defined in Section 7.8.

5. In cases where a final decision on gelatinous condition cannot be made in the thawed uncooked state, the disputed material is sectioned from the block and the gelatinous condition confirmed by cooking as defined in Section 7.7. or by using procedure in Section 7.6. to determine if greater than 86% moisture is present in any fillet. If cooking evaluation is inconclusive, then procedure in 7.6. would be used to make the exact determination of moisture content.

ANNEX B METHOD FOR THE DETERMINATION OF NET CONTENT OF FROZEN FISH BLOCKS COVERED BY GLAZE

Glazing is not used for Q.F. blocks of white fish. Only Q.F. blocks of herring, mackerel and other brown (fat) fish are glazed, which are destined for further processing (canning, smoking). For such blocks the following procedure may be applicable (tested with block frozen shrimps).

1. PRINCIPLE:

The pre-weighed glazed sample is immersed into a water bath by hand till all glaze is removed (as felt by fingers). As soon as the surface becomes rough, the still frozen sample is removed from the water bath and dried by use of a paper towel before estimating the net product content by repeated weighing. By this procedure thaw drip losses and/or re-freezing of adhering moisture can be avoided.

2. EQUIPMENT:

- Balance – sensitive to 1g.
- Water bath, preferably with adjustable temperature.
- Circular sieve with a diameter of 20cm and 1 – 3mm mesh apertures (ISO R 565)
- Paper or cloth towels with smooth surface.
- A freezed box should be available at the working place.

3. PREPARATION OF SAMPLES AND WATER BATH:

- The product temperature should be adjusted to –18℃/–20℃ to achieve standard deglazing conditions (especially necessary if a standard deglazing period shall be defined in case of regular shaped products).
- After sampling from the low temperature store remove, if present, external ice crystals or snow from the package with the frozen product.
- The water bath shall contain an amount of fresh potable water equal to about 10 times of the declared weight of the product; the temperature should be adjusted on about 15℃ to 35℃.

4. DETERMINATION OF GROSS – WEIGHT "A":

After removal of the package, the weight of the glazed product is determined: In case of single fish fillets, single weights are recorded ($A_1 - A_n$). The weighed samples are placed intermediately into the freezer box.

5. REMOVAL OF GLAZE:

The pre – weighed samples/sub – samples are transferred into the water bath and kept immersed by hand. The product may be carefully agitated, till no more glaze can be felt by the finger – tips on the surface of the product: change from slippery to rough. Needed time, depending on size/shape and glaze content of the product, 10 to 60 sec. (and more in case of higher glaze contents or if frozen together).

For block – frozen products in consumer packs (also for single glaze products, which are frozen together during storage) the following (preliminary) procedure may be applicable: The pre – weighed block or portion is transferred onto a suitable sized sieve and immersed into the water bath. By slight pressure of the fingers separating deglazed portions are removed fractionally. Short immersing is repeated, if glaze residues are still present.

6. DETERMINATION OF NET WEIGHT "B"

The deglazed sample/sub – sample, after removal of adhering water by use of a towel (without pressure) is immediately weighed. Single net – weights of sub – samples are summed up: $B_1 - B_n$.

7. DETERMINATION OF GLAZE – WEIGHT "C"

Gross weight "A" – Net weight "B" = Glaze weight "C"

8. CALCULATION OF PERCENTAGE PROPORTIONS:

% net content of the product "F" $= \frac{\text{“B”}}{\text{“A”}} \times 100$

% glaze – related to the gross weight of the product "G" $= \frac{\text{“C”}}{\text{“A”}} \times 100$

% glaze – related to the net weight of the product "H" $= \frac{\text{“C”}}{\text{“B”}} \times 100$

冻沾面包屑或挂浆鱼条（鱼棒）、鱼片和鱼块

CODEX STANDARD FOR QUICK FROZEN FISH STICKS (FISH FINGERS), FISH PORTIONS AND FISH FILLETS - BREADED OR IN BATTER

(CODEX STAN 166-1989, REV. 1-1995)

1 适用范围

本标准适用于从速冻鱼块上切下的或由碎鱼肉组成的速冻鱼条和鱼块以及自然状态的鱼片沾面包屑或裹面糊包（单独或混合）而制成，不用进行工业深加工即可直接供人类消费的生品或半熟品。

2 说明

2.1 产品定义

2.1.1 鱼条（鱼棒）是指包括裹衣在内，重量不小于20g，且不大于50g，形状上其长度不小于最大宽度的3倍的产品。每条的厚度不小于10mm。
2.1.2 鱼块包括裹衣，但与2.1.1条中规定不同，可以是任意形状、重量或尺寸。
2.1.3 鱼条或鱼块可由同一品种的鱼制成，也可由感官特性相似的品种的鱼混合后制成。
2.1.4 鱼片是具有不同尺寸和形状的鱼的切片。它是通过对鱼体进行平行于鱼脊骨地剖切，并对这些鱼片进行接合、修补而制成的。这些鱼片可以是有皮或无皮的。

2.2 加工定义

产品经过适当预处理后，应在符合以下规定的条件下进行冻结加工：冻结应在适当的设备中进行，并使产品迅速通过最大冰晶生成温度带；速冻加工只有在产品的中心温度达到并稳定在-18℃或更低的温度时才算完成；产品在运输、贮存、分销过程中应保持在深度冻结状态，以保证产品质量。

在保证质量控制条件下，允许按规定要求对半成速冻产品再次进行速冻加工或者进行工业再包装。

2.3 产品介绍

产品介绍应符合下列要求：
2.3.1 达到本标准的所有要求；
2.3.2 在产品标签中对产品进行详细描述，以免引起混淆或误导消费者。

3 基本成分及质量因素

3.1 原料

3.1.1 鱼

速冻沾面包屑或面糊包衣的鱼条、鱼块、鱼片应由鱼片、碎鱼肉或其混合物制备而成，原料应为品质新鲜的可食用品种，能作为鲜品出售供人们消费。

3.1.2 裹衣

裹衣及使用的所有其他配料均应具有食品级的质量，并且符合所有相应法规标准的规定。

3.1.3 烹调用脂肪（油）

烹调用脂肪（油）应适合人们消费，并能用来达到预期的成品品质（参看第 4 部分）。

3.2 成品

产品应符合本标准要求，根据第 9 章进行成品批次检查时，其质量应符合第 8 章的规定。检查方法应符合第 7 章的规定。

3.3 腐败

被测样品单位的组胺平均含量不应超过 10mg/100g。仅适用于鲱科（*Clupeidae*）、鲭科（*Scombridae*）、秋刀鱼科（*Scomberesocidae*）、鯥科（*Pomatomidae*）以及鲯鳅科（*Coryphaenedae*）等科的鱼类。

4 食品添加剂

只允许使用下列添加剂。

<table>
<tr><th colspan="2">添加剂</th><th>成品中的最高含量</th></tr>
<tr><td colspan="3">只适用于鱼片和碎鱼肉</td></tr>
<tr><td rowspan="9">保水剂</td><td>339（i）磷酸二氢钠
340（i）磷酸二氢钾
450（iii）焦磷酸钠
450（v）焦磷酸钾
451（i）三聚磷酸钠
451（ii）三聚磷酸钾
452（i）多聚磷酸钠
452（v）多聚磷酸钙</td><td>≤10g/kg，以 P_2O_5 计，
单用或混用
（包括天然磷酸盐）</td></tr>
<tr><td>401 褐藻酸钠</td><td>GMP</td></tr>
<tr><td rowspan="2">抗氧化剂</td><td>301 抗坏血酸（维生素 C）
301 抗坏血酸钠
303 抗坏血酸钾</td><td>GMP</td></tr>
<tr><td>304 抗坏血酸棕榈酸脂</td><td>1g/kg</td></tr>
</table>

续表

添加剂		成品中的最高含量
只适用于碎鱼肉		
调酸剂	330 柠檬酸	GMP
	331 柠檬酸钠	
	332 柠檬酸钾	
	412 瓜尔豆胶	
	410 角豆胶（刺槐豆胶）	
	440 果胶	
增稠剂	466 羧甲基纤维素钠盐	GMP
	415 黄原胶	
	407 卡拉胶及其 Na，K，NH_4 盐（包括红藻胶）	
	407a 经热加工处理的麒麟菜（PES）	
	461 甲基纤维素	
用于面包屑和面包糊		
发酵剂	341（i）磷酸一氢钙	≤1g/kg，以 P_2O_5 计，单用或混用
	341（ii）磷酸二钙	
	541 磷酸钠铝，碱性及酸性	
	500 碳酸钠	GMP
	501 碳酸钾	
	502 碳酸铵	
增味剂	621 谷氨酸钠（味精）	GMP
	622 谷氨酸钾	
色素	160b 胭脂树籽红（萃取物）	以类胡萝卜素表示 20mg/kg
	150a 焦糖色 I（纯）	GMP
	160a β－胡萝卜素（合成）	100mg/kg 单用或混用
	160e β－阿朴－8’－胡萝卜醛	
增稠剂	412 瓜尔豆胶	GMP
	410 角豆胶（刺槐豆胶）	
	440 果胶	
	466 羧甲基纤维素钠	
	415 黄原胶	
	407 卡拉胶及其 Na,K,NH_4 盐(包括红藻胶)	
	407 经热加工处理的麒麟菜（PES ）	
	461 甲基纤维素	
	401 褐藻酸钠	
	463 羟基丙基纤维素	
	464 羟基丙基甲基纤维素	
	465 甲基乙基纤维素	
乳化剂	471 甘油-脂肪酸酯	GMP
	322 卵磷脂	

续表

添加剂		成品中的最高含量
改性淀粉	1401 酸处理淀粉 1402 碱处理淀粉 1404 氧化淀粉 1410 磷酸单淀粉 1412 磷酸二淀粉酯化三偏磷酸钠；磷酸二淀粉酯化氢氧化磷 1414 磷酸乙酰化二淀粉 1413 磷酸化磷酸二淀粉 1420 醋酸淀粉酯化乙酸酐 1421 醋酸淀粉酯化乙烯基乙酸酯 1422 乙酰化二淀粉己二酸酯 1440 羧丙基淀粉 1442 磷酸羧丙基淀粉	GMP

5 卫生及处理

5.1 成品中不能含有任何危害人体健康的外来杂质

5.2 应用 CAC 规定的相应抽样及检测方法进行检验时：

（1）产品中微生物含量和由微生物产生的危害人体健康的有害物质的含量应该符合 CAC 标准的规定；

（2）组胺含量不应超过 20mg/100g，这仅适用于鲱科（Clupeidae）、鲭科（Scombridae）、秋刀鱼科（Scomberesocidae）、鯥科（Pomatomidae）以及鲯鳅科（Coryphaenedae）等科的鱼类；

（3）其他危害人体健康的有害物质的含量也不得超过 CAC 标准的规定。

5.3 建议本标准中涉及产品制备和处理的条款应符合《推荐性国际规范——食品卫生操作总则》（CAC/RCP1 - 1969，Rev. 3 - 1997）中的相关章节和以下相关标准：

（1）《推荐性国际操作规范—冻鱼》（CAC/RCP16 - 1978）；

（2）《推荐性国际操作规范—冻面糊包裹的和（或）沾面包屑的水产品》（CAC/RCP 35 - 1985）；

（3）《推荐性国际操作规范—机械分离法制备碎鱼肉》（CAC/ RCP 35 - 1985）；

（4）《推荐性国际操作规范—速冻食品的加工和处理》（CAC/RCP8 - 1976）。

6 标签

除了应符合《预包装食品标签通用标准》（CODEX STAN1 - 1985，Rev. 3 - 1999）的第 2，3，7 和 8 条要求外，还应遵守以下细节规定：

6.1 食品名称

6.1.1 标签上产品的名称应为“裹面包屑”和（或）“挂浆”，“鱼条”、“鱼块”或“鱼片”，或按出售该食品的国家的法律、习惯等实际情况使用其他不会引起混淆和误导消费者的名称。

6.1.2　标签上还应注明鱼的种类或混合品种的种类。

6.1.3　在产品标签上应标明“速冻”，除非在有的国家“冻”即代表速冻。

6.1.4　标签上还应按出售该食品的国家的法律和习惯，以不会混淆和误导消费者的方式，标明产品是由碎鱼肉、鱼片或二者混合制成的。

6.1.5　标签应注明产品必须在运输、贮藏、分销过程中保持的条件，以保证其质量。

6.2　贮藏说明

标签应注明产品须在 -18℃或更低的温度下贮藏。

6.3　非零售包装的标签

上述要求除了食品名称，批号、制造或分装厂名、地址、贮藏条件等都应在包装上和附文中出现，但批号、制造或分装厂名、地址也可用由一个鉴别性标志代替，如果这个鉴别性标志能在辅助文件中说明清楚。

7　抽样、检验和分析

7.1　取样

（1）产品批次检验用样品的抽样方法应符合 FAO/WHO 的食品法规《预包装食品的抽样方案》（AQL -6.5）（CODEX STAN 233 -1969）。预包装产品的样品单位是一整箱。对散装产品来说样品单位应至少是 1kg 的鱼条、鱼块或鱼片。

（2）对需检测净重批次的抽样，抽样计划应以 CAC 的相关严格标准为依据。

7.2　净含量的测定

代表每批次样品的每个完整的初级包装净重（不包括包装材料）应在冷冻状态下测定。

7.3　感官与物理检验

产品的感官与物理指标必须由经过此类检验培训的人员进行检验，并依据本标准 7.4 至 7.7 和附录 A 以及《实验室中水产品感官评价指南》（CAC/GL 31 -1999）所述程序进行。

7.4　鱼体中心温度的评估

按照 AOAC 方法 996.15 进行评估。

7.5　凝胶状态的检验

按照 AOAC 方法 983.18“肉及肉制品中水分含量测定时抽样的准备工作”和 950.46“肉中的水分含量”（方法 A）。

7.6　鱼片与碎鱼肉比例的评估

见附录 B。

7.7　蒸煮方法

冷冻样品应在感官评价之前，根据包装上说明进行蒸煮。如果没有说明，或没有说明中的蒸煮器具，则应根据下列可行的方法对冷冻样品进行蒸煮：

采用 AOAC 方法 976.16 进行；蒸煮使产品内部温度达到 65～70℃；蒸煮时间随产品大小和采用的温度而不同；如果要测定蒸煮时间，则另取样品加热，使用温度测量装置测定中心温度。

7.8 组胺检测

按照 AOAC 方法 977.13 进行检测。

8 缺陷界定

当样品呈现下列任何一项特征时，则认定其有“缺陷”。

8.1 外来杂质（熟）

样品单位中存在的任何不是来自于鱼体的物质（除包装材料外）。这些物质虽不会对人体健康造成危害，但其不用放大即可轻易辨别，或采用某些方法（包括放大）可以确定其存在。这表明不符合良好操作和卫生惯例。

8.2 鱼刺（熟）（在指明无刺的包装中）

每千克产品中检测出长度等于或大于 10mm 的鱼刺或者直径大于或等于 1mm 的鱼刺；鱼刺的长度等于或小于 5mm，且直径不大于 2mm 则该产品不被认定为有缺陷。如果鱼刺根部（与脊椎连接处）的宽度小于或等于 2mm，或其可用手指甲轻松地剥除，则可以忽略。

8.3 气味/风味（熟）

样品散发的持久、明显、令人厌恶的由腐败、酸败或饵料引起的气味或风味。

8.4 鱼肉异常

样品单位鱼肉内部出现类似凝胶状态，并伴有任何单片鱼片中水分达 86% 以上，或按重量计 5% 以上的样品被寄生虫感染导致质地呈现糊状。

9 批次验收

当满足以下条件时，可以认为此批次产品符合本标准的要求：

（1）根据本标准第 8 章中规定分类，缺陷总数不超过《预包装食品的抽样方案》（AQL－6.5）（CODEX STAN 233－1969）中相应抽样方案规定的可接受数（c）；

（2）所有样品单位中鱼肉的平均百分数不小于冻品重量的 50%；

（3）所有样品平均净重不少于标示量，在任何一个包装单位中没有过度的重量短缺；

（4）符合本标准第 4、5、6 款中对食品添加剂、卫生和处理以及标签的要求。

附录 A 感官及物理检验

用于感官及物理检验的样品与进行其他检验的样品不应为同一个样品。

1. 按本标准 7.2 条规定完成净重的测定。
2. 根据 7.4 条的规定对样品单位中的一组产品进行鱼体内部检测。

3. 如果需要，对鱼片及碎鱼肉的比例进行评估。

4. 对样品中的其他组进行蒸煮，检验其气味、风味、质地、外来杂质和鱼刺。

5. 对在解冻未蒸煮状态下无法最终确定其凝胶状态的产品，则应从产品中截取可疑部分，按7.7条规定的蒸煮方法，或用7.5测定在任何一个产品单位中是否含有超过86%的水分，即可确认其凝胶情况。如果无法根据蒸煮后的评估得到结论，则按照7.5条中规定测定截取部分的水分含量。

附录B　鱼片与碎鱼肉比例的评估（西欧水产技术员协会－WEFTA方法）

a）设备

天平：精确至0.1g。

圆形滤网：直径200mm，2.5或2.8网眼（ISO），软橡胶刃（或钝）的抹刀，叉子，适当大小的盘子，不透水的塑料袋。

b）样品制备

鱼块（条）：按所需的量尽量取鱼块肉，鱼肉内部样品约为200g（2kg）。如果产品沾面包屑或面糊，先根据7.4规定的方法去掉外层包衣。

c）测定冻鱼样品的重量“A”

在冷冻状态下称量单独的鱼块或去掉外层包衣的内部鱼块。将小鱼块组合成一个约200g的样品亚单位（如10个鱼条，每个约20g）。记录每个亚单位的重量“A”。将预先称重的亚单位装入防水的袋子中。

d）解冻

将袋子浸入水浴，轻微搅动进行解冻，水温约20℃，最高不能超过35℃。

e）沥液

解冻完成后（持续时间20～30min），取出每个样品单位，每次一个，在一个预先称重的圆形滤网上沥干2min，流出的液体（解冻的汁液），滤网倾斜角度为17°～20°。沥液完成后用纸巾去掉滤网底部附着的液体。

f）测量沥液后鱼肉样品的重量“B”以及解冻汁液的重量“C”

沥液后鱼肉样品的重量“B”：滤网加上鱼重减去滤网重。

解冻汁液的重量：A与B的重量差即为流出的液体的重量。

g）分离

将沥液后的鱼肉放在盘中，用叉子固定住鱼片，用软橡胶刃的抹刀刮掉碎鱼肉。

CODEX STANDARD FOR QUICK FROZEN FISH STICKS (FISH FINGERS), FISH PORTIONS AND FISH FILLETS - BREADED OR IN BATTER

(CODEX STAN 166 - 1989, REV. 1 - 1995)

1 SCOPE

This standard applies to quick frozen fish sticks (fish fingers) and fish portions cut from quick frozen fish flesh blocks, or formed from fish flesh, and to natural fish fillets, breaded or batter coatings, singly or in combination, raw or partially cooked and offered for direct human consumption without further industrial processing.

2 DESCRIPTION

2.1 PRODUCT DEFINITION

2.1.1 A fish stick (fish finger) is the product including the coating weighing not less than 20 g and not more than 50 g shaped so that the length is not less than three times the greatest width. Each stick shall be not less than 10 mm thick.

2.1.2 A fish portion including the coating, other than products under 2.1.1, may be of any shape, weight or size.

2.1.3 Fish sticks or portions may be prepared from a single species of fish or from a mixture of species with similar sensory properties.

2.1.4 Fillets are slices of fish of irregular size and shape which are removed from the carcass by cuts made parallel to the back bone and pieces of such fillets, with or without the skin.

2.2 PROCESS DEFINITION

The product after any suitable preparation shall be subjected to a freezing process and shall comply with the conditions laid down hereafter. The freezing process shall be carried out in appropriate equipment in such a way that the range of temperature of maximum crystallization is passed quickly. The quick freezing process shall not be regarded as complete unless and until the product temperature has reached -18℃ or colder at the thermal centre after thermal stabilization. The product shall be kept deep frozen so as to maintain the quality during transportation, storage and distribution.

Industrial repacking or further industrial processing of intermediate quick frozen material under controlled conditions which maintains the quality of the product, followed by the re-application of the quick freezing process, is permitted.

2.3 PRESENTATION

Any presentation of the product shall be permitted provided that it:

2.3.1 meets all the requirements of the standard, and

2.3.2 is adequately described on the label to avoid confusing or misleading the consumer.

3 ESSENTIAL COMPOSITION AND QUALITY FACTORS

3.1 RAW MATERIAL

3.1.1 Fish

Quick frozen breaded or battered fish sticks (fish fingers) breaded or battered fish portions and breaded or battered fillets shall be prepared from fish fillets or minced fish flesh, or mixtures thereof, of edible species which are of a quality such as to be sold fresh for human consumption.

3.1.2 Coating

The coating and all ingredients used therein shall be of food grade quality and conform to all applicable Codex standards.

3.1.3 Frying fat (oil)

A fat (oil) used in the cooking operation shall be suitable for human consumption and for the desired final product characteristic (see also Section 4).

3.2 FINAL PRODUCT

Products shall meet the requirements of this standard when lots examined in accordance with Section 9 comply with the provisions set out in Section 8. Products shall be examined by the methods given in Section 7.

3.3 DECOMPOSITION

The products shall not contain more than 10 mg/100 g of histamine based on the average of the sample unit tested. This shall apply only to species of Clupeidae, Scombridae, Scombresocidae, Pomatomidae and Coryphaenedae families.

4 FOOD ADDITIVES

Only the use of the following additives is permitted.

ADDITIVE	MAXIMUM LEVEL IN THE FINAL PRODUCT
For Fish Fillets and Minced Fish Flesh Only	
Moisture/Water Retention Agents	

continued

ADDITIVE	MAXIMUM LEVEL IN THE FINAL PRODUCT
339 (i) Monosodium orthophosphate 340 (i) Monopotassium orthophosphate 450 (iii) Tetrasodium diphosphate 450 (v) Tetrapotassium diphosphate 451 (i) Pentasodium triphosphate 451 (ii) Pentapotassium triphosphate 452 (i) Sodium polyphosphate 452 (iv) Calcium, polyphosphates	10 g/kg expressed as P_2O_5, singly or in combination (includes natural phosphate)
401 Sodium alginate	GMP
Antioxidants	
300 Ascorbic acid 301 Sodium ascorbate 303 Potassium ascorbate	GMP
304 Ascorbyl palmitate	1 g/kg
In Addition, for Minced Fish Flesh Only	
Acidity Regulator	
330 Citric acid 331 Sodium citrate 332 Potassium citrate	GMP
Thickeners	
412 Guar gum 410 Carob bean (Locust bean) gum 440 Pectins 466 Sodium carboxymethyl cellulose 415 Xanthan gum 407 Carrageenan and its Na, K, NH_4 salts (including Furcelleran) 407a Processed Euchema seaweed (PES) 461 Methyl cellulose	GMP
Food Additives for Breaded or Batter Coatings	
Leavening Agents	
341 (i) Monocalcium orthophosphate 341 (ii) Dicalcium orthophosphate 541 Sodium aluminium phosphate, basic and acidic	1 g/kg expressed as P_2O_5, singly or in combination
500 Sodium carbonates 501 Potassium carbonates 503 Ammonium carbonates	GMP
Flavour Enhancers	
621 Monosodium glutamate 622 Monopotassium glutamate	GMP

continued

ADDITIVE	MAXIMUM LEVEL IN THE FINAL PRODUCT
Colours	
160b Annatto extracts	20 mg/kg expressed as bixin
150a Caramel I (plain)	GMP
160a (i) β – carotene (Synthetic) 160e β – apo – carotenal	100 mg/kg singly or in combination
Thickeners	
412 Guar gum 410 Carob bean (Locust bean) gum 440 Pectins 466 Sodium carboxymethyl cellulose 415 Xanthan gum 407 Carrageenan and its Na, K, NH_4 salts (including Furcelleran) 407a Processed *Euchema* Seaweed (PES) 461 Methyl cellulose 401 Sodium alginate 463 Hydroxypropyl cellulose 464 Hydroxypropyl methylcellulose 465 Methylethylcellulose	GMP
Emulsifiers	
471 Monoglycerides of fatty acids 322 Lecithins	GMP
Modified Starches	
1401 Acid treated starches 1402 Alkaline treated starches 1404 Oxidized starches 1410 Monostarch phosphate 1412 Distarch phosphate esterified with sodium trimetaphosphate; esterified with phosphorus oxychloride 1414 Acetylated distarch phosphate 1413 Phosphated distarch phosphate 1420 Starch acetate esterified with acetic anhydride 1421 Starch acetate esterified with vinyl acetate 1422 Acetylated distarch adipate 1440 Hydroxypropyl starch 1442 Hydroxypropyl starch phosphate	GMP

5 HYGIENE AND HANDLING

5.1 The final product shall be free from any foreign material that poses a threat to human health.

5.2 When tested by appropriate methods of sampling and examination prescribed by the Codex Alimentarius Commission, the product:

(i) shall be free from microorganisms or substances originating from microorganisms in amounts which may present a hazard to health in accordance with standards established by the Codex Alimentarius Commission;

(ii) shall not contain histamine that exceeds 20 mg/100 g. This applies only to species of *Clupeidae*, *Scombridae*, *Scombresocidae*, *Pomatomidae* and *Coryphaenedae* families;

(iii) shall not contain any other substance in amounts which may present a hazard to health in accordance with standards established by the Codex Alimentarius Commission.

5.3 It is recommended that the products covered by the provisions of this standard be prepared and handled in accordance with the appropriate sections of the Recommended International Code of Practice – General Principles of Food Hygiene (CAC/RCP 1 – 1969) and the following relevant Codes:

(i) the Recommended International Code of Practice for Frozen Fish (CAC/RCP 16 – 1978);

(ii) the Recommended International Code of Practice for Frozen Battered and/or Breaded Fishery Products (CAC/RCP 35 – 1985);

(iii) the Recommended International Code of Practice for Minced Fish Prepared by Mechanical Separation (CAC/RCP 27 – 1983);

(iv) the Recommended International Code of Practice for the Processing and Handling of Quick Frozen Foods (CAC/RCP 8 – 1976).

6 LABELLING

In addition to Sections 2, 3, 7 and 8 of the Codex General Standard for the Labelling of Prepackaged Foods (CODEX STAN 1 – 1985) the following specific provisions apply:

6.1 THE NAME OF THE FOOD

6.1.1 The name of the food to be declared on the label shall be "breaded" and/or "battered", "fish sticks" (fish fingers), "fish portions", or "fillets" as appropriate or other specific names used in accordance with the law and custom of the country in which the food is sold and in a manner so as not to confuse or mislead the consumer.

6.1.2 The label shall include reference to the species or mixture of species.

6.1.3 In addition there shall appear on the label either the term "quick frozen" or the term "frozen" whichever is customarily used in the country in which the food is sold, to describe a product subjected to the freezing processes as defined in subsection 2.2.

6.1.4 The label shall show whether the products are prepared from minced fish flesh, fish fillets or a mixture of both in accordance with the law and custom of the country in which the food is sold and in a manner so as not to confuse or mislead the consumer.

6.1.5 The label shall state that the product should be maintained under conditions that will maintain the quality during transportation, storage and distribution.

6.2 STORAGE INSTRUCTIONS

The label shall include terms to indicate that the product shall be stored at a temperature of −18℃ or colder.

6.3 LABELLING OF NON-RETAIL CONTAINERS

Information specified above shall be given either on the container or in accompanying documents, except that the name of the food, lot identification, and the name and address of the manufacturers or packer, as well as storage instructions, shall always appear on the container. However, lot identification, and the name and address may be replaced by an identification mark, provided that such a mark is clearly identifiable with the accompanying documents.

7 SAMPLING, EXAMINATION AND ANALYSIS

7.1 SAMPLING

(i) Sampling of lots for examination of the product shall be in accordance with the FAO/WHO Codex Alimentarius Sampling Plans for Prepackaged Foods (AQL-6.5) (CODEX STAN 233-1969). For prepackaged goods the sample unit is the entire container. For products packed in bulk the sample unit is at least 1 kg of fish sticks (fish finger), fish portions or fillets.

(ii) Sampling of lots for examination of net weight shall be carried out in accordance with an appropriate sampling plan meeting the criteria established by the Codex Alimentarius Comission.

7.2 DETERMINATION OF NET WEIGHT

The net weight (exclusive of packaging material) is determined on each whole primary container of each sample representing a lot and shall be determined in the frozen state.

7.3 SENSORY AND PHYSICAL EXAMINATION

Samples taken for sensory and physical examination shall be assessed by persons trained in such examination and in accordance with procedures elaborated in Sections 7.4 through 7.7, Annex A and the *Guidelines for the Sensory Evaluation of Fish and Shellfish in Laboratories* (*CAC/GL 31-1999*).

7.4 ESTIMATION OF FISH CORE

According to AOAC Method 996.15.

7.5 DETERMINATION OF GELATINOUS CONDITIONS

According to the AOAC Methods – "Moisture in Meat and Meat Products, Preparation of Sample Procedure"; 983.18 and "Moisture in Meat" (Method A); 950.46.

7.6 ESTIMATION OF PROPORTION OF FISH FILLETS AND MINCED FISH FLESH See Annex B

7.7 COOKING METHODS

The frozen sample shall be cooked prior to sensory assessment according to the cooking instructions on the package. When such instructions are not given, or equipment to cook the sample according to the instructions is not obtainable, the frozen sample shall be cooked according to the applicable method (s) given below:

Use procedure 976.16 of the AOAC It is based on heating product to an internal temperature of 65 ~ 70℃. Cooking times vary according to size of product and equipment used. If determining cooking time, cook extra samples, using a temperature measuring device to determine internal temperature.

7.8 DETERMINATION OF HISTAMINE

According to the AOAC Methods 977.13.

8 DEFINITION OF DEFECTIVES

The sample unit shall be considered defective when it exhibits any of the properties defined below:

8.1 FOREIGN MATTER (COOKED STATE)

The presence in the sample unit of any matter which has not been derived from fish (excluding packing material), does not pose a threat to human health, and is readily recognized without magnification or is present at a level determined by any method including magnification that indicates non - compliance with good manufacturing and sanitation practices.

8.2 BONES (COOKED STATE) (IN PACKS DESIGNATED BONELESS)

More than one bone per kg greater or equal to 10 mm in length, or greater or equal to 1 mm in diameter; a bone less than or equal to 5 mm in length, is not considered a defect if its diameter is not more than 2 mm. The foot of a bone (where it has been attached to the vertebra) shall be disregarded if its width is less than or equal to 2 mm, or if it can easily be stripped off with a fingernail.

8.3 ODOUR AND FLAVOUR (COOKED STATE)

A sample unit affected by persistent and distinct objectionable odour and flavours indicative of decomposition, or rancidity or of feed.

8.4 FLESH ABNORMALITIES

Objectionable textural characteristics such as gelatinous conditions of the fish core together with greater than 86% moisture found in any individual fillet or sample unit with pasty texture resulting from parasites affecting more than 5% of the sample unit by weight.

9 LOT ACCEPTANCE

A lot shall be considered as meeting the requirements of this standard when:

(i) the total number of defectives as classified according to Section 8 does not exceed the acceptance number (c) of the appropriate sampling plan in the Sampling Plans for Prepackaged Foods (AQL – 6. 5) (CODEX STAN 233 – 1969) ;

(ii) the average percent fish flesh of all sample units is not less than 50% of the frozen weight;

(iii) the average net weight of all sample units is not less than the declared weight, provided there is no unreasonable shortage in any container; and

(iv) the Food Additives, Hygiene and Labelling requirements of Sections 4, 5 and 6 are met.

ANNEX A SENSORY AND PHYSICAL EXAMINATION

The sample used for sensory evaluation should not be the same as that used for other examinations.

1. Complete net weight determination, according to defined procedures in Section 7. 2.

2. Complete fish core determination on one set of the sample units according to defined procedures in Section 7. 4.

3. Complete the estimation of the proportion of fillets and minced flesh, if required.

4. Cook the other set of sample units and examine for odour, flavour, texture, foreign matter, and bones.

5. In cases where a final decision on gelatinous conditions cannot be made in the thawed uncooked state, the disputed material is sectioned from the product and gelatinous condition confirmed by cooking as defined in Section 7. 7 or by using the procedure in Section 7. 5 to determine if greater than 86% moisture is present in any product unit. If a cooking evaluation is inconclusive, then procedure in 7. 5 would be used to make the exact determination of moisture content.

ANNEX B ESTIMATION OF PROPORTION OF FISH FILLETS AND MINCED FISH FLESH (West European Fish Technologists Association – WEFTA Method)

a) Equipment

Balance, sensitive to 0. 1 g Circular sieve – 200mm diameter, 2. 5 or 2. 8 mesh opening (ISO) soft rubber edge (or blunt) spatula, forks, suitable sized plates, water tight plastic bags.

b) Preparation of Samples

Fish Portions/Sticks: Take as many portions as needed to provide a fish core sample of about 200g (2kg) . If breaded and/or battered firrst strip coating according to the method describer in section 7. 4.

c) Detemination of Weights “A” of the Frozen Fish Samples

Weight the single fish portions/decoated fish cores while they are still frozen. Smaller portions are combined to a sample sub – units of about 200g (e. g. 10) . fish sticks of about 20g each) . Record the weight “A” n of the sub – units. Place the pre – weighed sample sub – units into water tight bags.

d) Thawing

Thaw the samples by immersing the bags into a gently agitated water bath of about 20℃, but not more than 35℃.

e) Draining

After thawing has been completed (duration about 20 ~ 30min.) take each sample unit, one at a time, and drain the exuded fluid (thaw drip) for 2 minutes on a pre - weighed circular sieve incluned at an angle of 17 ~ 20 degrees. Remove adhering drip from the bottom of the sieve by use of a paper towel when draining is completed.

f) Determination of weight "B" of the Drained Fish Sample and Weight "C" of the Thaw Drip Determine the weight of the drained fish sample "B" - sieve plus fish minus sieve weight. The difference of "A" - "B" is the weight of exuded fluid - thaw drip.

g) Separation

Place the drained fish core on a plate and separate the minced flesh from the fillet using a fork to hold the fillet flesh and a soft, rubber edge spatula to scrape off the minced flesh.

盐渍和盐干鳕鱼

CODEX STANDARD FOR SALTED FISH AND DRIED SALTED FISH OF THE GADIDAE FAMILY OF FISHES

(CODEX STAN 167 - 1989, REV. 2 - 2005)

1 范围

本标准适用于盐腌制的及盐腌制后干制的鳕科鱼类。其腌制可以是将鱼全部饱和盐腌（重腌）或是不饱和盐腌（产品含盐量不少于产品总重的12%），产品不需工业深加工即可直接用于消费。

2 说明

2.1 产品定义

盐腌鱼产品取自：

（a）各种鳕科鱼；

（b）已放血、去除内脏、去头、劈开或切片、清洗并腌制的鱼；

（c）咸鱼干是咸鱼再干制而成的。

2.2 加工定义

产品应按照2.2.1中的一种方法进行腌制，并按照2.2.2中的一种或两种方法进行干制，产品的外观应符合2.3中所述的不同形式。

2.2.1 腌制

（1）干法腌制（堆腌）：将鱼与适量食盐充分混合并以可沥干剩余盐水的方式将其堆垛。

（2）湿法腌制（腌渍）：将鱼与适量食盐充分混合，并放入水密容器中，使之处于从鱼肉组织中析出的水形成的盐水溶液（即卤水）下面。也可向容器中添加盐水。随后将鱼从容器中移出并堆垛，以沥干盐水。

（3）盐水注射：将盐水直接注入鱼肉中，这种方法可以作为重腌加工中的一部分。

2.2.2 干制

（1）自然干制：露天风干。

（2）人工干制：将鱼置于温度和湿度可控的循环空气环境中晾干。

2.3 产品介绍

2.3.1 鲞片（Split fish）：沿鱼体主要长度从背部或腹部纵向劈（剖）开，并去除前段脊骨（约三分之二）。

2.3.2 含脊骨鲞片（Split fish with entire backbone）：纵劈（剖），但不去除脊骨。

2.3.3 鱼片：平行于鱼脊骨剖取两边鱼肉并切片，去除鱼鳍及主要的脊骨及骨刺，有时连腹壁肉也一同去除。

2.3.4 产品介绍的任何其他形式，应符合下列要求：

（1）与本标准中的所述有明显的区别；

（2）达到本标准的所有要求；

（3）在产品标签中对产品进行详细描述，以免引起混淆或误导消费者。

2.3.5 不同品种鱼的产品不能混装。

3 基本成分及质量因素

3.1 鱼

盐腌鱼应由适于人类消费的完好、健全的鱼加工制成。

3.2 盐

用来生产盐腌鱼的盐必须是清洁的。不含有任何外来杂质及其他外来结晶体。不含有肉眼可观察到的污秽、油渍、污水或其他外来污染并且符合《盐腌鱼操作规范》（CAC/REP. 26－1979）附录1（supplement 1）的要求。

3.3 成品

产品应符合本标准要求，根据第9章进行成品批次检查时，其质量应符合第8章的规定。检查方法应符合第7章的规定。

4 食品添加剂

只允许使用下列食品添加剂：

添加剂		成品中的最高含量
防腐剂	200 山梨酸 201 山梨酸钠 202 山梨酸钾	200mg/kg，以山梨酸计，单用或混用

5 卫生及处理

5.1 成品中不能含有任何危害人体健康的外来杂质

5.2 应用CAC规定的抽样及检测方法进行检验时：

（1）产品中微生物含量和由微生物产生的危害人体健康的有害物质的含量不能超过CAC标准的规定。

（2）其他危害人体健康的有害物质的含量也不能超过CAC标准的规定。

5.3 建议本标准中涉及产品预处理的条款应符合《推荐性国际规范—食品卫生操作总则》（CAC/RCP1－1969，Rev. 3－1997）中的相关章节和以下相关标准：

（1）《推荐性国际操作规范—鲜鱼》（CAC/RCP 9－1976）；

(2)《推荐性国际操作规范—冻鱼》(CAC/RCP 16－1978);

(3)《推荐性国际操作规范—盐腌鱼》(CAC/RCP 26－1979);

(4)《国际操作规范建议草案—鱼类和其他水产品》中水产养殖品的相关章节（正在完善中)[1]。

6 标签

除了应符合《预包装食品标签通用标准》（CODEX STAN 1－1985，Rev. 3－1999）的要求外，还应遵守以下规定:

6.1 食品名称

6.1.1 产品的名称可以在标签上注明是“咸鱼（salted fish)”、“湿咸鱼（wet salted fish)”、“咸鱼片（salted fillet)”、“咸鱼干（dried salted fish)”或“挪威盐干鳕（klippfish)”等其他符合出售该产品的国家的法律、习惯等实际情况的名称。另外在标签中还应与产品名称一起注明原料鱼的种名。

6.1.2 如果产品采用“鲞片”（2.3.1条）以外的其他外观形式，则应根据2.3.2和2.3.3在标签中结合食品名称注明其外观。标签上还应在与产品名称紧靠位置出现关于外观形式的词语，这些描述性词语应恰如其分地说明产品外观，并确保它们不会引起混淆或误导消费者。

6.1.3 术语“挪威盐干鳕（klippfish)”只能用于描述晾干前盐饱和度已达到95%的盐腌制鳕鱼干产品。

6.1.4 术语“湿咸鱼（wet salted fish)”只能用于盐饱和鱼产品。

6.2 非零售包装的标签

上述要求应既在包装上又在附文中出现，除了食品名称、批号、制造或分装厂名、地址外，还包括贮藏条件。批号、制造或分装厂名、地址也可用同一证明标志代替，只要证明标志能在辅助文件中说明清楚。

7 抽样、检验和分析

7.1 抽样

(1）对产品进行抽样检查应遵循《FAO/WHO预包装食品抽样检测标准》(AQL－6.5)（CAC/RM 42－1969）的要求，一个抽样单位应该是一个初级包装，若为散装的，每条鱼就是一个抽样单位。

(2）抽样的净重检测应严格遵守《FAO/WHO净重检测的抽样设计》的要求。

7.2 感官和物理检测

感官和物理检测的样品应该由专业人员进行评估，这些人员要经过专门的培训，检测要遵循附件A中描述的步骤，同时要符合《鱼贝类实验室感观评价指南》的要求。

[1] 国际操作规范的建议草案，在定稿后将替代所有现行的《国际操作规范建议草案——鱼类和其他水产品》。

7.3 净重的测定

一批样品中每个样品单位的净重（除去包装材料和额外的盐）都应该进行测定。

7.4 鱼样品的准备

（1）在取样品边缘之前，鱼体表面黏附的盐的结晶应通过不用水的干刷方式除去。

（2）鱼样品的准备应该按照 AOAC 937.07 的要求进行，这些样品是用来测定鱼体的盐度和水分含量，应选用鱼体的可食部分进行试验。

（3）检测至少要做两个平行样。

7.5 盐度的检测

7.5.1 原理

用水从预先称重的样品中提取盐，样品除去蛋白质后，所有氯化物的含量通过标准硝酸银溶液滴定的方法测定，最终结果以氯化钠计算。

7.5.2 仪器和试剂

—刷子

—锋利的刀或锯

—天平（精确到 0.01 克）

—校准的容量瓶（250ml）

—三角瓶

—电动匀浆机

—磁力搅拌器

—三角漏斗（快流速）

—移液管

—漏斗

—量筒

—铁氰化钾（$K_4Fe(CN)_6 \cdot 3H_2O$，15% w/v）

—硫酸锌（$ZnSO_4 \cdot 63H_2O$，30% w/v）

—氢氧化钠（NaOH，0.1N，0.41% w/v）

—标准硝酸银溶液（$AgNO_3$，0.1N，1.6987% w/v）

—铬酸钾（K_2CrO_4，5% w/v）

—1% 酚酞（乙醇溶解）

—双蒸水或去离子水

7.5.3 步骤

（1）称取 5g 匀浆后的样品于 250ml 容量瓶中，加入约 100ml 双蒸水，用力振摇；

（2）一边振摇一边加入 5ml 铁氰化钾溶液和 5ml 硫酸锌溶液；

（3）双蒸水定容至刻度线；

（4）再次振摇后，静置沉淀，用三角漏斗过滤；

（5）将滤液转移到三角瓶中，加入两滴酚酞，逐滴加入氢氧化钠溶液至滤液呈现微红色，然后用双蒸水稀释至约 100ml；

（6）加入约 1ml 铬酸钾后，在连续搅拌的情况下用标准硝酸银溶液滴定，滴定终点时颜色发生微弱但清楚可见的变化。振荡后这种微弱的红褐色没有消失。为便于观察颜色的变化，建议滴定时在白色的背景下进行；

（7）同时做空白实验；

（8）可以使用电位计或色度计等仪器来确定滴定终点。

7.5.4 结果计算

计算公式中将用到以下符号：

A = 滤液体积（ml）

C = 硝酸银的当量浓度（N）

V = 扣除空白后，达到滴定终点时消耗硝酸银的体积（ml）

W = 样品质量（g）

通过下列公式计算样品的盐度：

盐度（%） = （V × C × 58.45 × 250 × 100）/（A × W × 1000）

结果保留到小数点后一位。

7.5.5 参考方法

参考方法应包括在测氯之前根据上面（处理步骤 2 和步骤 4）描述的方法于 550℃ 的高温熔炉里将样品完全灰化。

7.5.6 注意

使用给定的方程时，检测到的所有氯均以氯化钠计。然而，通过这种方法不可能计算钠，因为碱和碱土元素以其他氯化物的形式存在。

除了氯之外的其他卤元素在鱼和盐中的量十分微少。

沉淀蛋白（2）的步骤对避免误导结果十分重要。

7.6 水分含量测定

（1）标准中要求的盐饱和度（以百分比表示）的测定，应该与 AOAC 950.46 部分一致（风干（a））。

（2）当商业贸易中需要对干的或湿的盐制鳕鱼全鱼进行水分含量测定时，取样的方法应与附件 B 中的“采用剖面法测定全鱼的水分含量”一致。

8 缺陷的定义

8.1 当样品呈现下列任何一项特征时，则认定其有“缺陷”

8.1.1 外来杂质

样品单位中存在的任何不是来自于鱼体的物质。这些物质虽不会对人体健康造成危害，但其不用放大即可轻易辨别，或采用某些方法（包括放大）可以确定其存在。出现外来杂质表明不符合良好操作和卫生惯例。

8.1.2 气味/风味

样品散发的持久、明显、令人厌恶的由腐败、酸败或饵料引起的气味或风味。

8.1.3 粉红色

表明存在嗜盐细菌（halophilic bacteria）。

8.1.4 外观

鱼肉组织全面下降，表现为超过鱼体表面积 2/3 处出现大量裂纹，或是鱼体被毁伤、撕裂或打碎到鲞片被分成两块或多块但其皮肤仍然相连的程度。

8.2 当样品单位中等于或多于30%出现下列任何一项缺陷时，则认定其有“缺陷”

8.2.1 嗜盐霉菌（halophytic mould）

明显出现嗜盐霉菌群落的总面积超过鱼体背部一边总面积的1/3。

8.2.2 肝斑（liver stains）

肝斑是因留存肝脏造成的，超过鱼体背部一边总面积1/4发生明显的变色（黄色或橙黄色）。

8.2.3 严重损伤

超过鱼外表面积1/2出现严重损伤。

8.2.4 严重灼伤

鱼背部（带皮边）超过1/2的地方由于干燥时过度晾晒而变黏。

9 批次验收

当满足以下条件时，可以认为此批产品符合本标准的要求：

（1）根据本标准第8章中规定分类，缺陷总数不超过《预包装食品的抽样方案》（AQL－6.5）（CODEX STAN 233－1969）中相应抽样方案规定的可接受数（c）；

（2）所有样品单位平均净重不少于标示量，个体样品不得少于标示重量的95%；

（3）不符合本标准2.3中规定数量的样品总数不超过《预包装食品的抽样方案》（AQL－6.5）（CODEX STAN 233－1969）中相应抽样方案规定的可接受数（c）；

（4）符合本标准第4、5、6款中对食品添加剂、卫生和处理以及标签的要求。

附录A 感官及物理检验

1. 检测样品中的鱼是否完整。
2. 检测产品的性状情况。
3. 检测鱼的外来杂质、嗜盐细菌、嗜盐霉菌、肝斑、严重损伤、严重灼伤的情况及鱼肉组织。
4. 依据《实验室中水产品感官评价指南》（CAC/ GL 31－1999）对产品的气味以及品质进行评定。

附录B 整条鱼水分含量的测定——剖面法

1 原理

按描述的方法将鱼切成段，再将鱼段切成更小的块混合成试样。试样中水分含量用烘干法测定。试验和经验表明，这种方式采集的混合试样中的水分含量接近于鱼中的真实水分含量。

2 仪器

—软刷

—盆（钢、玻璃、陶瓷）

—剪刀

—带锯

—小刀

—天平（精确到1g）

—烘箱（103～105℃）

—干燥器

3 样品制备

刷掉鱼体表面盐粒。

鱼体称重精确到1g。

测量鱼尾裂至两侧耳骨顶端连线之间的长度作为鱼体长度。

4 步骤

（1）如附图所示进行取样。

①用刀子将湿的盐渍鱼切成段。

②用手锯将盐渍鱼和盐干鱼锯成段。

· 从耳骨间连线（图中的虚线所示部分）向下切出20mm长的鱼段；

· 再切出40mm长的鱼段；

· 从40mm鱼段的前端切下2mm的鱼片，进行采集（见7. “注意事项”）；

· 从上次切口再向下切出40mm长的鱼段；

· 再从该40mm鱼段的前端切下2mm的鱼片，进行采集；

· 如图所示，将整条鱼切成40mm的段，再从每段上切下2mm的鱼片；

· 采集图中所有Ⅱ，Ⅳ，Ⅵ，Ⅷ等偶数标记处2mm的鱼片制成混合试样。

（2）切分完鱼后，马上将2mm鱼片混合试样，用剪刀剪成更小的碎片置于已称量皿中。

（3）将装有样品的称量皿称重。

（4）将装有样品的称量皿放入103~105℃的烘箱内，烘干至恒重（18小时过夜）。

（5）将称量皿从烘箱中取出放入干燥器、冷却。

（6）将称量皿称重。

5 计算

计算公式中将用到下述符号：

W_1 = 烘干前样品鱼和称量皿的总重量（g）。

W_2 = 烘干后样品鱼和称量皿的总重量（g）。

WS = 称量皿的重量（g）。

鱼体中的水分含量的计算公式：

水分含量，g/100g = $100 \times (W_1 - W_2) / (W_1 - W_S)$。

计算结果整化到最相近的克数，并注明所试验鱼的长度和重量。

6 整鱼的控制分析

采用剖面法测得的整鱼的水分含量与整鱼烘干后测定的水分含量相比，结果相差不大。

7 注意事项

样品鱼在进行试验前必须贮存并密封于塑料袋中，从取样到分析的过程中样品鱼应在冷冻或冰箱中存放。

对鱼取样后必须尽快进行分析。

当鱼体水分含量高于50%时，切2mm鱼片可能比较困难，但应尽量接近2mm。

为尽量减少2mm鱼片的水分流失，鱼切成段后，立即对采集的样品进行称重非常重要。

应该做两个平行样。下图为取样步骤，将图中偶数标记的鱼段收集组成一个样品。

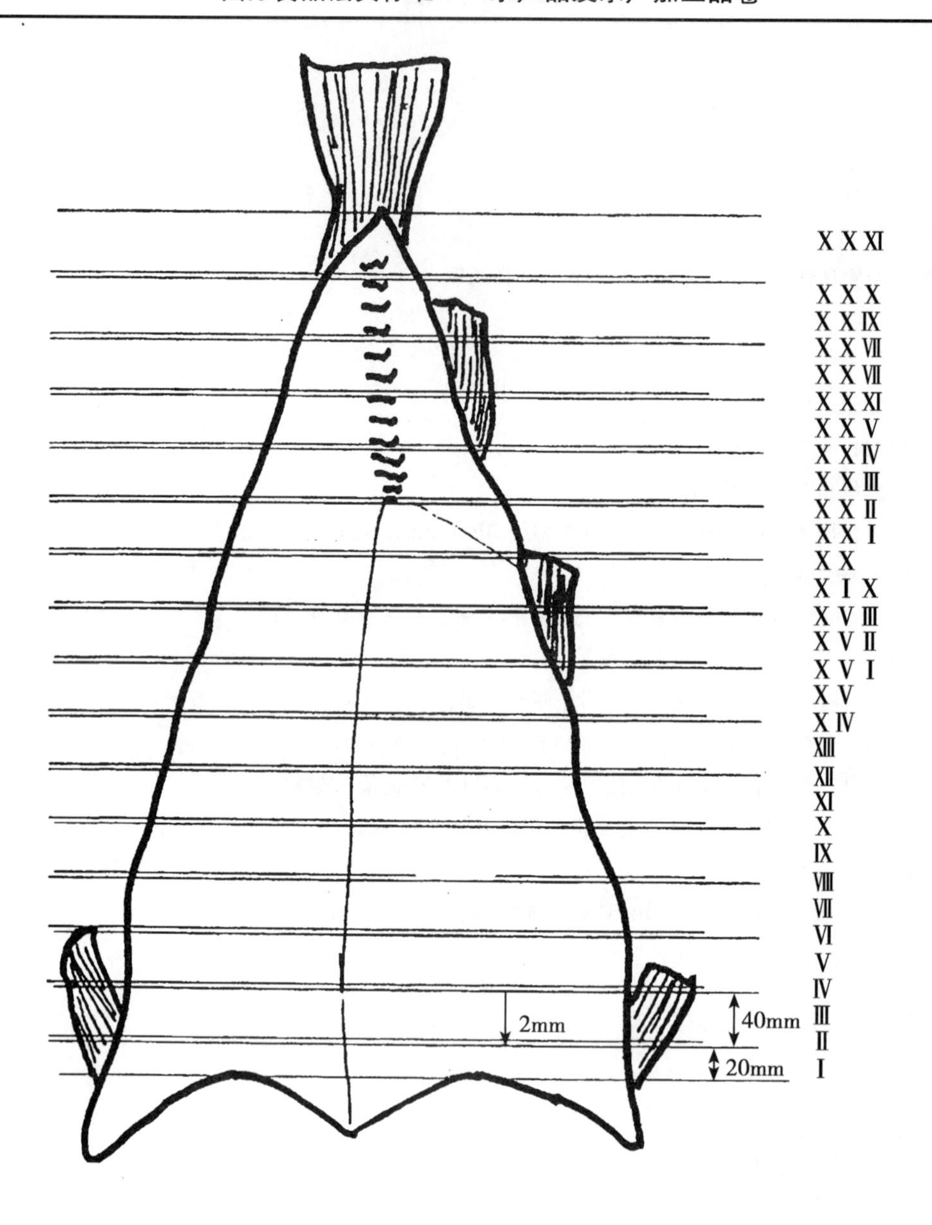
XXXI
XXX
XXIX
XXVII
XXVII
XXXI
XXV
XXIV
XXIII
XXII
XXI
XX
XIX
XVIII
XVII
XVI
XV
XIV
XIII
XII
XI
X
IX
VIII
VII
VI
V
IV
III
II
I
2mm
40mm
20mm

CODEX STANDARD FOR SALTED FISH AND DRIED SALTED FISH OF THE GADIDAE FAMILY OF FISHES

(CODEX STAN 167 – 1989, REV. 2 – 2005)

1 SCOPE

This standard applies to salted fish and dried salted fish of the Gadidae family which has been fully saturated with salt (heavy salted) or to salted fish which has been preserved by partial saturation to a salt content not less than 12% by weight of the salted fish which may be offered for consumption without further industrial processing.

2 DESCRIPTION

2.1 PRODUCT DEFINITION

Salted fish is the product obtained from fish:

(a) of the species belonging to the family *Gadidae*; and

(b) which has been bled, gutted, beheaded, split or filleted, washed, salted.

(c) dried salted fish is salted fish which have been dried.

2.2 PROCESS DEFINITION

The product shall be prepared by one of the salting processes defined in 2.2.1 and one or both of the drying processes defined in 2.2.2 and according to the different types of presentation as defined in 2.3.

2.2.1 Salting

(a) Dry Salting (kench curing) is the process of mixing fish with suitable food grade salt and stacking the fish in such a manner that the excess of the resulting brine drains away.

(b) Wet Salting (pickling) is the process whereby fish is mixed with suitable food grade salt and stored in watertight containers under the resultant brine (pickle) which forms by solution of salt in the water extracted from the fish tissue. Brine may be added to the container. The fish is subsequently removed from the container and stacked so that the brine drains away.

(c) Brine Injection is the process for directly injecting brine into the fish flesh and is permitted as a part of the heavy salting process.

2.2.2 Drying

(a) Natural Drying-the fish is dried by exposure to the open air; and

(b) Artificial Drying-the fish is dried in mechanically circulated air, the temperature and humidity of which may be controlled.

2.3 PRESENTATION

2.3.1 Split fish -split and with the major length of the anterior of the backbone removed (about two thirds).

2.3.2 Split fish with entire backbone-split with the whole of the backbone not removed.

2.3.3 Fillet-is cut from the fresh fish, strips of flesh is cut parallel to the central bone of the fish and from which fins, main bones and sometimes belly flap is removed.

2.3.4 Other presentation: any other presentation of the product shall be permitted provided that it

(i) is sufficiently distinctive from the other forms of presentation laid down in this Standard;

(ii) meets all other requirements of this Standard; and

(iii) is adequately described on the label to avoid confusing or misleading the consumer.

2.3.5 Individual containers shall contain only one form of presentation from only one species of fish.

3 ESSENTIAL COMPOSITION AND QUALITY FACTORS

3.1 FISH

Salted fish shall be prepared from sound and wholesome fish, fit for human consumption.

3.2 SALT

Salt used to produce salted fish shall be clean, free from foreign matter and foreign crystals, show no visible signs of contamination with dirt, oil, bilge or other extraneous materials and comply with the requirements laid down in supplement 1 to the Code of Practice for Salted Fish (CAC/RCP 26 - 1979).

3.3 FINAL PRODUCT

Products shall meet the requirements of this standard when lots examined in accordance with Section 9. comply with the provisions set out in Section 8. Products shall be examined by the methods given in Section 7.

4 FOOD ADDITIVES

Only the use of following additives is permitted.

	Additives	Maximum level in the Final Product
Preservatives	200 Sorbic acid 201 Sodium sorbate 202 Potassium sorbate	200mg/kg, Subgky or in combination expressed as sorbic acid

5 HYGIENE AND HANDLING

5.1 The final product shall be free from any foreign material that poses a threat to human health.

5.2 When tested by appropriate methods of sampling and examination prescribed by the Codex Alimentarius Commission, the product:

(i) shall be free from microorganisms or substances originating from microorganisms in amounts which may present a hazard to health in accordance with standards established by the Codex Alimentarius Commission;

(ii) shall not contain any other substance in amounts which may present a hazard to health in accordance with standards established by the Codex Alimentarius Commission.

5.3 It is recommended that the products covered by the provisions of this standard be prepared and handled in accordance with the appropriate sections of the Recommended International Code of Practice General Principles of Food Hygiene (CAC/RCP 1 – 1969, Rev. 3 – 1997) and the following relevant Codes:

(i) the Recommended International Code of Practice for Fresh Fish (CAC/RCP 9 – 1976);

(ii) the Recommended International Code of Practice for Frozen Fish (CAC/RCP 16 – 1978);

(iii) the Recommended International Code of Practice for Salted Fish (CAC/RCP 26 – 1979);

(iv) The sections on the Products of Aquaculture in the Proposed Draft International Code of Practice for Fish and Fishery Products (under elaboration)[1].

6 LABELLING

In addition to the provisions of the Codex General Standard for the Labelling of Prepackaged Foods (CODEX STAN 1 – 1985, Rev. 1 – 1991), the following specific provisions apply:

6.1 THE NAME OF THE FOOD

6.1.1 The name of the food to be declared on the label shall be "salted fish", "wet salted fish" or "salted fillet" "dried salted fish" or "klippfish" or other designations according to the law, custom or practice in the The Proposed Draft Code of Practice, when finalized, will replace all current Codes of Practice for Fish and Fishery Products country in which the product is to be distributed. In addition, there shall appear on the label in conjunction with the name of the product, the name of the species of fish from which the product is derived.

6.1.2 For forms of presentation other than those described in 2.3.1 "split fish", the form of presentation shall be declared in conjunction with the name of the product in accordance with sub-section 2.3.2 as appropriate. If the product is produced in accordance with sub-section 2.3.3, the label shall contain in close proximity to the name of the food, such additional words or phrases that will avoid misleading or confusing the consumer.

6.1.3 The term "klippfish" can only be used for dried salted fish which has been prepared from fish which has reached 95% salt saturation prior to drying.

6.1.4 The term "wet salted fish" can only be used for fish fully saturated with salt.

6.2 LABELLING OF NON-RETAIL CONTAINERS

Information specified above shall be given either on the container or in accompanying documents, except that the name of the food, lot identification, and the name and address of the manufacturer or packer

[1] The Proposed Draft Code of Practice, when finalized, will replace all current Codes of Practice for Fish and Fishery Products.

shall always appear on the container.

However, lot identification, and the name and address may be replaced by an identification mark, provided that such a mark is clearly identifiable with the accompanying documents.

7 SAMPLING, EXAMINATION AND ANALYSES

7.1 SAMPLING

(i) Sampling of lots for examination of the product shall be in accordance with the FAO/WHO Codex Alimentarius Sampling Plans for Prepackaged Foods (AQL-6.5) (CODEX STAN 233-1969). A sample unit shall be the primary container or where the product is in bulk, the individual fish is the sample unit.

(ii) Sampling for net weight shall be carried out in accordance with the FAO/WHO Sampling Plans for the Determination of Net Weight (under elaboration).

7.2 SENSORY AND PHYSICAL EXAMINATION

Samples taken for sensory and physical examination shall be assessed by persons trained in such examination and in accordance with procedures elaborated in Annex A and in accordance with *Guidelines for the Sensory Evaluation of Fish and Shellfish in Laboratories* (CAC/GL 31-1999).

7.3 DETERMINATION OF NET WEIGHT

The net weight (excluding packaging material and excess salt) of each sample unit in the sample lot shall be determined.

7.4 PREPARATION OF FISH SAMPLE

1. Before preparing of a sub-sample adhering salt crystals should be removed by brushing from the surface of the sample without using water.

2. The preparation of fish samples for the determination of salt content, and water content in order to calculate the % salt saturation of the fish should be carried out according to AOAC 937.07. The analysis should be on the edible portion of the fish.

3. Determination should be performed at least in duplicate.

7.5 DETERMINATION OF SALT CONTENT

1. Principle

The salt is extracted by water from the preweighed sample. After the precipitation of the proteins, the chloride concentration is determined by titration of an aliquot of the solution with a standardized silver nitrate solution (Mohr method) and calculated as sodium chloride.

2. Equipment and chemicals

—Brush

—Sharp knife or saw

—Balance, accurate to 0.01 g

—Calibrated volumetric flasks, 250 ml

—Erlenmeyer flasks

—Electric homoginizer

—Magnetic stirrer

—Folded paper filter, quick running

—Pipettes

—Funnel

—Burette

—Potassium hexacyano ferrate (II), $K_4Fe(CN)_6 \cdot 3H_2O$, 15% w/v (aq)

—Zinc sulphate, $ZnSO_4 \cdot 6H_2O$, 30% w/v (aq)

—Sodium hydroxide, NaOH, 0.1 N, 0.41% w/v (aq)

—Silver nitrate, $AgNO_3$, 0.1 N, 1.6987% w/v (aq), standardized

—Potassium chromate, K_2CrO_4 5% w/v (aq)

—Phenolphthalein, 1% in ethanol

—Distilled or deionized water

3. Procedure

(i) Five gram of homogenized subsample is weighted into a 250 ml volumetric flask and vigorously shaken with approximately 100 ml water.

(ii) Five millilitre of potassium hexacyano-ferrate solution and 5 ml of zinc sulphate solution are added, the flask is shaken.

(iii) Water is added to the graduation mark.

(iv) After shaking again and allowing to stand for precipitation, the flask content is filtered through a folded paper filter.

(v) An aliquot of the clear filtrate is transferred into an Erlenmeyer flask and two drops of phenolphthalein are added. Sodium hydroxide is added dropwise until the aliquot takes on a faint red colour.

The aliquot then diluted with water to approximately 100 ml.

(vi) After addition of approximately 1 ml potassium chromate solution, the diluted aliquot is titrated under constant stirring, with silver nitrate solution. Endpoint is indicated by a faint, but distinct, change in colour. This faint reddish-brown colour should persist after brisk shaking.

To recognize the colour change, it is advisable to carry out the titration against a white background.

(vii) Blank titration of reagents used should be done.

(viii) Endpoint determination can also be made by using instruments like potentiometer or colorimeter.

4. Calculation of results

In the equation of the calculation of results the following symbols are used:

A = volume of aliquot (ml)

C = concentration of silver nitrate solution in N

V = volume of silver nitrate solution in ml used to reach endpoint and corrected for blank value

W = sample weight (g)

The salt content in the sample is calculated by using the equation:

Salt concentration (%) = $(V \times C \times 58.45 \times 250 \times 100) / (A \times W \times 1000)$

Results should be reported with one figure after the decimal point.

5. Reference method

As reference method a method should be used which includes the complete ashing of the sample in a muffle furnace at 550℃ before chloride determination according to the method described above (leaving out steps

(ii) and (iv)).

6. Comments

By using the given equation all chloride determined is calculated as sodium chloride. However it is impossible to estimate sodium by this methodology, because other chlorides of the alkali and earth alkali elements are present which form the counterparts of chlorides.

The presence of natural halogens other than chloride in fish and salt is negligible.

A step, in which proteins are precipitated (ii), is essential to avoid misleading results.

7.6 DETERMINATION OF WATER CONTENT

i) Determination of % salt saturation as required by the standard, should be in accordance to AOAC 950.46. B (Airdrying (a)).

ii) Determination of water content in the whole fish, when needed in the commercial trade of klippfish and wet salted fish, the method of sampling the fish should be carried out according to the "Determination of Water Content in Whole Fish by Cross Section Method" defined in "Annex B".

8 DEFINITION OF DEFECTIVES

8.1 The sample unit shall be considered defective when it exhibits any of the properties defined below.

8.1.1 Foreign Matter

The presence in the sample unit of any matter which has not been derived from Gadidae fish, does not pose a threat to human health, and is readily recognized without magnification or is present at a level determined by any method including magnification that indicates non-compliance with good manufacturing and sanitation practices.

8.1.2 Odour

A fish affected by persistent and distinct objectionable odours indicative of decomposition (such as sour, putrid, etc.) or contamination by foreign substances (such as fuel oil, cleaning compounds, etc.).

8.1.3 Pink

Any visible evidence of red halophilic bacteria.

8.1.4 Appearance

Textural breakdown of the flesh which is characterized by extensive cracks on more than 2/3 of the surface area or which has been mutilated, torn or broken through to the extent that the split fish is divided into two or more pieces but still held together by skin.

8.2 The sample unit shall be considered defective when 30% or more of the fish in the sample unit are affected by any of the following defects.

8.2.1 Halophilic Mould (dun)

A fish showing an aggregate area of pronounced halophilic mould clusters on more than 1/3 of the total surface area of the face side.

8.2.2 Liver Stains

A pronounced yellow or yellowish orange discoloration caused by the presence of liver and affecting more than 1/4 of the total surface area of the face of the fish.

8. 2. 3 Intense Bruising

Any fish showing more than 1/2 of the face of the fish with intense bruising.

8. 2. 4 Severe Burning

A fish with more than 1/2 of the back (skin side) tacky or sticky due to overheating during drying.

9 LOT ACCEPTANCE

A lot shall be considered as meeting the requirements of this standard when:

(i) the total number of defectives as classified according to section 8 does not exceed the acceptance number (c) of the appropriate sampling plan in the Sampling Plans for Prepackaged Foods (AQL – 6. 5) (CODEX STAN 233 – 1969);

(ii) the average net weight of all sample units is not less than the declared weight, provided no individual container is less than 95% of the declared weight; and

(iii) the total number of sample units not meeting the form of presentation as defined in section 2. 3 does not exceed the acceptance number (c) of the appropriate sampling plan in the Sampling Plans for prepackaged Foods (AQL – 6. 5) (CODEX STAN 233 – 1969);

(iv) the Food Additives, Hygiene and Handling and Labelling requirements of Sections 4, 5 and 6 are met.

ANNEX A SENSORY AND PHYSICAL EXAMINATION

1. Examine every fish in the sample in its entirety.
2. Examine the product for the form of presentation.
3. Examine the fish for foreign matter, pink conditions, halophilic mould, liver stains, intense bruising, severe burning and texture.
4. Assess odour in accordance with the *Guidelines for the Sensory Evaluation of Fish and Shellfish in Laboratories* (*CAC/GL* 31 – 1999).

ANNEX B DETERMINATION OF WATER CONTENT IN WHOLE FISH BY CROSS SECTION METHOD

1 Principle

The fish is cut in sections as described in method. The sections are cut in smaller bits to a collected sample. The water content of the collected sample is determined by drying. Examinations and experience have shown that the water content of this collected sample is closed to the "true" water content of the fish.

2 Equipment

—Soft brush

—Basins (steel, glass, porcelain)

—Scissors

—Band saw

—Knife
—Weight, 1 g precision
—Oven. 103 ~ 105℃
—Desiccator

3 Preparation of sample

Salt particles on the surface of the fish are brushed away.

The weight of the fish is determined to 1 g accuracy.

The length of the fish is measured as the distance between the cleft in the tail and a line drawn between the tips of the earbones.

4 Procedure

(i) The sampling of the fish is described in the enclosed figure.

A) Wet salted fish is sliced in sections by knife.

B) Salted and dried salted fish is sliced in sections by band saw.

1) A section of 20mm measured from a line drawn between the earbones, dotted line on figure, is cut.

2) The next cut is a 40mm section.

3) A 2mm section is cut from the front part of the 40mm section and collected (see 7. Comments).

4) The next cut is a new cut of a 40mm section.

5) A 2mm section is cut from the front part of the 40mm section and collected.

6) The entire fish is cut in 40mm sections from which are cut 2mm sections (see enclosed figure).

7) All sections of 2mm, marked Ⅱ, Ⅳ, Ⅵ, Ⅷ in the figure, even numbers, are collected to a collected sample.

(ii) The 2mm sections in the collected sample are cut with scissors in smaller pieces directly in tared basins just after the fish is cut.

(iii) The basins containing the sample are weighted.

(iv) The basins containing the samples are put in the oven at 103 ~ 105℃ for drying to constant weight (18 hours over night).

(v) The basins are taken from the oven to a desiccator and cooled.

(vi) The basins are weighted.

5 Calculation of results

In the equation of the calculation of results the following symbols are used:

W_1 = Weight of fish and basins before drying, g.

W_2 = Weight of fish and basins after drying, g.

W_s = Weight of tared basins, g.

The water content in the fish is calculated by using the equation:

Water content, g/100g $= 100 \times (W_1 - W_2) / (W_1 - W_s)$

The result is reported to the nearest gram, together with the length and the weight of the analysed fish.

6 Control analysis of whole fish

The determination of water content in whole fish by cross section method appears to give the closest result compared to water content determined by the drying of the whole fish (ALINORM 03/18, Appendix IX).

7 Comments

Each sampled fish should be packed and sealed in a plastic bag before analysis. The samples should be stored under chilled or refrigerated conditions from the time of sampling to the time of analysis.

The analysis must be performed as soon as possible after the fish has been sampled.

It might be difficult to cut sections of 2mm when the fish has a water content above 50% but the section must be close to 2mm.

To minimise the loss of water from the 2mm sections it is important to weight the collected sample immediately after the fish is cut in sections.

Determination should be performed at least in duplicate.

FIGURE

Sampling procedure.

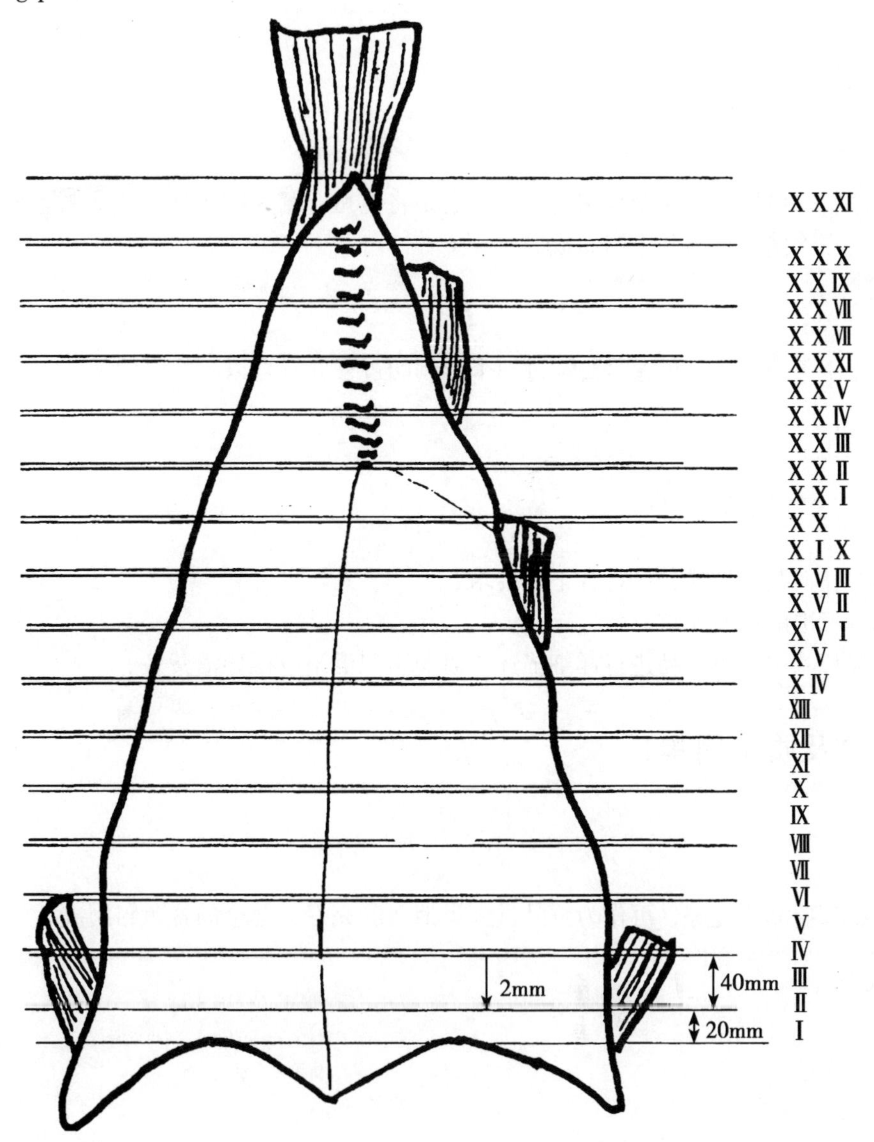

All section labelled by even numbers, II, IV, VI, VIII etc. are collected to constitute one sample.

鱼　翅

CODEX STANDARD FOR DRIED SHARK FINS

（*CODEX STAN 189 - 1993*）

1　范围

本标准适用于深加工用干鱼翅。

2　说明

2.1　产品定义

干鱼翅的原料是可供人类安全食用的鲨鱼。它是将鲨鱼背鳍和胸鳍从根部呈弧形割下，而尾鳍下叶直接切下，去除鳍上鱼肉并干制而成的。

2.2　加工定义

为达到3.2.4条的要求，鱼翅需要经过一个干制的过程并符合以下规定的条件。

2.3　产品介绍

2.3.1　干鱼翅可以带表皮或去皮

2.3.2　其他形式

产品介绍的任何其他形式，应符合下列要求：

（1）达到本标准的所有要求；

（2）在产品标签中应对产品进行详细描述，以免引起混淆或误导消费者。

3　基本成分及质量因素

3.1　鲨鱼

鲨鱼翅是由那些品质良好，可作为鲜品供人类消费的完好、健全的鲨鱼鳍制成。

3.2　其他成分

无。

3.3　成品

3.3.1　外观要求

产品不得含有任何外来杂质。

3.3.2　气味

产品不得带有令人厌恶的气味。

3.3.3　组织

干鱼翅不得有令人厌恶的组织特征。

3.3.4　水分含量

成品水分含量不得超过18%。

4　食品添加剂

不允许使用任何添加剂。

5　卫生及处理

5.1　成品中不能含有任何危害人体健康的外来杂质

5.2　应用CAC规定的抽样及检测方法进行检验时，应符合下列要求：

（i）产品中微生物含量和由微生物产生的危害人体健康的有害物质的含量不能超过CAC标准的规定。

（ii）其他危害人体健康的有害物质的含量也不能超过CAC标准的规定。

5.3　建议本标准中涉及产品预处理的条款应符合推荐性国际操作规范《CAC/RCP 1－1969，Rev. 3－1997 食品卫生总则》相关章节和相关标准推荐性国际操作规范《CAC/RCP 9－1976 鲜鱼》

6　标签

除了应符合《CODEX STAN 1－1985，Rev. 3－1999 预包装食品标签通用标准》的要求外，还应遵守以下规定：

6.1　食品名称

产品名称应为干鱼翅，或按出售该产品的国家的法律、习惯采用其他适当的名称。

6.1.1　标签上还应在与产品名称紧靠位置出现关于外观形式的词语，这些描述性词语应恰如其分地说明产品外观，并应确保它们不会引起混淆或误导消费者。

6.1.2　另外，还要在标签上注明鱼种名、鱼鳍的类型及其尺寸规格。

6.2　非零售包装的标签

上述要求应既在包装上又在附文中出现，除了食品名称、批号、制造或分装厂名、地址外，还应包括贮藏条件。批号、制造或分装厂名、地址也可用同一证明标志代替，证明标志要在辅助文件中说明清楚。

7　抽样、检验和分析

7.1　取样

（1）产品批次检验用样品的抽样方法应符合FAO/WHO的食品法典《CODEX STAN 233－1969

预包装食品的抽样方案》(AQL－6.5)。

(2) 对需检测净重批次的抽样，应按照 FAO/WHO《净重测定抽样计划》（正在制定中）进行。

7.2 感官与物理检验

产品的感官与物理指标须由经过此类检验培训的人进行检验，并依据本标准 7.3 条和附录 B（待述）以及《CAC/GL 31－1999 实验室中水产品感官评价指南》所述程序。

7.3 净重的测定

应测定抽样批次中各样品单位的净重（不包括包装材料）。

7.4 水分含量的测定

测定方法待述。

8 缺陷的定义

当样品单位未达到 3.3 中对成品的要求时，则认定其有“缺陷”。

8.1 外来杂质

样品单位中存在的任何不是来自于鱼体的物质。这些物质虽不会对人体健康造成危害，但用肉眼可直接辨别，或采用某些方法（包括放大）可以确定其存在。出现外来杂质表明不符合良好操作和卫生惯例。

8.2 气味

样品散发的持久、明显、令人厌恶的由腐败、酸败引起的气味或风味。

8.3 组织

鱼鳍组织全面下降，表现为鱼鳍变软。

8.4 水分

被检样品单位水分含量超过 18%。

9 批次验收

当满足以下条件时，可以认为此批次产品符合本标准的要求：

(1) 根据本标准第 8 章中规定分类，缺陷总数不超过《CODEX STAN 233－1969 预包装食品的抽样方案》(AQL－6.5) 中相应抽样方案规定的可接受数（c）；

(2) 不符合本标准 2.3 中规定数量的样品总数不超过《CODEX STAN 233－1969 预包装食品的抽样方案》(AQL－6.5) 中相应抽样方案规定的可接受数（c）；

(3) 所有样品单位平均净重不少于标示量，在任何一个包装单位中没有不合理的重量短缺；

(4) 符合本标准第 4、5、6 款中对食品添加剂、卫生和处理以及标签的要求。

CODEX STANDARD FOR DRIED SHARK FINS

(*CODEX STAN 189 - 1993*)

1 SCOPE

This Standard applies to dried shark fins intended for further processing.

2 DESCRIPTION

2. 1 PRODUCT DEFINITION

Dried shark fins are the dorsal and pectoral fins cut in the form of an arc and the lower lobe of the caudal fin cut straight, from which all flesh has been removed, and are cut from species of sharks which are safe for human consumption.

2. 2 PROCESS DEFINITION

The fins shall be subjected to a drying process so as to meet the requirements of Section 3. 2. 4 and shall comply with the conditions laid down hereafter.

2. 3 PRESENTATION

2. 3. 1 Dried shark fins may be presented with the skin on or as skinless.

2. 3. 2 *Other Forms of Presentation*

Any other presentation shall be permitted provided that it:

(i) meets all other requirements of this standard; and

(ii) is adequately described on the label to avoid confusing or misleading the consumer.

3 ESSENTIAL COMPOSITION AND QUALITY FACTORS

3. 1 SHARK

Dried shark fins shall be prepared from sound sharks which are of a quality fit to be sold fresh for human consumption.

3. 2 OTHER INGREDIENTS

None.

3.3 FINAL PRODUCT

3.3.1 Appearance

The final product shall be free from foreign material.

3.3.2 Odour

The product shall be free from objectionable odours.

3.3.3 Texture

The dried shark fins shall be free from objectionable textural characteristics.

3.3.4 Percentage of Moisture

The final product shall have a moisture content not exceeding 18%.

4 FOOD ADDITIVES

No additives are permitted.

5 HYGIENE AND HANDLING

5.1 The final product shall be free from any foreign material that poses a threat to human health.

5.2 When tested by appropriate methods of sampling and examination prescribed by the Codex Alimentarius Commission, the product:

(i) shall be free from microorganisms or substances originating from microorganisms in amounts which may present a hazard to health in accordance with standards established by the Codex Alimentarius Commission;

(ii) shall not contain any other substance in amounts which may present a hazard to health in accordance with standards established by the Codex Alimentarius Commission.

5.3 It is recommended that the product covered by the provisions of this standard be prepared and handled in accordance with the appropriate sections of the Recommended International Code of Practice-General Principles of Food Hygiene (CAC/RCP 1 - 1969, Rev. 3 - 1997) and the following relevant Code: Recommended International Code of Practice for Fresh Fish (CAC/RCP 9 - 1976).

6 LABELLING

In addition to the General Standard for the Labelling of Prepackaged Foods (CODEX STAN 1 - 1985, Rev. 3 - 1999), the following specific provisions shall apply:

6.1 NAME OF THE FOOD

The name of the product shall be "dried shark fins" or any other appropriate name in accordance with the law and custom of the country in which the product is to be distributed.

6.1.1 There shall appear on the label reference to the form of presentation in close proximity to the name of the product in such descriptive terms that will adequately and fully describe the nature of the presentation of the product to avoid misleading or confusing the consumer.

6.1.2 In addition to the specified labelling designations above, the name of the species, the type of fin, and its size shall also appear on the label.

6.2 LABELLING OF NON-RETAIL CONTAINERS

Information on the above provisions shall be given either on the container or in accompanying documents, except that the name of the product, lot identification, and the name and address of the manufacturer or packer, shall appear on the container.

However, lot identification, and the name and address of the manufacturer or packer may be replaced by an identification mark provided that such a mark is clearly identifiable with the accompanying documents.

7 SAMPLING, EXAMINATION AND ANALYSIS

7.1 SAMPLING

(i) Sampling of lots for examination of the product shall be in accordance with the Codex Sampling Plans of Prepackaged Foods (AQL - 6.5) (CAC/RM 42 - 1969);

(ii) The sampling of lots for examination of net weight shall be carried out according to the Codex Sampling Plans for the Determination of Net Weight (under elaboration).

7.2 SENSORY AND PHYSICAL EXAMINATION

Samples taken for sensory and physical examination shall be assessed by persons trained in such examination and in accordance with the procedures set out in Section 7.3, Annex B "Sensoric and Physical Examination"[1] and the *Guidelines for the Sensory Evaluation of Fish and Shellfish in Laboratories* (CAC/GL 31 - 1999).

7.3 DETERMINATION OF NET WEIGHT

The net weight (exclusive of packaging material) of each sample unit in the sample lot shall be determined.

7.4 DETERMINATION OF MOISTURE

[Method to be developed.]

8 CLASSIFICATION OF DEFECTIVES

A sample unit shall be considered defective when it fails to meet any of the following final product requirements referred to in Section 3.3.

8.1 FOREIGN MATTER

The presence in the sample unit of any matter which has not been derived from fish, does not pose a threat to human health, and is readily recognized without magnification or is present at a level determined

---

[1] To be developed.

by any method including magnification that indicates non-compliance with good manufacturing and sanitation practices.

8.2 ODOUR

A sample unit affected by persistent and distinct objectionable odours indicative of decomposition.

8.3 TEXTURE

Textural breakdown of the fin, indicative of decomposition, characterized by softness.

8.4 MOISTURE

The sample unit exceeds 18% moisture.

9 LOT ACCEPTANCE

A lot shall be considered as meeting the requirements of this Standard when:

(i) the total number of defectives as classified according to Section 8 does not exceed the acceptance number (c) of the appropriate sampling plan in the Sampling Plans for Prepackaged Foods (AQL-6.5) (CAC/RM 42-1969);

(ii) the average net weight of all sample units is not less than the declared weight, provided there is no unreasonable shortage in any container; and

(iii) the total number of sample units not meeting the form of presentation as defined in Section 2.3 does not exceed the acceptance number (c) of the appropriate sampling plan in the Sampling Plans for prepackaged Foods (AQL-6.5) (CAC/RM 42-1969);

(iv) the Food Additive, Hygiene and Handling and Labelling requirements of Sections 4, 5.1, 5.2 and 6 are met.

ANNEX A

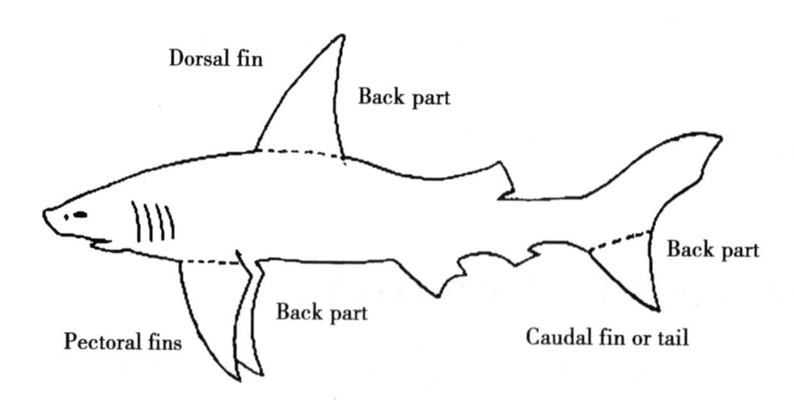

速冻鱼片

CODEX GENERAL STANDARD FOR QUICK FROZEN FISH FILLETS

(*CODEX STAN 190 – 1995*)

1　适用范围

本标准适用于符合下述定义的，不需深加工即可直接消费的速冻鱼片，不适用于标明还需进行深加工或用于其他工业目的的产品。

2　说明

2.1　产品定义

速冻鱼片是指从同种的适用于人类食用的鱼体上取下的大小形状不规则的鱼肉切片。为便于包装，切片时与脊椎骨平行切割。切割之后再根据 2.2 进行加工。

2.2　加工定义

产品经过适当预处理后，应在符合以下规定的条件下进行冻结加工：冻结应在合适的设备中进行，并使产品迅速通过最大冰晶生成温度带。速冻加工只有在产品的中心温度达到并稳定在 –18℃ (0 ℉) 或更低的温度时才算完成。产品在运输、贮存、分销过程中应保持在深度冻结状态，以保证产品质量。

在产品的加工和包装过程中应尽量减少脱水和氧化作用的影响。

在保证质量的受控条件下，允许按规定要求对速冻产品再次速冻加工，并按照被认可的操作进行再包装。

2.3　产品介绍

2.3.1　产品介绍应符合下列要求

（1）达到本标准的所有要求；

（2）在标签中对产品进行详细描述，以免引起混淆或误导消费者。

2.3.2　鱼片如果已经剔除了包括细刺在内的鱼骨，则为无刺鱼片

3　基本成分及质量因素

3.1　鱼

速冻鱼片应由品质良好、可作为鲜品供人类消费的个体完好的鱼制备。

3.2 冰衣

用于镀冰衣或制备镀冰衣所用溶液的水应达到饮用水或清洁海水的质量要求。饮用水应符合WHO最新版本的《国际饮用水质量规范》要求，清洁海水应是达到饮用水的微生物标准且不含异物的海水。

3.3 其他成分

所使用的其他成分应具有食品级的质量，并且符合所有相应法规标准的规定。

3.4 腐败

被测样品单位的组胺平均含量不应超过10mg/100g。这仅适用于鲱科（*Clupeidae*）、鲭科（*Scombridae*）、秋刀鱼科（*Scombresocidae*）、鲑科（*Pomatomidae*）以及鳍鳅科（*Coryphaenedae*）等科的鱼类。

3.5 成品

产品应符合本标准要求，根据第9章进行成品批次检查时，其质量应符合第8章的规定。检查方法应符合第7章的规定。

4 食品添加剂

添加剂		成品中的最高含量
保水剂	339（i）磷酸二氢钠 340（i）磷酸二氢钾 450（iii）焦磷酸钠 450（v）焦磷酸钾 451（i）三聚磷酸钠 451（ii）三聚磷酸钾 452（i）多聚磷酸钠 452（iv）多聚磷酸钙	≤10g/kg，以 P_2O_5 计，单用或混用（包括天然磷酸盐）
	401 褐藻酸钠	GMP
抗氧化剂	301 抗坏血酸钠 303 抗坏血酸钾	GMP

5 卫生及处理

5.1 成品中不能含有任何危害人体健康的外来杂质

5.2 应用CAC规定的抽样及检测方法对成品进行检验时，应遵循以下规定：

（1）产品中微生物含量和由微生物产生的危害人体健康的有害物质的含量不能超过CAC标准的规定；

（2）组胺含量不应超过20mg/100g，这仅适用于鲱科（*Clupeidae*）、鲭科（*Scombridae*）、秋刀鱼科（*Scombresocidae*）、鲑科（*Pomatomidae*）以及鳍鳅科（*Coryphaenedae*）等科的鱼类；

（3）其他危害人体健康的有害物质的含量也不能超过CAC标准的规定。

5.3 建议本标准中涉及产品预处理的条款应符合推荐性国际规范《CAC/RCP1 - 1969，Rev. 3 - 1997 食品卫生操作总则》中的相关章节和以下相关标准：

（1）推荐性国际操作规范《CAC/RCP 16－1978 冻鱼》；
（2）推荐性国际操作规范《CAC/RCP 8－1976 速冻食品的加工和处理》；
（3）《水产品国际操作规程草案》中水产养殖品的相关章节（正在完善中）❶。

6 标签

标签除应符合《CODEX STAN 1－1985，Rev. 3－1999 预包装食品标签通用标准》的要求外，还应遵循以下规定：

6.1 食品名称

6.1.1 在标签上产品的名称应按出售该食品的国家的法律、习惯和实际情况，注明"……片"或"……的片"。
6.1.2 标签上还应在产品名附近注明产品外观，这些术语应恰如其分地描述产品性状，以免引起混淆或误导消费者。
6.1.3 在产品标签上应标明"速冻"，除非在有的国家"冻"即代表"速冻"。
6.1.4 标签应注明产品必须在运输、贮藏、分销过程中保持的条件，以保证其质量。
6.1.5 如果产品用海水镀冰衣，则应予以说明。

6.2 净含量（镀冰衣产品）

对于镀冰衣的产品，其内容物净含量不包括冰衣重。

6.3 贮藏说明

标签应注明产品须在－18℃或更低的温度下贮藏。

6.4 非零售包装的标签

上述要求应既在包装上又在附文中出现，除了食品名称、批号、制造或分装厂名、地址外，还包括贮藏条件。但批号、制造或分装厂名、地址也可用同一证明标志代替，只要证明标志能在辅助文件中说明。

7 抽样、检验和分析

7.1 取样

（1）产品批次检验用样品的抽样方法应符合 FAO/WHO 的食品法典标准《CODEX STAN 233－1969 预包装食品的抽样方案》（AQL－6.5）。样品单位是初级包装，对单冻（IQF）产品至少以1kg 样品为单位。
（2）对需检测净重批次的抽样，抽样计划应以食品法典委员会的相关标准为依据。

7.2 感官与物理检验

产品的感官与物理指标须由经过此类检验培训的人员进行检验，并依据本标准 7.3 至 7.6 条和

❶ 操作规程建议草案，在定稿后将替代所有现行的《水产品操作规程》。

附录 A 以及《CAC/GL 31－1999 实验室中水产品感官评价指南》所述程序进行。

7.3　净重的测定

7.3.1　代表每批次的样品单位的净重（不包括包装材料）应在冷冻状态下测定

7.3.2　镀冰衣产品净重的测定

从低温贮藏状态下取出样品后，立即打开包装，将内容物放在冷水喷头下，用缓慢的水流喷洒内容物，并小心摇动样品，不要弄破样品；直到可以看到或用手感觉到全部冰衣已被去除时，停止喷洒；用纸巾除去附着的水并放在已知皮重的盘中称重。

7.4　无皮鱼片中寄生虫的检验程序（方法 1）

对整个抽样的检测可以通过以下非破坏性方法进行：将适当大小的解冻样品置于一 5mm 厚、透光率为 45% 的丙烯酸塑料薄片上，并在其上方 30cm 处，用 1500Lux（勒克司）的光源进行灯检。

7.5　凝胶状态的检验

按照 AOAC 方法 983.18“肉及肉制品中水分含量测定时抽样的准备工作”和 950.46“肉中的水分含量”（方法 A）。

7.6　蒸煮方法

蒸煮使产品内部温度达到 65～70℃。不能过度蒸煮，蒸煮时间随产品大小和采用的温度不同而不同。准确的蒸煮时间和条件应依据预先实验来确定。

烘焙：用铝箔包裹产品，并将其均匀放入扁平锅或浅平锅上。

蒸：用铝箔包裹产品，并将其置于带盖容器中沸水之上的金属架上。

袋煮：将产品放入可煮薄膜袋中加以密封，浸入沸水中煮。

微波：将产品放入适于微波加热的容器中，若用塑料袋，应检查确定塑料袋不会发出任何气味。根据设备说明加热。

7.7　组胺检测

依据 AOAC 方法 977.13 进行检测。

8　缺陷定义

当样品呈现以下任何一项特征时，则认定其有“缺陷”：

8.1　深度脱水

样品单位中表面积超过 10%，或按下表给出的包装大小，出现水分的过度损失，明显表现为鱼体表面呈现异常的白色或黄色，覆盖了肌肉本身的颜色并已渗透至表层以下，在不过分影响产品外观的情况下，不能轻易地用刀或其他利器刮掉，具体要求如下表。

包装大小	缺陷面积
a）200g	25cm^2
b）201～500g	50cm^2
c）501～5 000g	150cm^2

8.2 外来杂质

样品单位中存在的任何不是来自于鱼体的物质。这些物质虽不会对人体健康造成危害，但用肉眼可直接辨别，或采用某些方法（包括放大）可以确定其存在。含有外来杂质表明不符合良好操作和卫生习惯。

8.3 寄生虫

采用7.4方法，每千克样品中检测到2个或2个以上直径大于3mm的囊状寄生虫，或非囊状但长度大于10mm的寄生虫。

8.4 骨刺（在标明无刺的包装中）

每千克产品中检出一个以上长度为10mm或者直径大于或等于1mm的骨刺；骨刺的长度等于或小于5mm，如果其直径不大于2mm则该产品不被认定为缺陷。如果骨刺的根部（与脊椎骨的连接处）的宽度小于或等于2mm，或其可用手指甲轻松地剥除，则其可被忽略。

8.5 气味/风味

样品散发的持久、明显、令人厌恶的由腐败、酸败或饵料引起的气味或风味。

8.6 鱼肉异常

样品单位出现过量凝胶状态的鱼肉并伴有任何单片鱼片中水分达86%以上，或按重量计算5%以上的样品被寄生虫感染导致质地呈现糊状。

9 批次验收

当满足以下条件时，可以认为此批次产品符合本标准的要求：

（1）根据本标准第8章中规定分类，缺陷总数不超过《CODEX STAN 233－1969 预包装食品的抽样方案》（AQL－6.5）中相应抽样方案规定的可接受数（c）；

（2）所有样品单位平均净重不少于标示量，在任何一个包装单位中没有不合理的重量短缺；

（3）符合本标准第4、5、6款中对食品添加剂、卫生和处理以及标签的要求。

附录A 感官及物理检验

1. 净含量的测定按本标准7.3条规定执行（按要求去冰衣）。

2. 通过测定只能用小刀或其他利器除去的面积，检查冻鱼片中脱水的情况。测量样品单位的总表面积，计算受影响的面积百分比。

3. 解冻并且逐个地检测样品单位中每片鱼片有无外来杂质、寄生虫、骨刺（需要时）、臭味及鱼肉质地异常等缺陷。

4. 对解冻后未蒸煮状态下无法判定其气味的样品，则应从样品单位中截取一小块可疑部分（约200g），并立即使用7.6条中规定的某种蒸煮方法，确定其气味和风味。

5. 对在解冻后未蒸煮状态下无法最终确定其凝胶状态的产品，则应从产品中截取可疑部分，按7.6条规定的蒸煮方法，或用7.5测定在任何鱼片中是否含有超过86%的水分，即可确认其凝胶情况。如果无法根据蒸煮后的评估得到结论，则使用7.5条中的程序准确测定水分含量。

CODEX GENERAL STANDARD FOR QUICK FROZEN FISH FILLETS

(*CODEX STAN 190 - 1995*)

1 SCOPE

This standard applies to quick frozen fillets of fish as defined below and offered for direct consumption without further processing. It does not apply to products indicated as intended for further processing or for other industrial purposes.

2 DESCRIPTION

2.1 PRODUCT DEFINITION

Quick frozen fillets are slices of fish of irregular size and shape which are removed from the carcass of the same species of fish suitable for human consumption by cuts made parallel to the backbone and sections of such fillets cut so as to facilitate packing, and processed in accordance with the process definitions given in Section 2.2.

2.2 PROCESS DEFINITION

The product after any suitable preparation shall be subjected to a freezing process and shall comply with the conditions laid down hereafter. The freezing process shall be carried out in appropriate equipment in such a way that the range of temperature of maximum crystallization is passed quickly. The quick freezing process shall not be regarded as complete unless and until the product temperature has reached 18℃ (0℉) or colder at the thermal centre after thermal stabilization. The product shall be kept deep frozen.

so as to maintain the quality during transportation, storage and distribution.

These products shall be processed and packaged so as to minimize dehydration and oxidation.

The recognized practice of repacking quick frozen products under controlled conditions which will maintain the quality of the product, followed by the reapplication of the quick freezing process as defined, is permitted.

2.3 PRESENTATION

2.3.1 Any presentation of the product shall be permitted provided that it:

(a) meets all requirements of this standard, and

(b) is adequately described on the label to avoid confusing or misleading the consumer.

2.3.2 Fillets may be presented as boneless, provided that boning has been completed including the removal of pinbones.

3 ESSENTIAL COMPOSITION AND QUALITY FACTORS

3.1 FISH

Quick frozen fish fillets shall be prepared from sound fish which are of a quality fit to be sold fresh for human consumption.

3.2 GLAZING

If glazed, the water used for glazing or preparing glazing solutions shall be of potable quality or shall be clean sea-water. Potable water is fresh-water fit for human consumption. Standards of potability shall not be less than those contained in the latest edition of the WHO "International Guidelines for Drinking Water Quality". Clean sea-water is sea-water which meets the same microbiological standards as potable water and is free from objectionable substances.

3.3 OTHER INGREDIENTS

All other ingredients used shall be of food grade quality and conform to all applicable Codex standards.

3.4 DECOMPOSITION

The products shall not contain more than 10 mg/100 g of histamine based on the average of the sample unit tested. This shall apply only to species of *Clupeidae*, *Scombridae*, *Scombresocidae*, *Pomatomidae and Coryphaenedae families.*

3.5 FINAL PRODUCT

Products shall meet the requirements of this standard when lots examined in accordance with Section 9 comply with the provisions set out in Section 8. Products shall be examined by the methods given in Section 7.

4 FOOD ADDITIVES

	Additive	Maximum level in the final product
Moisture/Water Retention Agents	339 (i) Monosodium orthophosphate 340 (i) Monopotassium orthophosphate 450 (iii) Tetrasodium diphosphate 450 (v) Tetrapotassium diphosphate 451 (i) Pentasodium triphosphate 451 (ii) Pentapotassium triphosphate 452 (i) Sodium polyphosphate 452 (iv) Calcium, polyphosphates	≤10 g/kg expressed as P_2O_5, singly or in combination (includes natural phosphate)
	401 Sodium alginate	GMP
Antioxidants	301 Sodium ascorbate 303 Potassium ascorbate	GMP

5 HYGIENE AND HANDLING

5.1 The final product shall be free from any foreign material that poses a threat to human health.

5.2 When tested by appropriate methods of sampling and examination prescribed by the Codex Alimentarius Commission, the product:

(i) shall be free from microorganisms or substances originating from microorganisms in amounts which may present a hazard to health in accordance with standards established by the Codex Alimentarius Commission;

(ii) shall not contain histamine that exceeds 20 mg/100g. This applies only to species of *Clupeidae*, *Scombridae*, *Scombresocidae*, *Pomatomidae and Coryphaenedae families*;

(iii) shall not contain any other substance in amounts which may present a hazard to health in accordance with standards established by the Codex Alimentarius Commission.

5.3 It is recommended that the product covered by the provisions of this standard be prepared and handled in accordance with the appropriate sections of the Recommended International Code of Practice General Principles of Food Hygiene (CAC/RCP 1 - 1969) and the following relevant Codes:

(i) the Recommended International Code of Practice for Frozen Fish (CAC/RCP 16 - 1978);

(ii) The Recommended International Code of Practice for the Processing and Handling of Quick Frozen Foods (CAC/RCP 8 - 1976);

(iii) The sections on the Products of Aquaculture in the Proposed Draft International Code of Practice for Fish and Fishery Products (under elaboration)[1]

6 LABELLING

In addition to the General Standard for the Labelling of Prepackaged Foods (CODEX STAN 1 - 1985), the following specific provisions apply:

6.1 NAME OF THE FOOD

6.1.1 The name of the product as declared on the label shall be "……fillets" or "fillets of……" according to the law, custom or practice in the country in which the product is to be distributed.

6.1.2 There shall appear on the label reference to the form of presentation in close proximity to the name of the food in such additional words or phrases that will avoid misleading or confusing the consumer.

6.1.3 The term "quick frozen", shall also appear on the label, except that the term "frozen" may be applied in countries where this term is customarily used for describing the product processed in accordance with subsection 2.2 of this standard.

6.1.4 The label shall state that the product should be maintained under conditions that will maintain the quality during transportation, storage and distribution.

6.1.5 If the product has been glazed with sea-water, a statement to this effect shall be made.

---

[1] The Proposed Draft Code of Practice, when finalized, will replace all current Codes of Practice for Fish and Fishery Products.

6.2 NET CONTENTS (GLAZED PRODUCTS)

Where the food has been glazed the declaration of net contents of the food shall be exclusive of the glaze.

6.3 STORAGE INSTRUCTIONS

The label shall include terms to indicate that the product shall be stored at a temperature of -18℃ or colder.

6.4 LABELLING OF NON-RETAIL CONTAINERS

Information on the above provisions shall be given either on the container or in accompanying documents, except that the name of the product, lot identification, and the name and address of the manufacturer or packer as well as storage instructions, shall appear on the container.

However, lot identification, and the name and address of the manufacturer or packer may be replaced by an identification mark provided that such a mark is clearly identifiable with the accompanying documents.

7 SAMPLING, EXAMINATION AND ANALYSES

7.1 SAMPLING

(i) Sampling of lots for examination of the product shall be in accordance with the FAO/WHO Codex Alimentarius Sampling Plans for Prepackaged Foods (AQL-6.5) (CODEX STAN 233-1969). A sample unit is the primary container or for individually quick frozen products is at least a 1 kg portion of the sample unit.

(ii) Sampling of lots for examination of net weight shall be carried out in accordance with an appropriate sampling plan meeting the criteria established by the Codex Alimentarius Commission.

7.2 SENSORY AND PHYSICAL EXAMINATION

Samples taken for sensory and physical examination shall be assessed by persons trained in such examination and in accordance with procedures elaborated in Sections 7.3 through 7.6, Annex A and the *Guidelines for the Sensory Evaluation of Fish and Shellfish in Laboratories* (CAC/GL 31-1999).

7.3 DETERMINATION OF NET WEIGHT

7.3.1 The net weight (exclusive of packaging material) of each sample unit representing a lot shall be determined in the frozen state

7.3.2 Determination of Net Weight of Products Covered by Glaze

As soon as the package is removed from low temperature storage, open immediately and place the contents under a gentle spray of cold water. Agitate carefully so that the product is not broken. Spray until all ice glaze that can be seen or felt is removed. Remove adhering water by the use of paper towel and weight the product in a tared pan.

7.4 PROCEDURE FOR THE DETECTION OF PARASITES (TYPE 1 METHOD) IN SKINLESS FILLETS

The entire sample unit is examined non-destructively by placing appropriate portions of the thawed sample unit on a 5mm thick acryl sheet with 45% translucency and candled with a light source giving 1500lux 30 cm above the sheet.

7.5 DETERMINATION OF GELATINOUS CONDITION

According to the AOAC Methods- "Moisture in Meat and Meat Products, Preparation of Sample Procedure"; 983.18 and "Moisture in Meat" (Method A); 950.46.

7.6 COOKING METHODS

The following procedures are based on heating the product to an internal temperature of 65 ~ 70℃. The product must not be overcooked. Cooking times vary according to the size of the product and the temperatures used. The exact times and conditions of cooking for the products should be determined by prior experimentation.

Baking Procedure: Wrap the product in aluminum foil and place it evenly on a flat cookie sheet or shallow flat pan.

Steaming Procedure: Wrap the product in aluminum foil and place it on a wire rack suspended over boiling water in a covered container.

Boil-in-Bag Procedure: Place the product in a boilable film-type pouch and seal. Immerse the pouch in boiling water and cook.

Microwave Procedure: Enclose the product in a container suitable for microwave cooking. If plastic bags are used, check to ensure that no odour is imparted from the plastic bags. Cook according to equipment instructions.

7.7 DETERMINATION OF HISTAMINE

AOAC 977.13.

8 DEFINITION OF DEFECTIVES

A sample unit shall be considered as defective when it exhibits any of the properties defined below:

8.1 DEHYDRATION

Greater than 10% of the surface area of the sample unit or for pack sizes described below, exhibits excessive loss of moisture clearly shown as white or yellow abnormality on the surface, which masks the colour of the flesh and penetrates below the surface, and cannot be easily removed by scraping with a knife or other sharp instrument without unduly affecting the appearance of the product.

Pack Size	Defect Area
a) ≤200g units	≥ 25cm^2
b) 201 ~ 500g units	≥ 50cm^2
c) 501 ~ 5 000g units	≥ 150cm^2

8.2 FOREIGN MATTER

The presence in the sample unit of any matter, which has not been derived from fish, does not pose a threat to human health, and is readily recognized without magnification or is present at a level determined by any method including magnification that indicates non-compliance with good manufacturing and sanitation practices.

8.3 PARASITES

The presence of two or more parasites per kg of the sample unit detected by the method described in 7.4 with a capsular diameter greater than 3mm or a parasite not encapsulated and greater than 10mm in length.

8.4 BONES (IN PACKS DESIGNATED BONELESS)

More than one bone per kg of product greater or equal to 10mm in length, or greater or equal to 1mm in diameter; a bone less than or equal to 5mm in length, is not considered a defect if its diameter is not more than 2mm. The foot of a bone (where it has been attached to the vertebra) shall be disregarded if its width is less than or equal to 2mm, or if it can easily be stripped off with a fingernail.

8.5 ODOUR AND FLAVOUR

A sample unit affected by persistent and distinct objectionable odours or flavours characteristic of decomposition, rancidity or feed.

8.6 FLESH ABNORMALITIES

A sample unit affected by excessive gelatinous condition of the flesh together with greater than 86% moisture found in any individual fillet or a sample unit with pasty texture resulting from parasitic infestation affecting more than 5% of the sample unit by weight.

9 LOT ACCEPTANCE

A lot will be considered as meeting the requirements of this standard when:

(i) the total number of "defectives" as classified according to Section 8 does not exceed the acceptance number (c) of the appropriate sampling plan in the Sampling Plans for Prepackaged Foods (AQL – 6.5) – (CODEX STAN 233 – 1969);

(ii) the average net contents of all containers examined is not less than the declared weight, provided there is no unreasonable shortage in any containers;

(iii) the Food Additives, Hygiene and Handling and the Labelling requirements of Sections 4, 5 and 6 are met.

ANNEX A SENSORY AND PHYSICAL EXAMINATION

1. Complete net weight determination, according to defined procedures in Section 7.3 (de-glaze as required).

2. Examine the frozen fillets for the presence of dehydration by measuring those areas which can only be removed with a knife or other sharp instrument. Measure the total surface area of the sample unit, and calculate the percentage affected.

3. Thaw and individually examine each fillet in the sample unit for the presence of foreign matter, parasites, bone where applicable, odour, and flesh abnormality defects.

4. In cases where a final decision on odour cannot be made in the thawed uncooked sate, a small portion of the disputed material (approximately 200g) is sectioned from the sample unit and the odour and flavour confirmed without delay by using one of the cooking methods defined in Section 7. 6.

5. In cases where a final decision on gelatinous condition cannot be made in the thawed uncooked state, the disputed material is sectioned from the product and gelatinous condition confirmed by cooking as defined in Section 7. 6 or by using the procedure in Section 7. 5 to determine if greater than 86% moisture is present in any fillet. If a cooking evaluation is inconclusive, then the procedure in 7. 5 would be used to make the exact determination of moisture content.

速冻生（原条）鱿鱼

CODEX STANDARD FOR QUICK FROZEN RAW SQUID

(*CODEX STAN 191 - 1995*)

1 适用范围

本标准适用于符合下述定义的，不需深加工即可直接消费的完整或部分的速冻生鱿鱼。不适用于标明要进行深加工或其他工业目的的产品。

2 说明

2.1 产品定义

速冻鱿鱼的原料应取自以下“科”的属种：

（1）枪乌贼科（*Loliginidae*）；

（2）柔鱼科（*Ommastrepbidae*）。

2.2 加工定义

产品经过适当预处理后，应在以下规定的条件下进行冻结加工：冻结应在合适的设备中进行，并使产品迅速通过最大冰晶生成温度带。速冻加工只有在产品的中心温度达到并稳定在 -18℃（0℉）或更低的温度时才算完成。产品在运输、贮存、分销过程中应保持在深度冻结状态，以保证产品质量。

在保证质量的条件下，允许按规定要求对速冻产品再次进行速冻加工，并按照被认可的操作进行再包装。

在产品的加工和包装过程中应尽量减少脱水和氧化作用的影响。

2.3 产品介绍

2.3.1 产品介绍应符合下列要求：

（1）达到本标准的所有要求；

（2）在产品标签中对产品进行详细描述，避免引起混淆或误导消费者。

3 基本成分及质量因素

3.1 鱿鱼

速冻鱿鱼应由品质良好、可作为鲜品供人类消费的完好的鱿鱼制备。

3.2 冰衣

用于镀冰衣或制备镀冰衣所用溶液的水应达到饮用水或清洁海水的质量要求。饮用水应符合WHO最新版本的《国际饮用水质量规范》要求，清洁海水应是达到饮用水的微生物标准且不含异物的海水。

3.3 成品

产品应符合本标准要求，根据第9章进行成品批次检查时，其质量应符合第8章的规定。检查方法应符合第7章的规定。

4 食品添加剂

此类产品中不允许使用食品添加剂。

5 卫生及处理

5.1 成品中不能含有任何危害人体健康的外来杂质

5.2 应用CAC规定的抽样及检测方法进行检验时，应遵循以下规定：

（1）产品中微生物含量和由微生物产生的危害人体健康的有害物质的含量不能超过CAC标准的规定；

（2）其他危害人体健康的有害物质的含量也不能超过CAC标准的规定。

5.3 建议本标准中涉及产品预处理的条款应符合推荐性国际规范《CAC/RCP 1 - 1969，Rev. 3 - 1997 食品卫生操作总则》中的相关章节和以下相关标准：

（1）推荐性国际操作规范《CAC/RCP 16 - 1978 冻鱼》；

（2）推荐性国际操作规范《CAC/RCP 8 - 1976 速冻食品的加工和处理》；

（3）推荐性国际操作规范《CAC/RCP 37 - 1989 头足纲动物》。

6 标签

标签除应符合《CODEX STAN 1 - 1985，Rev. 3 - 1999 预包装食品标签通用标准》的要求外，还应遵循以下规定：

6.1 食品名称

6.1.1 产品名称应为“鱿鱼”，或按照出售该产品的国家的法律、习惯等实际情况使用其他名称。

6.1.2 标签上还应在与产品名称附近注明产品性状形式，这些术语应恰当充分地反应产品性状，以免引起混淆或误导消费者。

6.1.3 在产品标签上应标明“速冻”，除非在有的国家“冻”即代表“速冻”。

6.1.4 标签应注明产品必须在运输、贮藏、分销过程中保持的条件，以保证其质量。

6.1.5 如果产品用海水镀冰衣，则应予以说明。

6.2 净含量（镀冰衣产品）

对于镀冰衣的产品，其内容物净含量不包括冰衣重。

6.3 贮藏说明

标签应注明产品须在 -18℃或更低的温度下贮藏。

6.4 非零售包装的标签

上述要求应既在包装上又在附文中出现，除了食品名称、批号、制造或分装厂名、地址外，还包括贮藏条件。但批号、制造或分装厂名、地址也可用同一证明标志代替，只要证明标志能在辅助文件中说明清楚。

7 抽样、检验和分析

7.1 取样

（1）产品批次检验用样品的抽样方法应符合 FAO/WHO 的食品法典《CODEX STAN 233 - 1969 预包装食品的抽样方案》（AQL - 6.5）。样品单位是初级包装，对单冻（IQF）产品至少以 1kg 样品为单位。

（2）对需检测净重批次的抽样，抽样计划应以食品法典委员会的相关标准为依据。

7.2 感官与物理检验

产品的感官与物理指标须由经过此类检验培训的人员进行检验，并依据本标准 7.3 至 7.5 条和附录 A 以及《CAC/GL 31 - 1999 实验室中水产品感官评价指南》所述程序进行。

7.3 净重的测定

7.3.1 未镀冰衣的产品净重的测定

代表每批次的样品单位净重（不包括包装材料）应在冷冻状态下测定。

7.3.2 镀冰衣的产品净重的测定

（待述）

7.4 解冻步骤

解冻时将样品装入薄膜袋中，浸入室温（温度不高于 35℃）的水中，不时用手轻捏袋子，至袋中无硬块和冰晶时为止，但不要损坏鱿鱼的质地。

7.5 蒸煮方法

蒸煮使产品内部温度达到 65 ~ 70℃。不能过度蒸煮，蒸煮时间随产品大小和采用的温度而不同。准确的蒸煮时间和条件应依据预先实验来确定。

烘焙：用铝箔包裹产品，并将其均匀放入扁平锅或浅平锅上。

蒸：用铝箔包裹产品，并将其置于带盖容器中沸水之上的金属架上。

袋煮：将产品放入可煮薄膜袋中加以密封，浸入沸水中煮。

微波：将产品放入适于微波加热的容器中，若用塑料袋，应检查确定塑料袋不会发出任何气味。根据设备说明加热。

8 缺陷的定义

当样品呈现下列任何一项特征时，则认定其有“缺陷”。

8.1 深度脱水

样品中表面积超过10%出现的水分过度损失。明显表现为鱼体表面呈现异常的白色或黄色，覆盖了肌肉本身的颜色，并已渗透至表层以下，在不过分影响产品外观的情况下，不能轻易地用刀或其他利器刮掉。

8.2 外来杂质

样品单位中存在的任何不是来自于鱿鱼体的物质。这些物质虽不会对人体健康造成危害，但用肉眼可直接辨别，或采用某些方法（包括放大）可以确定其存在。出现外来杂质表明不符合良好操作和卫生惯例。

8.3 气味/风味

样品散发的持久、明显、令人厌恶的由腐败、酸败或饵料引起的气味或风味。

8.4 品质下降

鱿鱼肉质地全面下降，组织呈现糊状、膏状或出现鱼肉与鱼骨分离等特征而显示出腐败。

9 批次验收

当满足以下条件时，可以认为此批次产品符合最后的质量要求：

（1）根据本标准第8章中规定分类，缺陷总数不超过《CODEX STAN 233－1969 预包装食品的抽样方案》（AQL－6.5）中相应抽样方案规定的可接受数（c）；

（2）所有样品单位平均净重不少于标示量，在任何一个包装单位中没有不合理的重量短缺；

（3）符合本标准第4、5、6款中对食品添加剂、卫生和处理以及标签的要求。

附录A 感官及物理检验

1. 净含量的测定按本标准7.3条规定执行（按要求去冰衣）。
2. 通过测定只能用小刀或其他利器除去的面积，检查冻结的样品中脱水的情况。测量样品单位的总表面积，计算受影响的面积百分比。
3. 按本标准7.4条的规定解冻，并分别检查每件样品单位中每条鱿鱼的外来杂质及变色等缺陷。
4. 根据第8章标准检查每条鱿鱼，在平行于肌肉的表面切一刀，鉴定露出的表面，以检验其气味。
5. 对在解冻后未蒸煮状态下无法最终判定其气味的样品，则应从样品单位中截取一小块可疑部分，并立即使用7.5条中限定的某一种蒸煮方法，确定其气味和风味。

CODEX STANDARD FOR QUICK FROZEN RAW SQUID

(*CODEX STAN 191 - 1995*)

1 SCOPE

This standard applies to quick frozen raw squid and parts of raw squid, as defined below and offered for direct consumption without further processing. It does not apply to products indicated as intended for further processing or for other industrial purpose.

2 DESCRIPTION

2.1 Product Definition

Quick frozen squid and parts of squid are obtained from squid species of the following families:

(i) *Loliginidae*;

(ii) *Ommastrephidae.*

2.2 Process Definition

The product after any suitable preparation shall be subjected to a freezing process and shall comply with the conditions laid down hereafter. The freezing process shall be carried out in appropriate equipment in such a way that the range of temperature of maximum crystallization is passed quickly. The quick freezing process shall not be regarded as complete unless and until the product temperature has reached -18℃ or colder at the thermal centre after thermal stabilization. The product shall be kept deep frozen so as to maintain the quality during transportation, storage and distribution.

Industrial repacking of intermediate quick frozen material under controlled conditions which maintain the quality of the product, followed by the reapplication of the quick freezing process as defined above is permitted.

Quick frozen squid and parts of squid shall be processed and packaged so as to minimize dehydration and oxidation.

2.3 Presentation

Any presentation of the product shall be permitted provided that it:

(i) meets all the requirements of this standard, and

(ii) is adequately described on the label to avoid confusing or misleading the consumer.

3 ESSENTIAL COMPOSITION AND QUALITY FACTORS

3.1 Squid

Quick frozen squid shall be prepared from sound squid which are of a quality fit to be sold fresh for human consumption.

3.2 Glazing

If glazed, the water used for glazing or preparing glazing solutions shall be of potable quality or shall be clean sea-water. Potable water is fresh-water fit for human consumption. Standards of potability shall not be less than those contained in the latest edition of the WHO "International Guidelines for Drinking Water Quality". Clean sea-water is sea-water which meets the same microbiological standards as potable water and is free from objectionable substances.

3.3 Final Product

Products shall meet the reguirements of this standard when lots examined in accordance with Section 9 comply with the provisions set out in Section 8. Products shall be examined by the methods given in Section 7.

4 FOOD ADDITIVES

No food additives are permitted in these products.

5 HYGIENE AND HANDLING

5.1 The final product shall be free from any foreign material that poses a threat to human health.

5.2 When tested by appropriate methods of sampling and examination prescribed by the Codex Alimentarius Commission, the product:

(i) shall be free from microorganisms or substances originating from microorganisms in amounts which may present a hazard to health in accordance with standards established by the CAC; and

(ii) shall not contain any other substance in amounts which may present a hazard to health in accordance with standards established by the Codex Alimentarius Commission.

5.3 It is recommended that the product covered by the provisions of this standard be prepared and handled in accordance with the appropriate sections of the Recommended International Code of Practice -General Principles of Food Hygiene (CAC/RCP 1 – 1969, Rev. 3 – 1997) and the following relevant Codes:

(i) the Recommended International Code of Practice for Frozen Fish (CAC/RCP 16 – 1978);

(ii) The Recommended International Code of Practice for the Processing and Handling of Quick Frozen Foods (CAC/RCP 8 – 1976);

(iii) the Recommended International Code of Practice for Cephalopods (CAC/RCP 37 – 1989).

6 LABELLING

In addition to the provisions of the Codex General Standard for the Labelling of Prepackaged Foods (CODEX STAN 1 - 1985) the following specific provisions apply:

6.1 The Name of The Food

6.1.1 The name of the product shall be "squid", or another name according to the law, custom or practice in the country in which the product is to be distributed.

6.1.2 There shall appear on the label reference to the presentation, in close proximity to the name of the food in such additional words or phrases that will avoid misleading or confusing the consumer.

6.1.3 In addition, the labelling shall show the term "frozen", or "quick frozen" whichever is customarily used in the country in which the product is distributed, to describe a product subjected to the freezing process described in sub-section 2.2.

6.1.4 The label shall state that the product should be maintained under conditions that will maintain the quality during transportation, storage and distribution.

6.1.5 If the product has been glazed with sea-water, a statement to this effect shall be made.

6.2 Net Contents (Glazed Products)

Where the food has been glazed, the declaration of net contents of the food shall be exclusive of the glaze.

6.3 Storage Instructions

The label shall include terms to indicate that the product shall be stored at a temperature of -18℃ or colder.

6.4 Labelling of Non-Retail Containers

Information specified above shall be given either on the container or in accompanying documents, except that the name of the food, lot identification, and the name and address of the manufacturer or packer as well as storage instructions shall always appear on the container.

However, lot identification, and the name and address may be replaced by an identification mark, provided that such a mark is clearly identifiable with the accompanying documents.

7 SAMPLING, EXAMINATION AND ANALYSES

7.1 Sampling

7.1.1 Sampling of lots for examination of the product shall be in accordance with the FAO/WHO Codex Alimentarius Sampling Plans for Prepackaged Foods (AQL - 6.5) CAC/RM 42 - 1977. Sampling of lots composed of blocks shall be in accordance with the sampling plan developed for quick frozen fish blocks (reference to be provided). The sample unit is the primary container or for individually quick frozen products is at least 1 kg portion of the sample unit.

7.1.2 Sampling of lots for examination of net weight shall be carried out in accordance with an appropriate sampling plan meeting the criteria established by the CAC.

7.2 Sensory and Physical Examination

Samples taken for sensory and physical examination shall be assessed by persons trained in such examination and in accordance with procedures elaborated in Sections 7.3 through 7.5, Annex A and the *Guidelines for the Sensory Evaluation of Fish and Shellfish in Laboratories* (CAC/GL 31 - 1999).

7.3 Determination of Net Weight

7.3.1 Determination of Net Weight of Product not Covered by Glaze The net weight (exclusive of packaging material) of each sample unit representing a lot shall be determined in the frozen state

7.3.2 Determination of Net Weight of Products Covered by Glaze (to be elaborated)

7.4 Procedure for Thawing

The sample unit is thawed by enclosing it in a film-type bag and immersing in water at room temperature (not higher than 35℃). The complete thawing of the product is determined by gently squeezing the bag occasionally so as not to damage the texture of the squid until no hard core of ice crystals are left.

7.5 Cooking Methods

The following procedures are based on heating the product to an internal temperature of 65 ~ 70℃.

Cooking times vary according to the size of the product and the temperatures used. The exact times and conditions of cooking for the product should be determined by prior experimentation.

Baking Procedure: Wrap the product in aluminum foil and place it evenly on a flat cookie sheet or shallow flat pan.

Steaming Procedure: Wrap the product in aluminum foil and place it on a wire rack suspended over boiling water in a covered container.

Boil-In-Bag Procedure: Place the product into a boilable film-type pouch and seal. Immerse the pouch into boiling water and cook.

Microwave Procedure: Enclose the product in a container suitable for microwave cooking. If plastic bags are used, check to ensure that no odour is imparted from the plastic bags. Cook according to equipment instructions.

8 DEFINTION OF DEFECTIVES

The sample unit shall be considered defective when it exhibit any of the properties defined below.

8.1 Deep Dehydration

Greater than 10% of the surface area of the sample unit exhibits excessive loss of moisture clearly shown as white or yellow abnormality on the surface which masks the colour of the flesh and penetrates below the surface, and cannot be easily removed by scraping with a knife or other sharp instrument without unduly affecting the appearance of the squid.

8. 2 Foreign Matter

The presence in the sample unit of any matter which has not been derived from squid (excluding packing material), does not pose a threat to human health, and is readily recognized without magnification or is present at a level determined by any method including magnification that indicates non-compliance with good manufacturing and sanitation practices.

8. 3 Odour and Flavour

A sample unit affected by persistent and distinct objectional odours or flavours indicative of decomposition, which may be characterized also by light pinkish to red colour.

8. 4 Texture

Textural breakdown of the flesh, indicative of decomposition, characterized by muscle structure which is mushy or paste-like.

9 LOT ACCEPTANCE

A lot shall be considered as meeting the requirements of this standard when:

(i) the total number of defectives as classified according to Section 8 does not exceed the acceptance number (c) of the appropriate sampling plan in the Sampling plans for Prepackaged Foods (AQL - 6. 5) (CAC/RM 42 - 1977);

(ii) the average net weight of all sample units is not less than the declared weight, provided there is no unreasonable shortage in any container;

(iii) the Food Additives, Hygiene and Labelling requirements of Sections 4, 5. 1, 5. 2 and 6 are met.

ANNEX A SENSORY AND PHYSICAL EXAMINATION

1. Complete net weight determination, according to defined procedures in Section 7. 3 (de-glaze as required) .

2. Examine the frozen squid for the presence of deep dehydration by measuring those areas which can only be removed with a knife or other sharp instrument. Measure the total surface area of the sample unit, and determine the percentage affected using the following formula;

area affected × 100% = % affected by deep dehydration total surface area

3. Thaw and individually examine each squid in the sample unit for the presence of foreign matter and colour.

4. Examine each squid using the criteria outlined in Section 8. Flesh odours are examined by making a cut parallel to the surface of the flesh so that the exposed surface can be evaluated.

5. In cases where a final decision on odour and texture can not be made in the thawed uncooked state, a portion of the sample unit is sectioned off and the odour, flavour and texture confirmed without delay by using one of the cooking methods defined in Section 7. 5.

海淡水鱼类、甲壳类以及软体动物类制成的脆片标准

STANDARD FOR CRACKERS FROM MARINE AND FRESHWATER FISH, CRUSTACEAN AND MOLLUSCAN SHELLFISH

(*CODEX STAN 222 - 2001*)

1 范围

本标准适用于海水和淡水鱼类、甲壳类以及软体动物类等制成的脆片。不包括油炸即食的和人工调味的鱼类、甲壳类及软体动物类脆片。

2 说明

2.1 产品定义

该产品是一种传统食品，是由海水（包括红色肉和白色肉品种）和淡水鱼类、甲壳类（包括对虾类和虾类）以及软体动物类（包括鱿鱼、墨鱼、牡蛎、蛤、贻贝以及扇贝等）（见3.1）的鲜肉或冷冻碎肉，加上其他一些成分（见3.2）制成的。

2.2 加工定义

2.2.1 产品的生产需经过成分混合，成型、蒸煮 、冷却、切片及干燥等过程

2.2.2 产品需包装在适当的包装材料中，包装应防水且不透气

产品的加工和包装应使氧化作用的影响降低到最小程度。

2.3 实际操作

根据推荐性国际操作规范《CAC/RCP 9 - 1976 鲜鱼》，在收获新鲜的海水鱼、淡水鱼、甲壳类及软体类后应立即通过冷却或加冰的方式，使其温度达到0℃（32℉）或0℃以下，并尽快在此低温状态下将其保存起来。在加工前使其保持在适当的温度下，以防止损坏及细菌滋生。

3 基本成分及质量因素

3.1 原料

新鲜的海水鱼类、淡水鱼类、甲壳类及软体动物类指的是：刚捕获的、冷却或冻结的海淡水鱼类、甲壳类及软体动物类。冷冻碎肉指的是：进行过适当的加工处理的冷却或冻结的海淡水鱼类、

甲壳类及软体动物类。海淡水鱼、甲壳类及软体动物类应在外观、色泽及气味上具有新鲜的特征。

3.2 其他成分

所使用的其他成分应具有食品级的质量，并且符合所有相应法规标准的规定。

3.3 选择性添加的成分

产品可含白砂糖及其他适用的调味品。

3.4 成品

3.4.1 成品需具有统一的规格、形状、色泽、厚度及组织

3.4.2 成品需符合表1所列的要求

表1 海淡水鱼类、甲壳类及软体动物类脆片的要求

特征	等级	海水淡水鱼	甲壳类及双壳贝类
粗蛋白含量百分比 w/w（含氮量×6.25）	Ⅰ	12	8
	Ⅱ	8	5
	Ⅲ	5	2
水分含量百分比 w/w	Ⅰ		
	Ⅱ	8~14	8~14
	Ⅲ		

4 食品添加剂（表2）

表2 食品添加剂

添加剂		成品中的最高含量
螯合剂	452 多聚磷酸盐	5g/kg，以 P_2O_5 计，单用或混用
增味剂	621 谷氨酸钠（味精）	GMP

5 卫生及处理

5.1 建议本标准中涉及产品预处理的条款应符合推荐性国际规范《CAC/RCP 1－1969，Rev. 3－1997 食品卫生操作总则》中的相关章节

5.2 产品应符合所有依据《CAC/GL 21－1997 食品微生物学标准的制定和使用原则》制定的微生物标准

6 标签

标签除了应符合《CODEX STAN 1 1985，Rcv. 3 1999 预包装食品标签通用标准》的要求外，还应遵守以下规定：

6.1 食品名称

由海淡水鱼制成的产品可叫做“鱼片”，由各种甲壳类、软体动物类制成的产品可以用该品种的常用名命名，例如，“ 虾片 ”或“鱿鱼片”。

6.2 等级

需要时，按照表1的标准，在包装上标明等级。

6.3 附加要求

商标上应标明出厂日期及保质期，并注明烹饪方法。

7 抽样、检验和分析

7.1 取样

产品批次检验用样品的抽样方法应符合FAO/WHO的食品法典《CODEX STAN 233－1969 预包装食品的抽样方案》（AQL－6.5）。

7.2 粗蛋白测定

根据AOAC 920.87或960.52进行测定。

7.3 水分含量的测定

根据AOAC 950.46B进行测定（空气干燥）。

7.4 感官与物理检验

产品的感官与物理指标须由经过此类检验培训的人员进行检验，并依据附录A进行。

8 缺陷的定义

当样品单位呈现下列任何一项特征时，则认定其有“缺陷”：

8.1 外来杂质

样品单位中存在的任何不是来自于3.1、3.2、3.3所述的原料中的物质。这些物质虽不会对人体健康造成危害，但用肉眼可直接辨别，或采用某些方法（包括放大）可以确定其存在。出现外来杂质表明不符合良好操作和卫生惯例。

8.2 气味/口味

非油炸脆片、油炸脆片散发的受持久、明显、令人厌恶由腐败（腐烂等）或外来物质（如燃料、洗涤剂等）污染而引起的气味或风味。

8.3 骨刺

超过25%样品单位的脆片中出现一根以上直径大于3mm 、长度大于5mm的骨刺。

8.4 变色

超过10%的样品单位的脆片表面变黑、变白及变黄，变色表明滋生霉菌或真菌。

9 接受的条件

当满足以下条件时，可以认为此批次产品符合本标准的要求：

（1）根据本标准第8章中规定分类，缺陷总数不超过《CODEX STAN 233 - 1969 预包装食品的抽样方案》（AQL - 6.5）中相应抽样方案规定的可接受数（c）；

（2）所有样品单位平均净重不少于标示量，个体样品不得少于标示重量的95%；

（3）符合本标准第4、5、2.2、6款节中对食品添加剂、卫生和处理以及标签的要求。

附录A 感官及物理检验

用于感官及物理检验的抽样样品应不同于用于其他检测的样品。

1. 检测样品单位中的外来杂质、骨刺和变色情况。
2. 依据《CAC/ GL 31 - 1999 实验室中水产品感官评价指南》对非蒸煮产品样品的气味进行评定。
3. 依据《CAC/ GL 31 - 1999 实验室中水产品感官评价指南》对蒸煮产品样品的风味进行评定。
4. 依据样品厚度，将样品在190℃新鲜油中炸20 ~ 60 秒。

STANDARD FOR CRACKERS FROM MARINE AND FRESHWATER FISH, CRUSTACEAN AND MOLLUSCAN SHELLFISH

(*CODEX STAN 222 - 2001*)

1 SCOPE

This standard shall apply to crackers prepared from marine and freshwater fish, crustacean and molluscan shellfish. It does not include ready - to - eat fried as well as artificially flavoured fish, crustacean and molluscan shellfish crackers.

2 DESCRIPTION

2.1 PRODUCT DEFINITION

The product is a traditional food made from fresh fish or frozen minced flesh of either marine (including both the red meat and white meat species) or freshwater fish, crustacean (including prawns and shrimps) and molluscan shellfish (including squids, cuttlefish, oysters, clams, mussels and cockles) as described in section 3.1 and other ingredients as described in section 3.2.

2.2 PROCESS DEFINITION

2.2.1 The product shall be prepared by mixing all the ingredients, forming, cooking, cooling, slicing and drying.

2.2.2 The product shall be packed in a suitable packaging material which is moisture proof and gas impermeable. It shall be processed and packaged so as to minimize oxidation.

2.3 HANDLING PRACTICE

Fresh marine and freshwater fish, crustacean and molluscan shellfish shall be preserved immediately after harvesting by chilling or icing to bring its temperature down to 0℃ (32℉) as quickly as possible as specified in the Recommended International Code of Practice for Fresh Fish (CAC/RCP 9 - 1976) and kept at an adequate temperature to prevent spoilage and bacterial growth prior to processing.

3 ESSENTIAL COMPOSITION AND QUALITY FACTORS

3.1 RAW MATERIAL

Fresh marine and freshwater fish, crustacean and molluscan shellfish shall mean freshly caught, chilled or frozen marine and freshwater fish, crustacean and molluscan shellfish. Frozen minced flesh shall mean freshly caught, chilled or frozen marine and freshwater fish, crustacean and molluscan shellfish which has been appropriately processed. The marine and freshwater fish, crustacean and molluscan shellfish shall have a characteristic fresh appearance, colour and odour.

3.2 OTHER INGREDIENTS

Other ingredients shall be of food grade quality and conform to all applicable Codex Standards.

3.3 OPTIONAL INGREDIENTS

The product may contain sugar as well as suitable spices.

3.4 FINAL PRODUCT

3.4.1 The product shall display a uniform size, shape, colour, thickness and texture

3.4.2 The product shall comply with the requirements prescribed in Table 1

TABLE 1 REQUIREMENTS FOR CRACKERS FROM MARINE AND FRESHWATER FISH, CRUSTACEAN AND MOLLUSCAN SHELLFISH

Characteristics	Grade	Fish	Crustacean and Molluscan Shellfish
Crude protein (N × 6.25), percent w/w min.	Ⅰ	12	8
	Ⅱ	8	5
	Ⅲ	5	2
Moisture content, percent w/w	Ⅰ Ⅱ Ⅲ	8 to 14	8 to 14

4 FOOD ADDITIVES

Additives		Maximum Level in the Final Product
Sequestrants	452 Polyphosphates	5g/kg expressed as P_2O_5, single or in combination
Flavour enhancers	621 Monosodium glutamate	Limited by GMP

5 HYGIENE

5.1 It is recommended that the product covered by the provisions of this standard be prepared and handled in accordance with the appropriate sections of the Recommended International Code of Practice General Principles of Food Hygiene (CAC/RCP 1 – 1985, Rev 2 – 1997), and the Recommended International

Code of Practice for Fresh Fish (CAC/RCP 9 - 1976)

5.2 The products should comply with any microbiological criteria established in accordance with the Principles for the Establishment and Application of Microbiological Criteria for Foods (CAC/GL 211997)

6 LABELLING

In addition to the provisions of the Codex General Standard for the Labelling of Prepackaged Foods (CODEX STAN 1 - 1985, Rev. 1 - 1991), the following specific provisions apply:

6.1 THE NAME OF THE FOOD

The name of the product from marine and freshwater fish shall be "Fish Crackers" and those from crustacean and molluscan shellfish shall depict the common name of the species, like "Prawn Crackers" or "Squid Crackers".

6.2 GRADES

When declared by grade, the package shall declare the grade as prescribed in Table 1.

6.3 ADDITIONAL REQUIREMENTS

The package shall bear clear directions for keeping the product from the time it is purchased from the retailer to the time of its use and directions for cooking.

7 SAMPLING, EXAMINATION AND ANALYSIS

7.1 SAMPLING

Sampling of lots for examination of the products shall be in accordance with the FAO/WHO Codex Alimentarius Sampling Plans for Prepackaged Foods (1969) (AQL - 6.5) (CODEX STAN 233 - 1969).

7.2 DETERMINATION OF CRUDE PROTEIN

According to AOAC 920.87 or 960.52.

7.3 DETERMINATION OF MOISTURE

According to AOAC 950.46B (air drying).

7.4 SENSORY AND PHYSICAL EXAMINATION

Samples taken for sensory and physical examination shall be assessed by persons trained in such examination and in accordance with Annex A.

8 DEFINITION OF DEFECTIVES

The sample unit shall be considered defective when it exhibits any of the properties defined below:

8. 1 FOREIGN MATTER

The presence in the sample unit of any matter which has not been derived from materials specified in section 3. 1, 3. 2, 3. 3, does not pose a threat to human health and is readily recognized without magnification that indicates non – compliance with good manufacturing and sanitation practices.

8. 2 ODOUR AND FLAVOUR

Unfried crackers affected by persistent and distinct objectionable odours and fried crackers affected by persistent and distinct objectionable flavours indicative of decomposition (such as putrid), or contamination by foreign substances (such as fuel oil and cleaning compound).

8. 3 BONES

Crackers with more than one bone greater than 3mm in diameter and 5mm in length that affects more than 25% of the sample unit.

8. 4 DISCOLOURATION

Pronounced black, whitish or yellowish discolouration indicative of mould or fungal growth on the surface of crackers that affects more than 10% of the sample unit.

9 LOT ACCEPTANCE

A lot shall be considered as meeting the requirements of this standard when:

1. the total number of defectives as classified according to Section 8 does not exceed the acceptable number of the appropriate sampling plan in the Sampling Plans for Prepackaged Foods (1969) (AQL6. 5) (CODEX STAN 233 – 1969).

2. the average net weight of all sample units is not less than the declared weight, provided no individual container is less than 95% of the declared weight; and

3. the Food Additives, Hygiene, Packing and Labelling requirements of Section 4, 5, 2. 2 and 6 are met.

ANNEX A SENSORY AND PHYSICAL EXAMINATION

The sample used for sensory evaluation should not be same as that used for other examination.

1. Examine the sample unit for foreign matter, bones and discolouration.

2. Assess the odour in the uncooked sample in accordance with the Guidelines for the Sensory Evaluation of Fish and Shellfish In Laboratories (CAC/GL 31 – 1999).

3. Assess the flavour in cooked sample in accordance with the Guidelines for the Sensory Evaluation of Fish and Shellfish In Laboratories (CAC/GL 31 – 1999).

4. The sample shall be deep – fried in fresh cooking oil at 190℃ for 20 – 60 seconds as appropriate to the thickness of the crackers.

煮盐干鳀鱼

CODEX STANDARD FOR BOILED DRIED SALTED ANCHOVIES

(*CODEX STAN 236－2003*)

1 范围

本标准适用于经过盐腌、煮熟、干制加工的所有鳀科（Engraulidae）商品鱼种类。本产品是为食用前煮熟而制作。本标准不包括在盐水中经过酶催熟的产品。

2 说明

2.1 产品说明

本产品由新鲜鳀科鱼类制得。原料鱼的获得见3.1的说明。

2.2 加工说明

2.2.1 本产品应采用盐水或清洁海水清洗鲜鱼，经盐腌、用盐水或清洁海水煮后，再进行干燥制得。干燥加工指晒干或人工干燥。

2.2.2 本产品应该用防潮且不透气的合适的包装材料包装。加工和包装措施应使氧化程度降至最小。

2.3 预处理

捕获后未立即加工的新鲜鳀鱼应在卫生条件下处理，以保证运输期间、贮藏期及加工时的质量，建议鲜鱼应适当冷却或加冰使温度尽快降至0℃（32℉），以符合《推荐性鲜鱼操作国际规范》（CAC/RCP 9－1979）的规定，并保持在适宜的温度，防止发生加工前的变质，组胺形成、腐败和细菌生长。干燥过程应尽可能短，以防止肉毒梭菌毒素的形成。

3 主要成分和质量指标

3.1 生原料

3.1.1 鱼

本产品的原料鱼应为清洁、完整的、具有新鲜鱼特有的色泽、气味的鱼。

3.1.2 盐

盐即氯化钠，其质量要求依照《国际推荐咸鱼操作规范》（CAC/CRP 26－1979）5.4.2的规定。

3.2 成品

3.2.1 产品应符合本标准的要求，当按照第9款进行每批抽查检验时，产品应符合第8章的规定。

产品检验应按照第7章给出的方法进行。

3.2.2 本产品应遵守表1所规定的要求。

表1 盐干鳀鱼的要求

特 征	要 求
氯化钠，重量百分比，最大值（d.b）	15
水分活度（aw）最大值	0.75
酸不溶性灰分，重量百分比，最大值（d.b）	1.5

3.3 破损

3.3.1 破损指鱼体（包括鳍和鳞）受损。破损率指试样中的破损鱼数量占试样总数的百分比。

3.3.2 按3.3.1定义的破损率不应超过3.5规定的限量。

3.4 腐败

根据所测检样的平均值，本产品的组胺含量不应超过10mg/100g。

3.5 规格

参见附录A。

4 食品添加剂

本标准所涉及的产品不允许使用食品添加剂。

5 卫生和处理

5.1 建议本标准条款涉及的产品的加工和处理应按照《推荐性国际操作规范－食品卫生总则》（CAC/RCP 1－1969，Rev 2－1997）、《推荐性鲜鱼操作国际规范》（CAC/RCP 9－1976）和《推荐性咸鱼操作国际规范》（CAC/RCP 26－1979）中的相应章节规定

5.2 本产品应符合根据《食品微生物标准的建立和应用原则》（CAC/GL 21－1997）所确定的微生物指标

5.3 样品中的组胺含量不得超过20mg/100g

5.4 本产品中其他物质的含量应符合食品法典委员会的标准要求，不应对人体健康构成危害

6 标签

标签除应遵守《预包装食品标签通用标准》（CODEX STAN 1－1985，Rev 1－1991）的条款外，还应符合下列特定条款：

6.1 食品名称

产品名称应为“煮盐干鳀鱼”，此外，应按照销售国的法律和习俗标注该产品的商品名，产品命名方式不应使消费者产生误解。

6.2 产品等级和规格

标注产品等级和规格时，参照附件A中的表的规定。

6.3 学名

在贸易文件中应注明鱼的学名。

6.4 附加要求

包装上应明确说明产品的保藏、从零售商购买的时间至其使用的时间以及蒸煮说明。

7 抽样，检验与分析

7.1 抽样

各批次产品检验抽样应按照FAO/WHO食品法典《预包装食品抽样方案》（1969）（AQL－6.5）（CODEX STAN 233－1969）的要求进行。

7.2 氯化钠的测定

参照AOAC 937.09（以氯表示氯化钠）。

7.3 组胺的测定

参照AOAC 977.13。

7.4 感官和物理检验

样品的感官和物理检验应由经过这方面检验的培训的人员执行，并按照附件B的规定评定。

8 缺陷界定

当样品表现出以下任何特征时，即认为是“缺陷”：

8.1 异物

样品单位中存在的非鳀鱼类自身产生的任何物质，对健康不会构成威胁，肉眼容易辨认，或者其含量可通过包括放大在内的任何方法测定，不符合良好生产和卫生操作的规定。

8.2 破损

大面积的鱼类组织破损，其特征是在样品单位中鱼体部分裂开、或破裂、或撕成两段、或两段以上者占25%以上。

8.3 气味

样品散发出持久而明显的腐败味（如腐烂）或油脂酸败味。

8.4 红变

在样品单位中，鱼体表面出现明显的红色嗜盐菌占鱼体表面25%以上。

8.5 霉变

鱼体具明显霉菌生长的总面积占样品单位的25%以上。

9 批次验收

当符合下列规定时，可认为此批次产品符合本标准要求：

（1）按照第 8 章所列的缺损总数不超过《预包装食品抽样方案》（1969）（AQL－6.5）（CODEX STAN 233－1969）中相应抽样方案的认可数（c）；

（2）单个包装不少于标示重量的 95%，且所有样品单位的平均净重应不少于标示重量；

（3）卫生、包装和标签符合第 4、5 和 6 款的要求。

附件 A

1 规格

根据产品（全鱼）的体长划分规格

规格	体长
小	小于 3.5cm
中	3.5～6.5cm
大	大于 6.5cm

2 分级

各种规格的盐干鳀鱼均可按以下规定分为两级：

特　　征	级　　别	
	A	B
破损率	小于 5%	小于 15%
颜色(颜色的比较必须在同一品种之间进行)	带白色、蓝色或黄色(鱼的品种特征)	变色
气味	无恶臭或酸腐味	无恶臭或酸腐味

附件 B　感官和物理检验

用于感官检验的样品与用于其他检验的样品不同。

1. 对样品单位中的每尾鱼进行异物，破损，变红状况和发霉等项目的检验。

2. 按照《鱼类和贝壳类实验室感官评定指南》（CAC/GL 31－1999）来评定未煮样品的气味。

3. 按照《鱼类和贝壳类实验室感官评定指南》（CAC/GL 31－1999）来评定煮熟样品的风味。

样品在评定之前应按照包装上的说明进行烹煮。如果包装中没有给出这种说明，样品应在 190℃ 的新鲜烹调油中炸 1～2min，以达到适当的规格。

CODEX STANDARD FOR BOILED DRIED SALTED ANCHOVIES

(*CODEX STAN 236 – 2003*)

1 SCOPE

This standard shall apply to all commercial species of fish belonging to the family Engraulidae that have been salted, boiled and dried. This product is intended for cooking before consumption. This Standard does not cover products which have undergone an enzymatic maturation in brine.

2 DESCRIPTION

2.1 PRODUCT DEFINITION

The product shall be prepared from fresh fish of the family Engraulidae obtained from the raw material described in Section 3.1.

2.2 PROCESS DEFINITION

2.2.1 The product shall be prepared by washing fresh fish in brine or clean sea water and salting by boiling in brine or clean sea water and drying. The drying process shall mean sundrying or artificial drying.
2.2.2 The product shall be packed in a suitable packaging material which is moisture proof and gas impermeable. It shall be processed and packaged so as to minimize oxidation.

2.3 HANDLING PRACTICE

Fresh anchovies that are not processed immediately after harvesting shall be handled under such hygienic conditions as will maintain the quality during transportation and storage up to and including the time of processing. It is recommended that the fish shall be properly chilled or iced to bring its temperature down to 0℃ (32℉) as quickly as possible as specified in the "Recommended International Code of Practice for Fresh Fish" (CAC/RCP 9 – 1976) and kept at an adequate temperature to prevent deterioration, histamine formation, spoilage and bacterial growth prior to processing. The drying process must be sufficiently short to preclude the formation of Clostridium botulinum toxin.

3 ESSENTIAL COMPOSITION AND QUALITY FACTORS

3.1 RAW MATERIAL

3.1.1 Fish

The product shall be prepared from clean, sound fish which have characteristic fresh appearance, colour and odour.

3.1.2 Salt

Salt shall mean sodium chloride of suitable quality as specified in sub-section 5.4.2 of the "Recommended International Code of Practice for Salted Fish" (CAC/CRP 26 – 1979).

3.2 FINAL PRODUCT

3.2.1 Products shall meet the requirements of this standard when lots examined in accordance with Section 9 comply with the provisions set out in Section 8. Products shall be examined by the methods given in Section 7

3.2.2 The product shall comply with the requirements prescribed in Table 1

Table 1 Requirements for Dried Salted Anchovies

Characteristics	Requirement
Sodium chloride, percent by weight, max (d. b.)	15
Water activity (aw), max	0.75
Acid insoluble ash, percent by weight, max. (d. b.)	1.5

3.3 BREAKAGE

3.3.1 Breakage shall mean fish (excluding fins and scales) which is not intact. The percentage of breakage is determined by the number of broken fish over the total number of fish in the test sample

3.3.2 The percent breakage defined in section 3.3.1 shall not exceed the limits specified in section 3.5

3.4 DECOMPOSITION

The products shall not contain more than 10 mg/100g of histamine based on the average of the sample unit tested.

3.5 SIZE CLASSIFICATION

According to Annex A.

4 FOOD ADDITIVES

No food additives are permitted in these products.

5 HYGIENE AND HANDLING

5.1 It is recommended that the product covered by the provisions of this standard be prepared and handled in accordance with the appropriate sections of the Recommended International Code of Practice-General Principles of Food Hygiene (CAC/RCP 1 - 1969, Rev 2 - 1997), and the Recommended International Code of Practice for Fresh Fish (CAC/RCP 9 - 1976) and Salted Fish Code (CAC/RCP 26 - 1979).

5.2 The products should comply with any microbiological criteria established in accordance with the Principles for the Establishment and Application of Microbiological Criteria for Foods (CAC/GL 21 - 1997).

5.3 No sample unit shall contain histamine that exceeds 20 mg/100g.

5.4 The product shall not contain any other substance in amounts which may present a hazard to health of in accordance with standards established by the Codex Alimentarius Commission.

6 LABELLING

In addition to the provisions of the Codex General Standard for the Labelling of Prepackaged Foods (CODEX STAN 1 - 1985, Rev. 1 - 1991), the following specific provisions apply:

6.1 THE NAME OF THE FOOD

The name of the product shall be "Boiled Dried Salted Anchovies" in addition the common name of the fish shall be declared in accordance with the law and custom of the country in which the product is sold, in a manner not mislead the consumer.

6.2 GRADE AND SIZE OF PRODUCT

If the grade and size of fish is declared the table of Annex A should be applied.

6.3 SCIENTIFIC NAMES

The scientific names of the fish shall be declared on trade documents.

6.4 ADDITIONAL REQUIREMENTS

The package shall bear clear directions for keeping the product from the time they are purchased from the retailer to the time of their use and directions for cooking.

7 SAMPLING, EXAMINATION AND ANALYSIS

7.1 SAMPLING

Sampling of lots for examination of the products shall be in accordance with the FAO/WHO Codex Alimentarius Sampling Plans for Prepackaged Foods (1969) (AQL - 6.5) (CODEX STAN 233 - 1969).

7.2 DETERMINATION OF SODIUM CHLORIDE

According to AOAC 937.09 (chloride expressed as sodium chloride).

7.3 DETERMINATION OF HISTAMINE

According to AOAC 977.13.

7.4 SENSORY AND PHYSICAL EXAMINATION

Samples taken for sensory and physical examination shall be assessed by persons trained in such examination and in accordance with Annex B.

8 DEFINITION OF DEFECTIVES

The sample unit shall be considered defective when it exhibits any of the properties defined below:

8.1 FOREIGN MATTER

The presence in the sample unit of any matter, which has not been derived from the Engraulidae family, and does not pose a threat to human health, and is readily recognized without magnification or is present at a level determined by any method including magnification that indicates non-compliance with good manufacturing and sanitation practices.

8.2 BREAKAGE

Extensive textural breakdown of the fish which is characterized by the body part being split or broken or torn into two or more pieces in more than 25% of the fish in the sample unit.

8.3 ODOUR AND FLAVOUR

A sample unit affected by persistent and distinct objectionable odours and flavours indicative of decomposition (such as putrid) or rancidity.

8.4 PINK

Any visible evidence of red halophilic bacteria on the surface of the fish in more than 25% of the fish in the sample unit.

8.5 MOULD GROWTH

Fish with an aggregate area of pronounced mould growth in more than 25% the sample unit.

9 LOT ACCEPTANCE

A lot shall be considered as meeting the requirements of this standard when:

1. the total number of defectives as classified according to Section 8 does not exceed the acceptable number of the appropriate sampling plan in the Sampling Plans for Prepackaged Foods (1969) (AQL – 6.5) (CODEX STAN 233 – 1969).
2. the average net weight of all sample units is not less than the declared weight, provided no individual container is less than 95% of the declared weight; and
3. the Hygiene, Packing and Labelling requirements of Section 4, 5 and 6 are met.

ANNEX A

1 SIZING

Size shall be determined by the length of the product (whole fish) .

Size Designation	Length
Small less than	3. 5 cm
Medium	3. 5 ~ 6. 5 cm
Big greater than	6. 5 cm

2 GRADING

Each size of dried salted anchovies shall be classified into two grades as defined below:

Characteristics	Grade	
	A	B
Breakage	Less than 5%	Less than 15%
Colour (comparison of colour must be among the same species of fish)	Whitish or bluish or yellowish (characteristic of species)	Off colour
Odour	No foul or rancid smell	No foul or rancid smell

ANNEX B SENSORY AND PHYSICAL EXAMINATION

The sample used for sensory evaluation should not be same as that used for other examination.

1. Examine every fish in the sample unit for foreign matter, breakage, pink condition and mould growth.

2. Assess the odour in uncooked sample in accordance with the Guidelines for the Sensory Evaluation of Fish and Shellfish In Laboratories (CAC/GL 31 - 1999) .

3. Assess the flavour in cooked sample in accordance with the Guidelines for the Sensory Evaluation of Fish and Shellfish In Laboratories (CAC/GL 31 - 1999) .

The sample shall be cooked prior to assessment according to the cooking instructions on the package. When such instructions are not given, the sample shall be deep fried in fresh cooking oil at 190℃ for 1 ~ 2minutes as appropriate to the size.

盐渍大西洋鲱和盐渍黍鲱鱼

STANDARD FOR SALTED ATLANTIC HERRING AND SALTED SPRAT

(*CODEX STAN 244 - 2004*)

1　范围

本标准适用于盐渍大西洋鲱（*Clupea harengus*）和黍鲱鱼（*Sprattus sprattus*）[1]。通过添加酶制剂、酸或人造酶生产的鱼制品不适用于本标准。

2　说明

2.1　产品定义

产品由新鲜或冷冻的鱼制成，盐渍的鱼包括整鱼、去头鱼、去头和内脏、去鳃和内脏或鱼片（带皮或去皮）。可以添加香料、糖和其他可选配料。产品的消费国可以允许产品不去内脏，也可以要求加工之前或之后去除内脏，因为即使遵守了良好操作，在对梭状肉毒杆菌（*clostridium botulinum*）的控制中，出错的余地仍很小，并且后果是严重的。产品可用于人们的直接消费或进一步加工。

2.2　加工定义

原料鱼经过适当的预处理后进入盐渍过程，并应符合下述条件：必须对盐渍过程，包括温度和时间，进行充分控制，以防梭状肉毒杆菌的繁殖，或者在盐渍之前先去除内脏。

2.2.1　腌制

腌制就是将原料鱼与适当比例的食盐、糖类及其他可选配料混合，或添加适当浓度的盐水溶液。腌制应在不漏水的容器（桶等）中进行。

2.2.2　盐渍鱼类型

2.2.2.1　极轻腌鱼

在水箱中，鱼肉里的盐含量高于1g/100g，且低于或等于4g/100g。

2.2.2.2　轻腌鱼

在水箱中，鱼肉里的盐含量高于4g/100g，且低于或等于10g/100g。

2.2.2.3　中腌鱼

在水箱中，鱼肉里的盐含量高于10g/100g，且低于或等于20g/100g。

2.2.2.4　重腌鱼

在水箱中，鱼肉里的盐含量高于20g/100g。

---

[1] 针对该标准，鱼包括了大西洋鲱和黍鲱。

2.2.3 贮存温度

产品应当冷冻或冷藏，贮存时间和温度应确保产品的安全性和质量符合第3章和第5章的要求。极轻腌鱼必须在加工后进行冷冻。

2.3 表现形式

在符合下述两个条件的情况下，允许产品呈任何表现形式：

2.3.1 符合本标准的所有要求

2.3.2 在标签上进行恰当的描述，以防消费者误解或误导消费者

3 主要成分和质量因素

3.1 鱼

盐渍大西洋鲱和盐渍黍鲱鱼需要由结实和完整的鱼制成，其质量应适于新鲜销售，供人类适当处理后消费。鱼肉应没有明显感染寄生虫。

3.2 盐和其他成分

所用的盐和其他配料应当具有食品级质量，并符合所有适用的法典标准。

3.3 终产品

当根据第9章进行的批检验符合第8章所列规定时，产品应符合本标准的要求。产品参照第7章所列方法进行检验。

3.4 腐败

产品中的组胺含量，不应大于10mg组胺/100g鱼肉（取所检测的样品单元的平均值）。

4 食品添加剂

仅允许使用以下添加剂。

酸度调节剂	产品中的最大含量
300 抗坏血酸	符合GMP要求
330 柠檬酸抗氧化剂	符合GMP要求
200~203 山梨酸盐防腐剂	200mg/kg（以山梨酸表示）
210~213 苯甲酸盐	200mg/kg（以苯甲酸表示）

5 卫生和处理

5.1 建议本标准条款涉及的产品的预处理和加工应该依据《推荐性国际操作规范－食品卫生总则》（CAC/RCP 1－1985，Rev.3，1997）的适用条款，和以下其他相关法典文件如操作规范和卫生操作规范：

（1）《推荐性咸鱼操作国际规范》（CAC/RCP 26－1979）；

（2）《推荐性鲜鱼操作国际规范》（CAC/RCP 9－1976）；

（3）《推荐性冻鱼操作国际规范》（CAC/RCP 16－1978）。

5.2 产品应该遵照任何根据《食品微生物标准的建立和应用原则》（CAC/GL 21－1997）建立的微生物标准

5.3 产品中任何其他物质的含量应符合食品法典委员会建立的标准，不应对人体健康造成危害

5.4 寄生虫：

鱼肉不应含有活的线虫幼体。线虫的生存应当参照附件Ⅰ进行检测。假如确认有活的线虫，在产品根据附件Ⅱ进行处理之前，不应上市供人类消费。

5.5 组胺：

检样鱼肉中的组胺含量不应超过20mg/100g。

5.6 外来物质：

最终产品中不应含有对人类健康造成危胁的外来物质。

6 标签

检签除了《预包装食品标签通用标准》（CODEX STAN 1－1985，Rev. 1－1991）的规定外，还应符合以下具体条款：

6.1 食品名称

6.1.1 产品名称应按销售国家的法律和习惯，定为“……盐渍鲱鱼”或“……盐渍黍鲱鱼”，其命名方式不应误导消费者。

6.1.2 另外，标签应含有其他描述性术语，以避免误导和迷惑消费者。

6.2 非零售包装的标签

除了食品名称、批号、生产者、包装者或出口商的名称和地址以及贮存说明等应在包装上标识外，上述信息应在包装上或在附加文件中给出。

批号、名称和地址可用识别码代替，只要该识别码在附带文件中也同样清楚地标明。

7 抽样、检测和分析

7.1 容器抽样规则

（1）产品质量检验的抽样批次应根据FAO/WHO食品法典《预包装食品抽样规则》（AQL－6.5）（CODEX STAN 233－1969）。一条鱼或基本容器为一个抽样单位；

（2）用于样品净重检测的抽样应根据食品法典委员会标准所规定的相关的抽样规则进行；

（3）用于样品病原微生物和寄生虫检验的抽样应符合《食品微生物标准建立和应用原则》（CAC/GL 21－1997）；

（4）用于样品组胺检验的抽样应符合《抽样通用指南草案》（分析和取样方法委员会正在制定中）。

7.2 感官和鱼体检测

用于感官和物理检验的样品应由受过该项专门培训的人员进行评定。其检测需遵照7.3至7.8和附件中详述的程序以及《鱼类和贝类实验室感官评定指南》（CAC/GL 31－1999）来进行。

7.3 盐分含量的检测

盐分含量的检测需按照法典标准《盐渍和盐干鳕鱼》（CODEX STAN 167 - 1989，Rev. 1 - 1995）中的方法进行。

7.4 水分含量的检测

水分含量的检测需按照 AOAC 950. 46B（空气干燥法（air drying））进行。

7.5 线虫存活力的检测：见附件 I

7.6 组胺检测

参照 AOAC 977. 13。

7.7 净重测定

应测定每批样品中每件样品的净重。

从容器（桶）中取出鲱鱼并放入合适的滤网之中，晾干 5min，去掉附着的盐晶体。称重并计算净重。

8 缺陷定义

8.1 当样品表现出以下任何一项特征时，则认定其存在“缺陷”

8.1.1 外来杂质

在样品中出现了非鱼类自身产生的任何物质，对人类健康不构成威胁，而且肉眼容易辨认，或其含量可通过包括放大在内的任何方法测定，表明不符合良好生产和卫生操作。

8.1.2 寄生虫

通过常规的目视检测鱼肉，在样品的可食用部分中，轻易可见寄生虫（见附件Ⅲ）。

8.1.3 异味和气味/口味

鱼体有持久的、明显的、令人反感的表明腐败的异味或气味（如酸、腐烂、腥臭、腐臭、刺激性气味等）或受到外来物质（如燃油、清洁剂等）的污染。

9 批次验收

当符合下列规定时，可认为此批次产品符合本标准要求：

（1）按照第 8 章所列的缺损总数不超过第 7 章中相应抽样方案的认可数（c）；

（2）单个包装不少于标示重量的 95%，且所有样品单位的平均净重不少于标示重量；

（3）食品添加剂、卫生、处理和标签符合第 4、5 和 6 款的要求。

附录 I 线虫存活力的检验（根据参考文献 1 修改过的方法）

原理：

通过消化作用，将线虫从鱼片中分离，移至 0. 5% 胃蛋白酶消化溶液中，观测其存活力。消化条件相当于哺乳动物消化道中的条件，并保证线虫的存活。

仪器：

—分样筛（直径：14cm 或更大，网眼尺寸：0.5mm）；

—带恒温加热板的磁力搅拌器；

—常规实验室仪器。

化学试剂：

—胃蛋白酶 2000FIP - U/g；

—盐酸。

溶液：

A：溶于 0.063 M 盐酸的 0.5% （W/V） 胃蛋白酶

步骤：

将约 200g 的鱼片撕碎，置入装有 1L 胃蛋白酶溶液 A 的 2L 烧杯中。将混合液置于 37℃磁力搅拌器上，连续低速搅拌 1 ~ 2h。如果鱼肉没有溶解，将溶液倒入筛内，用水清洗。剩余的鱼肉定量放入烧杯中，添加 700ml 消化液 A，同时，在温和加热下（最大 37℃），再次搅拌混合液，直到没有大片鱼肉留下。

消化液倒入筛中，用水清洗筛内物质。

用小镊子小心地将线虫移入盛有新鲜胃蛋白酶溶液 A 的培养皿内，将培养皿置于对光底盘上，注意其温度不能超过 37℃。

当用解剖针轻触活线虫时，便能看到它的活动和自然反应。有时出现的卷曲线虫的一次松弛不作为存活力的显著征兆。线虫必须表现出自发活动。

注意：

当检查盐渍或加糖盐渍产品中的活线虫时，线虫的复苏时间可持续 2 小时以上。

备注：

还有其他一些检测线虫存活力的方法（如：参考文献 2、3）。

之所以选择上述方法，是因为该方法易于操作，并且在同一步骤中结合了线虫的分离和存活力检测。

参考文献

1. Anon. : Vorläufiger Probenahmeplan, Untersuchungsgang und Beurterilungsvorschlag für die amtliche überprüfung der Erfüllung der Vorschriften des § 2 Abs. 5 der Fisch-VO. Bundesgesundheitsblatt 12, 486 ~ 487 (1998) .

2. Leinemann, M. and Karl, H. : Untersuchungen zur Differenzierung lebender und toter Nematodenlarven (Anisakis sp) in Heringen und Heringserzeugnissen. Archiv Lebensmittelhygiene 39, 147 ~ 150 (1998) .

3. Priebe, K. , Jendrusch, H. and Haustedt, U. : Problematik unk Experimentaluntersuchungen zum Erlöschen der Einbohrpotenz von Anisakis Larven des Herings bei der Herstellung von Kaltmarinaden. Archiv Lebensmittelhygiene 24, 217 ~ 222 (1973) .

附录Ⅱ 充分杀死活线虫的处理步骤

例如，将产品全部冷冻至 - 20℃，不少于 24 小时；

盐分含量和贮存时间和恰当组合（待细化）；

或其他具有相同效用的方法（待细化）。

附录Ⅲ 可见寄生虫的检测

1. 在被分成 20 ~ 30mm（一般的咬痕尺寸）大的鱼片中出现易见的寄生虫时，即使鱼片上还有其他物质，也仅仅考虑通常的可食用部分。寄生虫的检测应在光照充足（可轻松阅读报纸）的房间内进行，无须放大。

2. 尽管已有第一段内容，但是在产品进行进一步加工时，对所有中间产品中寄生虫的检测可在较晚的阶段进行。

STANDARD FOR SALTED ATLANTIC HERRING AND SALTED SPRAT

(*CODEX STAN 244 - 2004*)

1 SCOPE

The standard applies to salted Atlantic herring (*Clupea harengus*) and sprat (*Sprattus sprattus*)[1]. Fish products produced by use of added natural or artificial enzymatic preparations, acids and/or artificial enzymes are not covered by this standard.

2 DESCRIPTION

2.1 PRODUCT DEFINITION

The product is prepared from fresh or frozen fish. The fish is salted as whole fish or as headed or nobbed or headed and gutted or gibbed or filleted (skin-on or skin-off) fish. Spices, sugar and other optional ingredients may be added. Countries where the product are to be consumed may allow this product in an uneviscerated state or may require evisceration , either before or after processing, since the margin of error in the control of Clostridium botulinum is small even when good practices are followed and the consequences are severe. The product is either intended for direct human consumption or for further processing.

2.2 PROCESS DEFINITION

The fish after any suitable preparation shall be subjected to a salting process and shall comply with the conditions laid down hereafter. The salting process including the temperature and time should be sufficiently controlled to prevent the development of *Clostridium botulinum* or fish should be eviscerated prior to brining.

2.2.1 Salting

Salting is the process of mixing fish with the appropriate amount of food grade salt, sugar spices and all optional ingredients and/or of adding the appropriate amount of salt-solution of the appropriate concentration. Salting is performed in watertight containers (barrels etc.).

2.2.2 Types of salted fish

2.2.2.1 Very lightly salted fish

The salt content in the fish muscle is above 1 g/100 g in water phase and below or equal to 4 g/100 g or less in water phase.

---

[1] For the purpose of the standard, fish includes herring and sprats.

2.2.2.2 Lightly salted fish

The salt content in the fish muscle is above 4 g/100 g in water phase and below or equal to 10 g salt/100 g in water phase.

2.2.2.3 Medium salted fish

The salt content in the fish muscle is above 10 g salt/100 g water phase and below or equal to 20 g salt/100 g in water phase.

2.2.2.4 Heavily salted fish

The salt content of the fish muscle is above 20 g salt /100 g in water phase.

2.2.3 Storage temperatures

The products shall be kept frozen or refrigerated at a time/temperature combination which ensures their safety and quality in conformity with Sections 3 and 5. Very lightly salted fish must be kept frozen after processing.

2.3 PRESENTATION

Any presentation of the product shall be permitted provided that it:

2.3.1 meets all requirements of this standard, and

2.3.2 is adequately described on the label to avoid confusing or misleading the consumer.

3 ESSENTIAL COMPOSITION AND QUALITY FACTORS

3.1 FISH

Salted Atlantic herring and salted sprats shall be prepared from sound and wholesome fish which are of a quality fit to be sold fresh for human consumption after appropriate preparation. Fish flesh shall not be obviously infested by parasites.

3.2 SALT AND OTHER INGREDIENTS

Salt and all other ingredients used shall be of food grade quality and conform to all applicable Codex standards.

3.3 FINAL PRODUCT

Products shall meet the requirements of this standard when lots examined in accordance with Section 9 comply with the provisions set out in Section 8. Products shall be examined by the methods given in Section 7.

3.4 DECOMPOSITION

The products shall not contain more than 10 mg of histamine per 100 g fish flesh based on the average of the sample unit tested.

4 FOOD ADDITIVES

Only the use of the following additives is permitted.

Acidity regulators	Maximum level in the final product
300 Ascorbic acid	GMP
330 Citric acid Antioxidants	GMP
200 ~ 203 Sorbates Preservatives	200mg/kg (expressed as sorbic acid)
210 ~ 213 Benzoates	200mg/kg (expressed as benzoic acid)

5 HYGIENE AND HANDLING

5.1 It is recommended that the products covered by the provisions of this standard be prepared and handled in accordance with the appropriate sections of the Recommended International Code of Practice-General Principles of Food Hygiene (CAC/RCP 1 - 1985, Rev. 3, 1997) and other relevant Codex texts such as codes of practice and codes of hygienic practice, as follows;

(i) the Recommended International Code of Practice for Salted Fish (CAC/RCP 26 - 1979);

(ii) the Recommended International Code of Practice for Fresh Fish (CAC/RCP 9 - 1976);

(iii) the Recommended International Code of Practice for Frozen Fish (CAC/RCP 16 - 1978).

5.2 The products should comply with any microbiological criteria established in accordance with the Principles for the Establishment and Application of Microbiological Criteria to Foods (CAC/GL 211997)

5.3 The product shall not contain any other substance in amounts which may present a hazard to health in accordance with standards established by the Codex Alimentarius Commission

5.4 PARASITES

Fish flesh shall not contain living larvae of nematodes. Viability of nematodes shall be examined according to Annex I. If living nematodes are confirmed, products must not be placed on the market for human consumption before they are treated in conformity with the methods laid down in Annex Ⅱ.

5.5 HISTAMINE

NO SAMPLE UNIT SHALL CONTAIN HISTAMINE THAT EXCEEDS 20MG PER 100G FISH MUSCLE.

5.6 FOREIGN MATERIAL

The final product shall be free from any foreign material that poses a threat to human health.

6 LABELLING

In addition to the provisions of the Codex General Standard for the Labelling of Prepackaged Foods (CODEX STAN 1 - 1985, Rev. 1 - 1991) the following specific provisions apply:

6.1 NAME OF THE FOOD

6.1.1 The name of the product shall be ... - salted herring or ... - salted sprat in accordance with the law and custom of the country in which the product is sold, in a manner not to mislead the consumer

6.1.2 In addition the label shall include other descriptive terms that will avoid misleading or confusing the consumer

6.2 LABELLING OF NON-RETAIL CONTAINERS

Information specified above should be given either on the container or in accompanying documents, except that the name of the food, lot identification, and the name of and address of the manufacturer or packer or importer as well as storage instructions shall always appear on the container.

However lot identification, and the name and address may be replaced by an identification mark, provided that such a mark is clearly identifiable with accompanying documents.

7 SAMPLING, EXAMINATION AND ANALYSIS

7.1 SAMPLING PLAN FOR CONTAINERS (BARRELS)

(i) Sampling of lots for examination of the product for quality shall be in accordance with the FAO/WHO Codex Alimentarius Sampling Plan for Prepackaged Foods (AQL – 6.5) (CODEX STAN 233 – 1969). A sample unit is the individual fish or the primary container;

(ii) Sampling of lots for examination of net weight shall be carried out in accordance with an appropriate sampling plan meeting the criteria established by the Codex Alimentarius Commission;

(iii) Sampling of lots for pathogenic microorganisms and parasites will be in accordance with the Principles for the Establishment and Application of Microbiological Criteria to Foods (CAC/GL 21 – 1997);

(iv) Sampling of lots for histamine will be in accordance with the Draft General Guidelines on Sampling (under development by the Committee on Methods of Analysis and Sampling)

7.2 SENSORY AND PHYSICAL EXAMINATION

Samples taken for sensory and physical examination shall be assessed by persons trained in such examination and in accordance with procedures elaborated in Section 7.3 through 7.8 and Annexes and in accordance with the Guidelines for the Sensory Evaluation of Fish and Shellfish in Laboratories (CAC/GL 31 – 1999).

7.3 DETERMINATION OF SALT CONTENT

Determination of salt content is performed according to the method in the Codex Standard for Salted Fish and Dried Salted Fish of *Gadidae* Family of Fishes-CODEX STAN 167 – 1989, Rev. 1 – 1995.

7.4 DETERMINATION OF WATER CONTENT

Determination of water content is performed according to AOAC 950.46B (air drying).

7.5 DETERMINATION OF THE VIABILITY OF NEMATODES: SEE ANNEX I

7.6 DETERMINATION OF HISTAMINE

AOAC 977.13.

7.7 DETERMINATION OF NET WEIGHT

The net weight (excluding packaging material) of each sample unit in the sample lot shall be determined.

Remove the herring from the container (barrel) and put it on an appropriate sieve. Allow to drain for 5 min and remove adhering salt crystals. Weigh the herring and calculate net weigh.

8 DEFINITION OF DEFECTIVES

8.1 The sample unit shall be considered as defective when it exhibits any of the properties defined below.

8.1.1 Foreign matter

The presence in the sample unit of any matter which has not been derived from fish, does not pose a threat to human health, and is readily recognized without magnification or is present at a level determined by any method including magnification that indicates non-compliance with good manufacturing and sanitation practices.

8.1.2 Parasites

The presence of readily visible parasites in a sample of the edible portion of the sample unit detected by normal visual inspection of the fish flesh (see Annex Ⅲ).

8.1.3 Odour and flavour/taste

Fish affected by persistent and distinct objectionable odours or flavours indicative of decomposition (such as sour, putrid, fishy, rancid, burning sensation, etc.) or contamination by foreign substances (such as fuel oil, cleaning compounds, etc.).

9 LOT ACCEPTANCE

A lot shall be considered as meeting the requirements of this standard when:

(i) the total number of defectives as classified according to Section 8 does not exceed the acceptance number (c) of the appropriate sampling plan in Section 7; and

(ii) the average net weight of all sample units is not less than the declared weight, provided no individual container is less than 95% of the declared weight; and

(iii) the Food Additives, Hygiene and Handling and Labelling requirements of Sections 4, 5 and 6 are met.

ANNEX Ⅰ

VIABILITY TEST FOR NEMATODES (modified method according to Reference 1)

Principle:

Nematodes are isolated from fish fillets by digestion, transferred into 0.5% Pepsin digestion solution and inspected visually for viability. Digestion conditions correspond to conditions found in the digestive tracts of mammals and guarantee the survival of nematodes.

Equipment:

—Stacked sieves (diameter: 14 cm or larger, mesh size: 0.5mm);

—Magnetic stirrer with thermostated heating plate;

—normal laboratory equipment.

Chemicals:

—Pepsin 2000 FIP-U / g;

—Hydrochloric acid.

Solution:

A: 0.5 % (w/v) Pepsin in 0.063 M HCl

Procedure:

Fillets of approximately 200 g are manually shredded and placed in a 2l beaker containing 1l Pepsin solution A. The mixture is heated on a magnet stirrer to 37℃ for 1 ~ 2 h under continuous slow stirring. If the flesh is not dissolved, the solution is poured through a sieve, washed with water and the remaining flesh is quantitatively replaced in the beaker. 700 ml digestion solution A is added and the mixture stirred again under gentle heating (max. 37℃) until there are no large pieces of flesh left.

The digestion solution is decanted through a sieve and the content of the sieve rinsed with water.

Nematodes are carefully transferred by means of small forceps into Petri dishes containing fresh Pepsin solution A. The dishes are placed on a candling dish, and care has to be taken not to exceed 37℃.

Viable nematodes show visible movements or spontaneous reactions when gently probed with dissecting needles. A single relaxation of coiled nematodes, which sometimes occurs, is not a clear sign of viability.

Nematodes must show spontaneous movement.

Attention:

When checking for viable nematodes in salted or sugar salted products, reanimation time of nematodes can last up to two hours and more.

Remarks:

Several other methods exist for the determination of viability of nematodes (e. g. ref. 2, 3).

The described method has been chosen because it is easy to perform and combines isolation of nematodes and viability test within one step.

References

1. Anon.: Vorläufiger Probenahmeplan, Untersuchungsgang und Beurteilungsvorschlag für die amtliche überprüfung der Erfüllung der Vorschriften des §2 Abs. 5 der Fisch-VO. Bundesgesundheitsblatt 12, 486 ~ 487 (1988).

2. Leinemann, M. and Karl, H.: Untersuchungen zur Differenzierung lebender und toter Nematodenlarven (Anisakis sp.) in Heringen und Heringserzeugnissen. Archiv Lebensmittelhygiene 39, 147 ~ 150 (1988).

3. Priebe, K., Jendrusch, H. and Haustedt, U.: Problematik und Experimentaluntersuchungen zum Erlöschen der Einbohrpotenz von Anisakis Larven des Herings bei der Herstellung von Kaltmarinaden. Archiv Lebensmittelhygiene 24, 217 ~ 222 (1973).

ANNEX Ⅱ Treatment procedures sufficient to kill living nematodes

—e. g. freezing to -20℃ for not less than 24 h in all parts of the product;

—the adequate combination of salt content and storage time (To be elaborated);

—or by other processes with the equivalent effect (To be elaborated).

ANNEX Ⅲ Determination of the presence of visible parasites

1. The presence of readily visible parasites in a sample unit that is broken into normal bite-size pieces 2030 mm of flesh by the thickness of the fillet. Only the normal edible portion is considered even if other material is included with the fillet. Examination should be done in an adequately lighted room (where a newspaper may be read easily), without magnification, for evidence of parasites.

2. Notwithstanding paragraph 1, the verification of the presence of parasites in intermediate entire fishery products in bulk intended for further processing could be carried out at a later stage.

水产及水产加工品操作规范

（*CAC/RCP 52－2003*，*REV. 2－2005*）

引言
使用方法
第一部分　范围

第二部分　定义
2.1　概述
2.2　水产养殖
2.4　鲜鱼、冻鱼和碎鱼肉
2.5　速冻碎鱼糜
2.12　鱼类和贝类罐头

第三部分　前提条件
3.1　渔船和捕捞船的设计和构造
3.2　加工车间的设计和构造
3.3　设备和用具的设计和制造
3.4　卫生控制程序
3.5　工作人员的健康与卫生
3.6　运输
3.7　产品可追溯性与召回
3.8　培训

第四部分　处理鲜鱼、鲜贝的总体注意事项
4.1　与鲜鱼、鲜贝有关的潜在危害
4.2　时间和温度控制
4.3　把鱼、贝的腐败降低到最低限度——处理工序

第五部分　危害分析和关键控制点（HACCP）以及缺陷作用点（DAP）分析
5.1　HACCP 原则
5.2　DAP 分析
5.3　应用
5.4　结论

第六部分 养殖产品

6.1 概述
6.2 危害和缺陷的名称
6.3 产品操作

第八部分 鲜鱼、冻鱼和碎鱼肉的加工

8.1 鱼的预处理
8.2 真空包装或气调包装鱼类的加工
8.3 冻鱼的加工
8.4 碎鱼肉的加工
8.5 包装、标签和配料

第九部分 冷冻鱼糜的加工

9.1 冷冻鱼糜危害和缺陷的一般说明
9.2 鱼的预处理
9.3 鱼肉分离工序
9.4 清洗和脱水工序
9.5 精滤
9.6 最后脱水工序
9.7 辅配料混和添加
9.8 包装和称量
9.9 冻结操作
9.10 脱盘
9.11 金属探测
9.12 装箱和加标签
9.13 冻藏
9.14 原料接收——包装材料和配料
9.15 原料贮存——包装材料和配料

第十部分 速冻裹面包屑或挂浆鱼加工

10.1 前提条件的追加项
10.2 危害和缺陷的名称
10.3 加工操作

第十四部分 虾类的加工

14.1 冷冻虾——总述
14.2 小虾的准备

第十五部分 头足类动物的加工

15.1 头足类动物的接收
15.2 头足类动物的储存
15.3 解冻的控制

15.4 剖割、取内脏和清洗
15.5 去皮，修整
15.6 添加剂的使用
15.7 分级/包装/标签
15.8 冷冻
15.9 包装、标签和成分 - 接收和储存

第十六部分 鱼类、贝类罐头的加工

16.1 通则——前提条件件的补充
16.2 危害和缺陷的识别
16.3 加工操作
16.4 预煮和其他处理方法

第十七部分 运输

17.1 鲜、冰和冻品
17.2 活鱼和获贝
17.3 罐装的与和贝
17.4 所有的产品

第十八部分 零售

18.1 鱼、贝的接收和它们的相应产品的零售——概述

引　言

《水产及水产加工品操作规范》是食品法典水产及水产加工品委员会将附录Ⅻ中所列的一系列法规合并而成的，另外，增加了水产养殖类和冷冻鱼糜的部分。所列法规主要从技术角度提供了水产品生产、贮存以及在船上和岸上处理过程中应注意的基本事项，也涉及水产品的批发和销售的有关事宜。

本操作规范结合《推荐性国际操作规范——食品卫生总则》（CAC/RCP 1 - 1969，Rev. 3 1997）及“附录：HACCP 体系及应用准则”（Codex Volume 1B）中提出的 HACCP 系统方法，对所列规范进行了进一步的修改。为了确保食用安全和满足相关的食品法典的产品标准，本规范中的“前提条件”涵盖了在鱼、贝类及其制品的生产过程中卫生方面的技术指导和必要要求。本规范还包括 HACCP 方法的应用指南。

在本规范中有一套与 HACCP 相似的系统方法——“缺陷行动控制点（DAP）分析”贯穿始终，它是用于判定生产过程是否符合食品法典产品标准中有关质量、成分以及标签的规定。然而，DAP 分析是可选择的。

食品法典水产及水产品加工委员会在第二十次会议上提议，应将已从食品（水产品）法典标准中删除的商业性缺陷（例如，工艺方面的缺陷），转换成相应的法典操作规范，以便产品交易中买卖双方可随意应用。该委员会进一步提议，应该在“最终产品说明”部分对此进行详细阐述。因此，在本规范中已将这些内容列入附录Ⅱ ~ Ⅺ。本规范采纳的 DAP 分析体系方法可作为控制产品缺陷的指导。

本规范将对从事水产品加工、生产、贮藏、批发、进出口和销售及相关的人员有所帮助，以确保符合食品法典标准（见附Ⅻ）的健康、卫生、安全的产品销往国内外市场。

使用方法

本规范旨在提供一个便于使用的、包含了详细的鱼和贝类加工管理系统背景资料和指导的文件。这一加工管理系统可与 GMP（良好操作规范）合并，同时也可以在没有实行 HACCP 的国家应用。除此之外，它还可以用于培训从事捕捞业和鱼、贝类加工业的有关人员。

这一国际规范在各国渔业的实际应用中，仍需要修改和完善，必须考虑到特定的地理环境和特殊的消费需求。因此，本规范并不想取代训练有素、经验丰富的专家们关于复杂技术和卫生问题提出的建议和指导，这些建议和指导可能是某一特定地区或特定渔业所特有的。事实上，本规范应作为这些建议和指导的补充。

本规范分为独立存在而又有内在联系的若干部分，为更好地建立 HACCP 或 DAP 计划，以下内容应予以适当的考虑：

（1）第二部分：定义

了解定义很重要，有助于对本规范全面的理解。

（2）第三部分：前提条件

对一个加工企业来讲，在应用 HACCP 或类似的系统方法之前，具备良好的卫生操作的硬件条件是很重要的。这一部分涵盖了一些基础工作，这些工作是应用危害和缺陷分析之前应达到的最低要求。

（3）第四部分：处理鲜鱼、鲜贝的总体注意事项

这部分提供了在制定 HACCP 或 DAP 计划时应考虑的潜在危害和缺陷的概况。它并不是一个详

尽的罗列，而是旨在帮助 HACCP 或 DAP 计划小组全面考虑鲜鱼、鲜贝和其他水产无脊椎动物可能存在的危害和缺陷因素，然后由小组确定与加工过程有关的显著危害和缺陷。

（4）第五部分：危害分析与关键控制点（HACCP）和缺陷行动控制点（DAP）分析

只有当第三部分中提到的基础工作令人满意地完成后，才能考虑应用第五部分中概述的原则。这部分以罐装金枪鱼的加工为例详细说明了 HACCP 原则在加工过程中的应用。

（5）第六和第七部分*：水产养殖类产品和软体动物类产品

涉及养殖和初加工的鱼类、甲壳类及软体动物类产品，而不是捕获自野生的。

尽管在第六至第十八部分的许多步骤中列出了潜在危害和潜在缺陷，但应该强调的是这只是为了引导和考虑其他更有可能发生的危害和缺陷。同时，这部分给出的模式是为了最大限度便于使用。因此，只有当这些“潜在的危害”或“潜在的缺陷”可能被引入到产品中或被控制的环节才会被列出，而不是在所有加工步骤都重复出现。

除此之外，还应强调的是，这些危害和缺陷以及它们后续的控制点或作用点，都具有产品和阶段特异性。因此，以第五部分为基础的全面的关键分析应针对每个具体操作加以完善。

（6）第八部分*：鲜鱼、冻鱼和碎鱼肉的加工过程

这部分是大多数后续部分的基础。它涉及原料鱼的处理到冷藏的全过程，列举了不同步骤可能出现的各种危害和缺陷并加以指导。这部分是所有加工过程（第九至第十六部分）的基础。第九至第十六部分是针对特定产品的加工过程给以补充指导。

（7）第九至第十六部分*：各种鱼类、贝类产品的加工过程

从事这些品种加工的人员应参阅相关部分的有关信息。

（8）第十七至第十八部分*：运输和销售

包括运输和销售的一般事宜。运输和销售原则适用于非特殊加工的大多数产品。这些部分应同其他加工步骤一样给予重视。

（9）在附录*中包括一些附加说明。

第一部分　范围

本操作规范适用于供人类食用的海、淡水鱼类、贝类和水生无脊椎类以及这些产品的养殖、捕获、处理、生产、加工、贮存、运输和销售。

第二部分　定义

本规范目的。

2.1　一般定义

术　语	定　义
生物毒素 Biotoxins	指天然存在于水产品中的毒素或因水生动物食用了有毒藻类而在体内富集的毒素，或这些生物体释放到水中的毒素
冷却 Chilling	指把鱼类、贝类的温度降到接近冰融化的温度的过程
净水 Clean Water	指来自任何水源有害物质、微生物污染和有毒浮游生物的含量被控制在一定范围内，不致影响鱼类、贝类及其制品的安全质量的水

续表

术 语	定 义
净化 Cleaning	指去除泥土、食物残渣、污物、油脂或其他令人不悦的东西
污染物 Contaminant	指生物的、化学的物质、异物或其他并非有意加入的、影响食品卫生安全或产品质量的物质
污染 Contamination	在鱼贝类以及其他产品中污染物的引入或出现
控制措施 Control Measure	指为预防或消除食品安全危害或将其降低到可以接受水平所采取的任何行动和活动在本规范中，它也应用在缺陷的预防或消除
纠偏行动 Corrective	指当 HACCP 监测结果表明偏离关键限值时所采取的任何行动。在本规范中，它也应用于 DAP 系统
关键控制点 Critical Control Piont (CCP)	指对危害进行控制的一个步骤，它对预防或消除一种食品卫生安全的危害或将其降低至可接受水平是非常关键的
关键限值 Critical Limit	指判定可接受和不可接受的界线，在本规范中，它也应用于 DAP 系统
判断树 Decision Tree	指对已确定可造成危害的每一加工步骤，判定该加工步骤是否是关键控制点的一系列问题。在本规范中，它也应用于 DAP 系统
腐败 Decomposition	指鱼类、贝类及其加工品的变质，包括组织破坏和散发出持续的令人不快的异味和臭味
质量缺陷 Defect	指产品在基本质量、组成和食品标签方面，不符合食品法规中产品标准的相映条款的情况
缺陷作用点（DAP） Defect Action Point	指可以实施控制的一个步骤，在此可将质量缺陷预防、消除或降低到可接受水平或消除欺诈风险
消毒 Disinfection	指用化学物质或物理方法，减少环境中的微生物，使不致危及食品卫生安全或质量
去头脏鱼 Dressed	指去头、除内脏后的鱼的其他部分
设备 Facility	指处理、加工、冷却、冻结、包装和贮存水产品的设施。在本规范中还包括渔船
鱼类 Fish	指任何冷血水生的脊椎动物。两栖类和爬行类不包括在内
危害 Hazard	指食品中潜存的可能影响人类身体健康的生物、化学和物理的物质或状态
危害分析 Hazard Analysis	指收集和评估有关危害的信息及导致危害存在的条件，并确定哪些对食品卫生安全的影响是显著的这样一个过程，该过程应在 HACCP 计划中写明
危害分析与关键控制点（HACCP） Hazard Analysis and CriticalControl Point	指确定、评估和控制显著的食品安全危害的系统
监测 Monitor	指实施一系列计划好的观察和测量控制参数的行动，用来评估某关键控制点是否在控制范围内。在本规范中，它也应用于 DAP 系统
饮用水 Potable Water	指新鲜的适合人类饮用的水。可饮用的标准应不低于世界卫生组织最新制定的“饮用水的国际标准”
前提条件 Pre – Requisite Programme	指应用 HACCP 体系之前必须具备的条件，以确保鱼类和贝类的加工设备、设施符合《食品卫生法规通则》、法典相关的规范及相关的食品安全法规

续表

术　语	定　义
原材料 Raw Material	指用于生产供人类食用的鱼类、贝类加工品的新鲜和冷冻的鱼类、贝类或其组成部分
冷却水 Refrigerated Water	指经过适当的制冷系统冷却后的净水
货架寿命 Shelf - Life	指在特定的贮存温度下，产品能够保持其微生物和化学稳定性以及感官质量的时间。它取决于对产品已确定的危害所采取的加热或其他处理方法、包装方法和可能的抑制因素
贝类 Shellfish	指那些通常被用来食用的水生软体动物类和甲壳类动物
步骤 Step	指食物链中的某一点、过程、操作或阶段，它包括原材料，从初级产品到最终消费
确认 Validation	指获得 HACCP 计划中的元素的有效证据
验证 Verification	指除了用监测以确定是否符合 HACCP 计划外，所应用的方法、程序、测试和其他评估。在本规范中，它也应用于 DAP 系统
全鱼（整鱼） Whole Fish(or Round Fish)	指捕获的未去内脏的鱼

2.2 水产业

术　语	定　义
水产养殖 Aquaculture	指对所有用于人类消费的水生动物，部分或全部生命周期的养殖，这些水生动物不包括哺乳动物、水生爬行动物和水生两栖动物，但是本规范第七部分涉及的品种除外。为简便起见，参考第 2.2 条和第六部分，在下文中我们将用“鱼类”来表示这里所说的“水生动物”
水产养殖设施 Aquaculture Establishment	指那些生产供人类消费用鱼类的经营场址，包括内部的基础设施和在同一管理下的周围环境
化学品 Chemicals	包括任何天然的或合成的可以影响鲜活鱼类及其病原体、养殖水体、养殖设备或养殖水域周围土质的物质
着色 Colouring	指通过在鱼用饵料中添加权威机构认可的天然或人工合成的成分或其他添加剂以获得具有特定色泽的目标体
病鱼 Diseased Fish	指鱼体出现了明显的病理学变化或呈现其他不正常症状，这些变化会影响食用的安全和质量
粗养 Extensive farming	指对鱼类养殖时的生长过程和养殖环境的不控制或不完全控制，这时鱼类的生长主要取决于养殖环境本身的营养情况
饲料添加剂 Feed Additives	指经许可在鱼类饲料中添加的除营养物质以外的其他化学品
水产养殖场 Fish farm	是一个水产养殖（陆基或水基）生产单位；通常包括存养设施（水箱、池塘、水沟 raceway、网箱等）、工厂（建筑物、仓库、加工设施）、维护设备和农具
鱼用饲料 Fish feed	指用于水产养殖鱼类的任何形式和成分的饲料

续表

术　语	定　义
良好水产养殖操作 Good Aquaculture (or Good Fish Farming) Practice	定义为在水产养殖环节，为生产符合相关食品法律法规要求的高质量产品所必需的操作
收获 Harvesting	将鱼从水中捞出的操作
精养 Intensive farming	指对鱼类养殖时的生长过程和养殖环境采取严格控制，这时鱼类的生长完全取决于外部供给的饲料
官方管辖机构 Official Agency Having Jurisdiction	指负责监控食品卫生和水产养殖卫生的官方权威机构或受政府指派的权威机构（有时称为主管机构）
农药 Pesticide	指用以防止、消灭、吸引、驱逐或控制所有害虫（包括在生产、贮藏、运输、销售以及食品、农产品和动物饲料的加工过程中出现的有害的植物或动物），或可能用于控制动物体外寄生虫的任何物质，该术语通常不包括肥料、动物和植物营养剂、食品添加剂和兽药
农药残留 Pesticide Residue	指食品、农产品或动物饲料中源于使用农药的特定物质。此术语还包括所有农药的派生物，诸如转化物、代谢物、反应物以及被认为有毒性的杂质等
残留 Residues	指由于施用或意外暴露而导致其于收获前留存在鱼体内的任何外来物质（包括其代谢物）
半密集养殖 Semi - intensive farming	指对鱼类生长过程和养殖环境进行部分控制的条件下的养殖鱼类，这时，鱼类的生长依赖于养殖环境本身的营养物和外部供给饲料
养殖密度 Stocking density	单位养殖面积或容积内所养殖的鱼的数量
兽药 Veterinary Drug	指所有的用于任何食用动物（包括产肉或产奶的动物、家禽、鱼类以及蜜蜂）的药物，无论是用于治疗、预防或诊断等目的还是用于改变其生理机能或习性
休药期 Withdrawal Time	指最后一次对某种鱼类使用某种渔药或使该种鱼类暴露于该渔药，同收获这些鱼之间必要的时间间隔，以此确保其用于人类消费的可食鱼肉中的兽药浓度不超过所允许的最高残留限量

2.4　鲜鱼、冻鱼和碎鱼肉

术　语	定　义
灯检 Candling	指鱼片通过一张半透明的桌子，从下方照明检查寄生生物及其他缺陷
脱水 Dehydration	通过蒸发使冻结的水产品失去水分，此情况在产品未经过适当的镀冰衣、包装而冷藏时可能出现。严重脱水会影响水产品外观和表面组织，通常被称为“冻斑”
鱼片 Fillet	指从鱼的胴体沿脊骨平行的方向将鱼切成的大小、形状不规则的薄片
冻结器 Freezer	指为冷冻鱼和其他食品设计的设备，通过快速降温使温度恒定后食品内部温度与贮存温度一致

续表

术　语	定　义
冻结过程 Freezing Process	指使用适当的设备，快速通过最大结晶温度范围的过程。除非在温度恒定后食品内部温度已经达到或低于 -18℃（0 ℉），否则不能认为冻结过程是完全的
冻藏设备 Frozen Storage Facility	指能够使鱼体温度保持在 -18℃（0 ℉）的设备
鲜鱼 Fresh Fish	指仅经过冷却而没经过其他处理的水产品
冻鱼 Frozen Fish	指经过充分冷冻的鱼。冷冻过程中使整个产品的温度降至能够保持其原有质量的温度，并一直保持，如在速冻有鳍鱼类（去脏和不去脏）的标准中所规定的其在运输、贮存和配送（包括最终出售）时的温度。因此，在本规范中，“冷冻”、“深度冷冻” 和 “速冻” 这些术语除特别说明外可看作同义词
镀冰衣 Glazing	指用净海水、饮用水或饮用水并适当添加经批准可使用的添加剂，喷洒或浸泡冷冻产品，在其表面形成一层冰保护层
碎鱼肉 Minced Fish	指去掉鱼皮和脊骨，并粉碎的鱼肉
气调包装（MAP） Modified Atmosphere Packaging	指鱼周围的气体组成与通常的气体不同的包装方法
分离 Separation	指为生产碎鱼肉而从鱼肉中基本去除鱼皮和脊骨的机械加工过程
分离器 Sparator	指用于分离的机械装置
鱼段 Steak	指近似垂直于鱼脊骨的角度进行切割，而得到的鱼的一段

2.5　速冻鱼糜

术　语	定　义
脱水 De - Watering	指从碎鱼肉中去除多余的漂洗水分
速冻鱼糜 Frozen Surimi	指需经进一步加工的鱼类蛋白产品。即已经去头、除内脏、清洗和机械分离鱼皮和脊骨后得到可食用的碎肉，再经过漂洗、精滤、脱水、与冷冻变性防止剂混合并冷冻
凝胶形成 Gel Forming	指鱼糜与盐混合、成形、加热形成有弹性的凝胶的能力。弹性是肌球蛋白
能力 Ability	拥有的特性，而肌球蛋白是肌原蛋白的主要成分的
肌原纤维蛋白 Myofibrillar Protein	骨骼肌肉蛋白的总称。例如，肌球蛋白和肌动蛋白
精滤 Refining	指用过滤器去除漂洗过的鱼肉中的小块鱼骨、肌腱、鱼鳞和带血的肉，使其不在最终产品中出现，从而浓缩了肌原纤维蛋白
鱼糜制品 Surimi - Based Products	指一系列在鱼肉中添加辅料和调味料而制成的产品，如 “鱼糜胶”、“模拟贝肉”

续表

术 语	定 义
水溶成分 Water - Soluble Component	指鱼肉中的任何水溶性蛋白、有机物质和无机盐
漂洗 Washing	指用旋转过滤器加冷水从碎鱼肉中去除血和水溶成分，从而提高其中肌原纤维蛋白的含量
漂洗过的鱼肉 Washed meat	指漂洗后再沥去水分的鱼肉

2.6 速冻裹衣产品

术 语	定 义
挂浆 Batter	由谷物产品，香料，食盐，糖和其他成分或添加剂组成的裹衣用的液体物质。有两种典型的挂浆：发酵型和非发酵型
裹面包 Breading	干的面包屑或其他来自谷物的含有着色剂和其他成分的物质，是最后裹衣用的。典型的裹面包屑的类型有：自由流式裹面包，粗略裹面包法和面粉型裹面包法
裹衣 Coating	把面浆或面包屑覆盖在水产品的表面
预干燥 Fro - frying	把裹面包和挂浆的水产品进行油浴的方法干燥，这样可以保持其内部还是冷冻的
切开 Sawing	用机器或人工把原料鱼分割成适合以后裹衣的规则的形状个体

2.10 虾类

术 语	定 义
去头 Dehead	指将虾头从虾或对虾整个身体上去掉
去肠虾 De - veined shrimp	指所有经过去壳并且背部切开去掉内脏后的虾
鲜虾 Fresh shrimp	指刚收获的、未经过贮藏处理的、或者是只经过冷藏保鲜的虾。不包括鲜煮后的虾
虾仁 Peeled shrimp	去掉头和外壳的虾
去头虾 Raw headless	去掉头部、保留外壳的生虾
虾 Shrimp	术语虾（shrimp）（包括经常被成为对虾“prawn”的种类）包括最近修订的 FAO 水产大纲第一卷 No. 125，世界虾类中列出的所有种类

2.11 头足类

术 语	定 义
剖割 Splitting	指沿着体侧切开头足类，使其形成一个肉片的过程

2.12 鱼类、贝类罐头

为满足本规范使用的目的，只收录了与罐头行业和第十三部分有关的主要术语的定义。要查找所有定义请参阅《推荐性国际法典 - 低酸和酸化低酸罐装食品卫生操作规范》（CAC/PRC 23 - 1979，Rev. 2（1993））。

术 语	定 义
罐装食品 Canned Food	指封入密封容器中的商业无菌食品
热加工食品的商业无菌 Commercial sterility of hermally processed food	指仅通过充分加热或同时经其他适当处理，使食品能够在非冷藏状态下免受微生物的侵袭，从而可以销售和贮存的条件
密封容器 Hermetically Sealed Container	指为保持其内容物在热加工过程中和其后免受微生物的侵袭而进行密封的容器
高压杀菌罐 Retort	指将装入密封容器中的食品进行热加工而设计的压力容器
预设程序（杀菌规范） Scheduled Process（or Sterilisation Schele）	指加工者为至少达到商业无菌标准而为某一特定产品和容器的规模选择的热加工程序
杀菌温度 Sterilisation Temperature	指热加工过程中，始终保持在预定程序中要求的温度
杀菌时间 Sterilisation time	指达到杀菌温度到冷却开始的一段时间
热加工 Thermal Process	指为获得商业无菌而限定了时间和温度的热处理过程
排气 Venting	指预定热加工之前，以先通入蒸汽的方法彻底去除杀菌罐中的气体

2.13 运输

2.14 零售

术 语	定 义
零售 Retail	指贮藏、预处理、包装、服务或其他形式的向消费者直接提供鱼、贝类及其产品供其处理后再食用的操作。包括水产品的自由市场，杂货店或商店的水产柜台，包装、冷藏或冷冻和（或）完整的服务
包装 Pakaged	指提前包装后放在冷藏或冷冻柜台供消费者直接取购
完全服务展示 Full Service Display	指售货员根据顾客的要求展示称重、包装冷藏的鱼类、甲壳类和贝类及其产品的服务过程

第三部分　前提条件

将 HACCP 应用于食品加工过程中的任一步骤之前，必须有良好卫生规范或主管当局要求的前提条件做支持。

前提条件的建立使HACCP计划小组可以把HACCP的应用重点放在有关的产品和选择的加工工艺上，而不必过多考虑那些来自周围环境的危害。前提条件对于特定加工企业或对不同的渔船都有其特殊性，并且需要通过监测和评估以保证其持续有效。

要想获得更多有助于水产加工车间或渔船制定前提条件的信息，请参阅《推荐性国际操作规范——食品卫生总则》（CAC/RCP 1－1969，Rev. 3 1997）及其附录《HACCP体系及应用准则》。

值得注意的是：以下列出的一些事项，例如，与损失有关的一些事项，是为了保证质量而非食品卫生安全。因此，以食品安全为取向的HACCP体系制定前提条件时，并不是必须的。

HACCP体系的这些原则同样也适用于缺陷行动控制点（DAP）体系。

3.1 捕捞船和收获船的设计和构造

由于不同地区的经济状况、环境、捕获的鱼、贝的种类各不相同，因此全世界各地有许多品种繁多的捕捞船和收获船。为了确保进一步加工和冷冻时鲜鱼和鲜贝的卫生和高质量，这部分强调了所有渔船都应最大限度的考虑清洁能力以及减少损失、污染和腐败等基本要求。

一般的捕捞船和用于捕获养殖的鱼类、贝类的收获船的设计及构造应参照如下注意事项：

3.1.1 易于清洁和消毒

（1）渔船设计和构造应尽可能减少内部的棱角和突起，以避免藏污纳垢；

（2）构造应便于充分排水；

（3）具有压力合适的，可供应充足的净水或饮用水（WHO饮用水质量指南，第二版，日内瓦，1993）的设施。

3.1.2 把污染降到最低

（1）操作区的所有表面应无毒、光滑、不渗水，并且状态良好，尽可能减少鱼黏液、血液、鳞和内脏的黏附以减少物理及微生物污染的风险；

（2）在需要的地方应提供足够的处理及清洗鱼类的设施，以及充足的冷却饮用水或净水；

（3）在需要的地方应提供足够的清洗和消毒设施；

（4）净水的入水口应放置在无污染的地方；

（5）供水和排水系统应能够满足使用高峰的需要；

（6）非饮用水管线应有明显的标志并与饮用水严格分开以避免交叉污染；

（7）防止包括船底污水、废气、燃油、润滑油、废水和其他固体或半固体废物在内的废弃物污染鱼获物；

（8）在需要的地方，应有将由不渗水的材料制成的可加盖密封的容器，并做明显的标志以放置水产品的内脏和废弃物；

（9）为防止水产品和干性材料（例如，包装材料）被下列物质污染，应提供充足的隔离设施：

（a）有毒有害物质；

（b）干燥的贮存材料和包装等；

（c）内脏和废弃物。

（10）提供足够便利的洗手和卫生间设施，并与水产品操作区严格分开；

（11）防止鸟类、昆虫或其他害虫及有害动物侵入。

3.1.3 尽量减少鱼类、贝类和其他水生无脊椎类的损失

（1）在操作区内，表面应尽可能减少棱角和突起；

（2）箱式和架式贮存区内，其设计应避免对渔获物施加过多压力；

（3）设计滑道和传送带时，应防止由长距离下滑或挤压而造成渔获物的物理损伤；

（4）捕捞渔具及其使用应尽量减少对渔获物的损伤，尽可能不使其变质。

3.1.4 尽量减少收获过程中对养殖水产品的损伤

使用围网、渔网或其他手段收获养殖水产品并将其活着运送到加工车间，应做到：

（1）慎重选择围网、拉网和定置网，把捕捞过程中对水产品的损伤降到最低；

（2）捕获区和所有捕捞、分类、分级、传送和运输鲜活水产品的装置应设计得易清洗、无污染，快捷有效地处理水产品避免机械损伤；

（3）运送活鱼和宰杀后的鱼的装置应使用不会锈蚀、不会传播有害物质且不会对产品产生机械损伤的材料；

（4）运送活鱼时，应尽量注意不要过于拥挤以免造成碰伤；

（5）活鱼在装箱和运输时，应注意一切对其有影响的条件（例如，CO_2、O_2、温度、氮气和废气物等）。

3.2 生产车间的设计和构造

生产车间应包括一个生产流程模式，设计该模式是为了防止潜在污染源，尽量减少可能导致产品质量进一步降低的加工延误，并防止成品和原材料之间的交叉污染。鱼、贝和其他水生脊椎类动物是易腐食品，故必须小心处理，及时冷却。因此，操作间的设计应方便快速加工和随后的贮存。

设计和建造操作间应考虑如下因素：

3.2.1 易于清洁和消毒

（1）墙面、隔板和地面应使用不透水的无毒材料；

（2）所有与水产品直接接触的表面应采用浅色、光滑、防锈、易清洗、不透水材料；

（3）墙面、隔板光滑表面的高度应符合操作要求；

（4）地面设计应使排水能充分通过且无积水；

（5）设计和建造天花板和顶棚结构时，应尽可能减少灰尘的黏附、水蒸气的凝结及灰尘脱落；

（6）建造窗户时，应尽可能减少灰尘的黏附，并在必要的地方安装活动的、可清洗的、防昆虫窗，如有必要可安装固定纱窗；

（7）门的表面应光滑，不吸水；

（8）地面和墙面之间的接缝应易于清洗（圆角接缝）。

3.2.2 尽量减少污染

（1）生产车间的平面布置应设计得可将交叉污染降到最低限度，这可以通过时间或空间的分隔来实现；

（2）操作区的所有表面应无毒、光滑、不渗水，并且状态良好，尽可能减少鱼黏液、血液、鳞和内脏的黏附以减少鱼体损伤和微生物污染的风险（请参见3.1.2第一条，并参看其原文）；

（3）直接接触水产品的工作台面应状态良好，经久耐用，易于保养。应采用光滑、不吸水的无毒材料，而且在正常操作条件下，不应和鱼、贝类原料及其制品以及消毒剂、清洁剂起化学反应；

（4）为处理和清洗水产品提供充分的设施和充足的冷饮用水；

（5）提供适当且足够的贮冰或制冰设施；

（6）顶灯应加罩或加其他防护，以防碎玻璃或其他材料的污染；

（7）须避免烟雾造成的交叉污染，通风设备应能够消除多余的蒸汽、烟雾和异味；

（8）在需要的地方应提供充分的清洗和消毒的设施；

（9）非饮用水管线应有明显标志并与饮用水管线严格分开以避免污染；

（10）供水和排水系统应能满足使用高峰的需要；

（11）尽可能减少固态、半固态、液态废弃物的积聚，以避免污染；

（12）在需要的地方，应有将由不渗水的材料制成的可加盖密封的容器，并做明显的标志以放置水产品的内脏和废弃物；

（13）为防止下列物质的污染，应提供充足的隔离设施：

—有毒有害物质；

—干燥的贮存材料和包装等；

—内脏和废弃物；

（14）提供充足便利的洗手和卫生间设施，并与水产品操作区严格分开；

（15）防止鸟类、昆虫或其他害虫及有害动物侵入；

（16）在适当的地方，应安装供水管道防返流装置。

3.2.3 提供充足的光源

向所有工作台面提供光源。

3.3 设备和用具的设计和建造

渔船上或生产车间中处理水产品的设备和用具因其性质和用途不同而各不相同。在使用中，它们要不断接触鱼、贝及其产品，应尽量减少残留物的黏附以防这些残留物成为污染源。

设备和用具的设计和构造应考虑如下因素：

3.3.1 易于清洗和消毒

（1）考虑到保养、清洗、消毒和监测的需要，设备应耐用、可移动或可拆卸；

（2）接触鱼、贝及其产品的设备、容器和用具的设计应考虑易于排水，并按照能充分清洁、消毒、保养从而避免污染的目的而制造；

（3）设计和建造设备和用具时，应尽可能减少棱角、突起、裂缝或缺口，防止灰尘黏附；

（4）提供充足的由官方机构批准使用的法定清洁用具和清洁剂。

3.3.2 尽量减少污染

（1）操作区所有设备的表面应无毒、光滑、不渗水，应尽量减少鱼黏液、血液、鳞和内脏的黏附并减少物理污染的机会；

（2）应尽量减少固态、半固态、液态废弃物的积聚以防止对水产品的污染；

（3）贮存容器和设备应易于充分排水；

（4）污水不能污染水产品。

3.3.3 尽量减少水产品的损失

（1）表面应尽可能减少棱角、突起；

（2）设计滑道和传送带时，应防止由长距离下滑或挤压而造成水产品的物理损伤；

（3）贮藏设备应避免水产品的挤压，适于达到尽量减少损失的目的。

3.4 卫生控制程序

应始终考虑到在渔船上的捕捞、处理或岸上车间中的生产加工对鱼、贝类及其产品的安全卫生及质量存在的潜在影响，特别是应考虑所有可能存在污染的环节和为保证生产出安全、卫生的产品所采取的特别措施。控制和监控的形式取决于加工的规模和特性。

卫生控制程序应实施如下措施：

（1）防止废弃物和碎屑的黏附；

（2）保护鱼类、贝类及其产品免受污染；

（3）以卫生方法处置废弃物；

（4）监控个人的健康卫生水平；

（5）监控害虫控制程序；

（6）监控清洁和消毒程序；

（7）监控水和冰供应的质量和安全卫生。

卫生控制计划应考虑如下因素：

3.4.1　持久的清洁和消毒计划

为确保定期地、适当地清洁渔船、加工车间及其设备的各部分，应起草一个长期的清洁和消毒计划。当渔船、加工车间及其设备发生变化时，应对该计划进行再评估。计划中应体现“离开时清洁”的原则。

典型的清洁和消毒应包括如下7个步骤：

（1）清洁前准备：需清洁的场地和设备的准备。包括把所有水产品移出操作区，保护精细部件和防止弄湿包装材料，用手或橡胶扫帚去除鱼的废弃物等；

（2）冲洗前准备：用水冲去大块松散的泥土；

（3）清洁：指去除掉泥土、食物残渣、污垢、油垢或其他脏东西；

（4）冲洗：用饮用水或净水冲洗，去除所有泥土和清洁剂残留；

（5）消毒：用法定机构批准使用的法定化学制剂和（或）加热以杀灭表面上的大多数微生物；

（6）再冲洗：用饮用水或净水进行最后冲洗，去除消毒剂残留；

（7）贮存：清洗、消毒后的设备、容器和用具应放到可防止污染的地方存放。

清洁有效性的检查：应适当监控清洁的有效性。

应当培训操作工人或清洁人员正确使用专用清洁用具和化学试剂，掌握设备的拆卸方法，了解与清洁有关的污染和危害。

3.4.2　清洁人员的指派

每个加工厂或渔船需要一个受过专门训练的工作人员负责加工车间或渔船及其设备的卫生工作。

3.4.3　基础设施、设备和用具的保养

（1）加工厂的建筑、材料、用具和一切设备包括排水系统应保持良好状态，整齐有序；

（2）设备、用具和其他加工厂或渔船的设施应保持干净清洁，维修良好；

（3）应制定仪器设备的保养、维修、调节和校准程序。这些程序应针对每种设备、使用方法、操作人员、保养频率等制定。

3.4.4　害虫控制体系

（1）应用良好卫生规范，杜绝滋生害虫的环境；

（2）害虫控制程序可以包括防止害虫侵入、消除隐匿场所、阻止大批侵入等内容，并应建立监控、侦查和根除体系；

（3）物理、化学和生物手段的使用须由有资格的专业人员去进行。

3.4.5　水、冰和蒸汽的供应

3.4.5.1　水

（1）在需要的地方，应提供充足的冷热饮用水或具有一定压力的净水；

（2）在必要的地方为避免污染应使用饮用水。

3.4.5.2　冰

（1）应以饮用水或净水制冰；

（2）必须保护冰块免受污染。

3.4.5.3　蒸汽

（1）为需要蒸汽的操作提供足够压力的、充足的蒸汽；

（2）应避免蒸汽直接接触鱼、贝及食品表面而对食品的卫生安全或质量构成威胁。

3.4.6 废弃物的管理

（1）鱼内脏和其他废弃物应定期清除出生产车间或渔船；

（2）盛放内脏和废弃物的容器应合理维护；

（3）摒弃船上废弃物时不能污染供水系统或加工中的产品。

3.5 工作人员的卫生与健康

工作人员必须保持适当的个人健康卫生，避免造成污染。

3.5.1 设施和设备

设施和设备应包括：适当而卫生的洗手和干手条件、合理设置和安排卫生间和更衣室。

3.5.2 个人卫生

（1）从事处理、加工或运输的工作人员不能有感染或开放性创伤，不能是传染病患者或带菌者；

（2）必要的地方应穿戴隔离衣、鞋、帽；

（3）车间中所有工作人员都应保持高度的个人清洁水平并采取必要的防护措施避免污染；

（4）加工区域内的全部工作人员必须洗手：在开始加工水产品前和再次进入加工区域时必须洗手。进过厕所后必须立即洗手；

（5）前处理和加工区域内严禁：抽烟、吐痰、咀嚼口香糖或进食、对着未加防护的食品咳嗽，打喷嚏、佩戴可能对食品安全卫生和质量造成影响的饰物如珠宝、手表、胸针或其他饰品。

3.6 运输

设计和建造运输车辆应注意：

（1）在适当的地方如内壁、地面和顶棚，应使用表面光滑、不渗水的防腐材料。地面应布设充分的排水管道；

（2）在需要的地方安装制冷设备，以保持冰鲜水产品在运输过程中保持在0℃左右，冻鱼、贝类及其产品保持在－18℃或更低（准备罐装用的盐水冻鱼，可以保持在－9℃或以下运输）；

（3）鲜活水产品因种类不同所耐受的温度不同；

（4）须提供保护措施以防止鱼、贝类及其产品遭受污染、日晒、风干或暴露在严酷的温度之下；

（5）使用机械方式制冷时，货物周围的冷气允许自由流动。

3.7 产品可追溯性和召回程序

经验证实，产品召回体系是必不可少的前提条件，因为没有一个程序是绝对安全的，产品的可追溯性（包括批量验证）对有效召回产品是必不可少的。

（1）管理者应保证合理有效的程序以达到追踪成品和从市场上快速召回任何一批水产品的目的；

（2）保留加工、生产和销售记录，且保留时间应超过食品的货架期；

（3）每个盛放待加工或待销售的鱼、贝及其制品的容器应有明显标志，以识别生产者或产品批次；

（4）一旦出现健康危害，同样生产条件下的产品对公共卫生会产生相似危害，必须召回，并应考虑公共卫生的警告；

（5）被召回的产品应在监督下销毁或用于供人类食用之外的其他用途或重新加工以确保食品

安全卫生。

3.8 培训

水产品卫生培训十分重要，所以工作人员必须认识到他们作为保护水产品免受污染和变质的角色和责任。操作者应全面掌握确保卫生地处理水产品的知识和技能。对那些使用强力化学制剂或有潜在危害的化学制品的工作人员必须进行安全操作技术指导。

每个水产品加工车间的工作人员都应接受有关 HACCP 体系制定、实施和程序控制方面良好的培训。HACCP 应用的人员培训是在水产品加工场所成功实现此计划的基础。此体系的实际应用，会因为每个参与者的积极参与所强化。管理者也应安排操作间的工作人员全程或阶段性的培训，使其了解与 HACCP 有关的原则。

第四部分 处理鲜鱼、鲜贝和其他无脊椎水产品的一般注意事项

当获知鱼、贝和其他无脊椎水产品有寄生虫、有害微生物、杀虫剂、兽药、毒素、腐烂或异物这些对人类健康有害的物质时，除非通过分拣和加工能把它们降到可接受水平，否则这些水产品是不被接受的。当发现不适合食用的产品时，应将其与其他水产品隔离，进行重新加工和其他适当处置。处理一切可食用的水产品时都应特别注意时间和温度的控制。

4.1 时间和温度的控制

温度是影响鱼、贝类中的微生物腐败和繁殖速率的最重要的因素。对于容易产生鲭鱼毒素的品种来说，时间和温度的控制可能是控制食品安全的最有效办法。因此，冷却的鲜鱼鱼片、贝类及它们的产品，应尽可能在0℃左右保存。

4.1.1 减少腐败——时间控制

要减少腐败，以下几点是很重要的：

（1）应尽可能早进行冷却；

（2）鲜鱼、贝类和其他水生无脊椎动物的冷却、加工和销售，应小心进行并减少延误。

4.1.2 减少腐败——温度控制

温度控制的地方应符合以下要求：

（1）恰当使用足够和适当的冰鲜、冷却或冷冻水系统，以保证鱼、贝类和其他水生无脊椎动物应尽可能保存在0℃左右；

（2）鱼、贝类和其他水生无脊椎动物应在薄冰和碎冰中贮存；

（3）活鱼、贝类的运输温度应与其品种适宜；

（4）冷却或冷冻水系统或冷藏系统应设计并维护良好，以提供最大负荷时足够的冷却或冻结能力；

（5）冷却水中的鱼的密度不能过大，以免降低工作效率；

（6）对冷却时间、冷却温度和均一性的监控应有规则的进行。

4.2 减少腐败——操作

不好的操作会加速鲜鱼、贝类和其他水生无脊椎动物腐败速率和增加收获后不必要的损失。操作损失能通过以下方法来减少：

（1）在鱼、贝类的运输、分类过程中应小心处理以避免刺破、毁损等物理损伤；

（2）贮存或运输活鱼、贝类时，应注意维持能够影响其健康的因素（如 CO_2、O_2、温度、含

氮废物等）；

（3）鱼、贝类不得被踏踩；

（4）用于贮存鱼、贝类的包装箱（盒）不应装得过满，也不应堆叠得太高；

（5）当鱼、贝类在甲板上时，为防止不必要的脱水，应尽可能减少会对产品带来不利影响的暴露；

（6）应尽可能使用碎冰，它能帮助减少鱼、贝类损伤并且增大冷却能力；

（7）在以冷冻水冷藏区域里，鱼的密度应加以控制以防止损伤。

第五部分　危害分析和关键控制点（HACCP）以及缺陷作用点（DAP）分析

危害分析和关键控制点（HACCP）是科学的体系，它旨在防止食品安全问题的发生，而不是反应最终产品是否合格。HACCP 体系是由识别特定危害和采取控制措施来完成的。一个有效的 HACCP 体系能减少对传统的最终产品检验的依赖。第五部分主要解释 HACCP 原则应用于水产养殖品和软体动物类及其生产和加工，但该规范只能提供此原则的使用指南和可能发生在各种水产品中的危害类型的建议。HACCP 计划，应与食品管理计划相融合，而且应尽可能简明完整地以文件形式表达出来。本节将演示一个可考虑用于 HACCP 计划推广应用的格式。

另外，第五部分描述了如何使用另一种相似的、涉及普遍原则的方法，它能作更广泛的应用，涵盖了食品法典标准中的对基本质量、成分和标签的规定，或其他非安全要求，即我们这里定义的缺陷作用点分析（Defect Action Point Analysis）。这种缺陷分析方法是一个可选择的、可达到同一目标的技术，图 5 - 1 简要描述了如何建立 HACCP 和缺陷分析体系。

HACCP 的实施原则是通过逻辑顺序更好地确定 HACCP 的实施方法（图 5 - 1），此流程图仅用作说明。对于工厂内实施的 HACCP 计划，每一个操作过程的流程图都必须要完整全面地绘制出来。

作为本规范相关部分的参考。

5.1　HACCP 原则

HACCP 体系由七项原则[1]组成，它们是：

原则 1——进行危害分析。

原则 2——确定关键控制点（CCP）。

原则 3——建立关键限值。

原则 4——建立 CCP 的监控体系。

原则 5——建立纠正措施。当监控表明特定的 CCP 失控时，施行之。

原则 6——建立验证程序以证实 HACCP 体系在有效运行。

原则 7——建立与七项原则及其应用相应的所有程序和记录的文件。

考虑 HACCP 时，必须遵循这些原则。

任何 HACCP 的相关事项均必须遵循这些原则。

HACCP 是一个重要管理工具，操作者可应用这一工具以保证食品加工过程的安全和高效。应当承认，为使 HACCP 有效，有必要进行个人培训。根据 HACCP 原则，使用者被要求罗列出任何产品类型在收获、卸货、运输、贮存或在加工过程中的任何一道程序或步骤中可能发生的危害。尤

---

[1] 《国际推荐的操作规范—食品卫生通则》（CAC/RCP1 - 1969，Rev. 3 - 1997），附件：HACCP 体系及其应用准则。

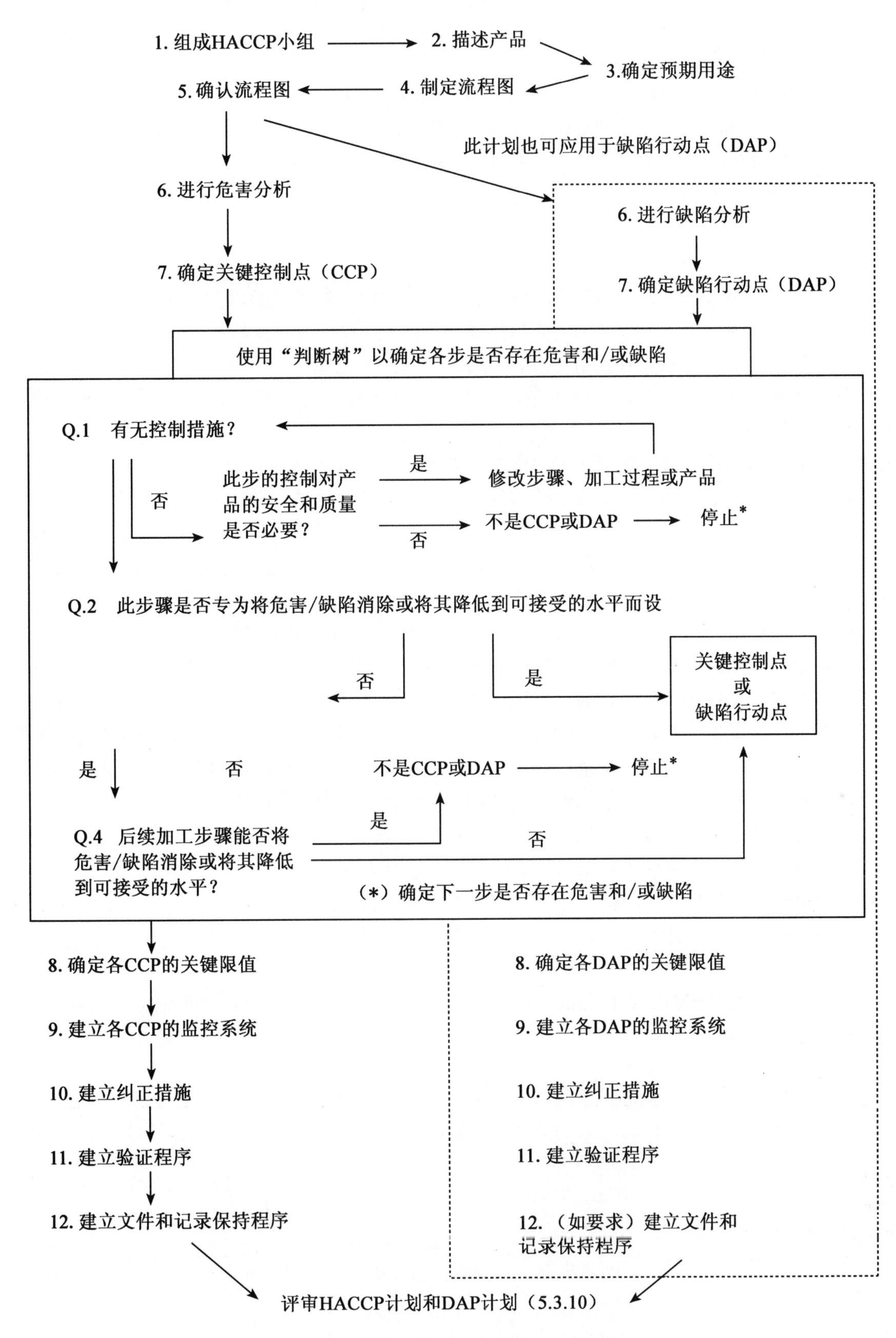

图 5－1 HACCP 和缺陷分析体系的建立

为重要的是 HACCP 原则被看作是反映操作风险的特殊基础。

5.2 缺陷作用点分析

由于该规范不仅针对与食品安全有关的危害，还包括其他与生产各方面有关的产品基本质量、成分和标识规定（这些在食品法典委员会制定的产品标准中已有描述）。该规范不仅包括对关键控制点（CCP）的描述，还包括对缺陷点（DAP）的描述。因此 HACCP 原则可以应用于确定 DAP，只要在各个步骤中用安全参数取代质量即可。

5.3 应用

任何一家水产养殖场、贝类和鱼类的加工厂，应确保满足相应的食品法典标准的规定。要达到此目的，依据该法规的描述，每家工厂应当实施基于 HACCP 原则的食品安全管理体系，同时至少应考虑与“缺陷”类似方法。HACCP 应用于培育、操作和加工链的每个环节之前必须有以良好卫生操作（参见第三部分）为基础的前提条件作支持。应当说明的是部分前提条件也可能在特定的过程中作为 CCP 或 DAP。

建立食品管理体系要明确职责和权力以及管理者、执行者和检查者的相互关系及其对该体系的影响。由多学科小组来完成对科学技术数据的收集、整理和评价是十分重要的。理想状态是该小组由相似水平的技术专家组成，这些专家具有产品和生产过程的详尽知识。例如，这类人员应包括加工厂的管理者、微生物学家、质量保证或质量控制专家，如有必要还应有其他人员，例如，买主、操作人员等。对于小规模加工企业，不可能建立这样的小组但有必要寻求外部咨询。

HACCP 计划的范围应当确定，并应当描述食品链中每个环节涉及的各类危害。

设计这个程序时，应当确定操作过程中的关键控制点，在该点处加工设备或产品应是受控的；满足规范或标准；具备关键控制点的监控频率和样品采集方案以及用于记录这些检验结果的监控系统和必要时的纠正措施。对于每个关键控制点，应提供显示其监控程序和纠正措施得到遵循的记录。记录应当保存以作为该工厂质量保证程序的证据。DAPS 的类似记录和程序，也要有必要程度的保存。建立与 HACCP 程序有关的识别、描述和记录方法，应作为 HACCP 程序的一部分。

验证活动包括所采用的方法、程序（复查/审核）和试验（除用于监控之外的）。还应确定：

（1）HACCP 或 DAP 计划达到预期结果的效果，即有效性；

（2）HACCP 或 DAP 计划的符合性，例如，复查/审核；

（3）HACCP 或 DAP 计划及其应用方法是否需要修改或重新确认。

5.3.1 描述产品

为了获得更多的对于被复查产品的了解和认识，应该对该产品做出详尽的描述。这样做有利于识别潜在危害或缺陷（表 5－1）。

表 5－1 原汁金枪鱼罐头的产品描述

产品名称 Product name（s）	目标	范例
	确定品种及加工方法	原汁金枪鱼罐头
原料来源 Source of raw material	描述鱼的来源	以围网捕自几内亚湾的黄鳍金枪鱼整条盐水冻结
最终产品重要特性 Important final product characteristics	列出影响产品安全和基本质量的特性，特别是那些有影响的微生物菌属	与食品法典标准：金枪鱼和鲣鱼罐头；低酸食品；罐密封性相一致
配料 Ingredients	列出加工过程中加入的每一种物料。只可以使用官方机构认可的法定配料	水、盐

续表

产品名称 Product name（s）	目标 确定品种及加工方法	范例 原汁金枪鱼罐头
包装 Packaging	列出所有包装材料。只可以使用官方机构认可的法定材料	镀铬钢制容器，容量212ml，总净重185g；鱼重150g；传统开启方式
最终产品如何使用 How the end product is to be used	规定成品将如何使用，特别是是否即食	即食
货架期（适时） Shelf life（is applicable）	规定按照说明的条件下存放，产品预期开始腐败的日期	3年
产品出售地 Where the product will be sold	指出预期的市场。这项信息应易于符合目标市场的法规和标准	国内零售市场
特别的标签说明 Special labeling instructions	列出所有安全贮存和食前准备的说明	“最好在标签显示的日期之前”
特别的销售控制 Special distribution control	列出所有安全分销产品的说明	无

5.3.2 流程图

对于危害和缺陷分析，有必要仔细检查产品和生产过程，并且制作生产流程图。流程图应尽可能简明。从原料的选择、加工到产品的分销、出售和消费者处理中的任何一步都应当通过一系列充分的技术数据（以避免含糊）清晰地概括出来。如果一个加工过程复杂到不能用单一的流程图表述，那么可以将其再细分为若干部分，前提是各部分之间的关系有明确的定义。这样有利于对各步骤编号和标注，从而方便查阅。一个准确、完整的结构流程图可以使多学科小组对加工程序具有清晰的认识。一旦CCP和DAP被确认，就可以将其具体体现在流程图上的相应工序中。图5－2为我们提供了金枪鱼罐头生产线流程图的实例，作为本规范相关部分的参考。不同加工过程的实例可见规范中图8－1、图9－1和图10－1。

5.3.3 进行危害和缺陷分析

危害分析的目的是确定在各个步骤中所有的对食品安全的危害，进而确定其重要性并且评定每一步骤对这些危害的控制措施是否有效。缺陷分析适用于潜在的质量缺陷，与危害分析的目的相同。

5.3.3.1 识别危害和缺陷

每个加工厂应当收集从初级生产、加工、制造、贮存、配送到消费全过程中与商业有关的可靠的科学技术数据。这些信息的集成和特性可以保证多学科小组能够确定和列出各个步骤中的所有危害（缺陷），这些危害（缺陷）很可能发生，如果缺少控制措施，就会生产出不受欢迎的食品。已经获知的与鱼类和贝类有关的潜在危害在附录1中有所描述。表5－2简述了鱼类和贝类收获前和收获过程中可能存在的安全危害，表5－3简述了鱼类和贝类收获后和深加工中存在的安全危害。

从工厂建设、设备使用、卫生操作（包括冰和水的使用）的角度而言，确定潜在危害和缺陷是十分重要的。这是前提条件所概括并用来表示过程中几乎任何点存在危害的。

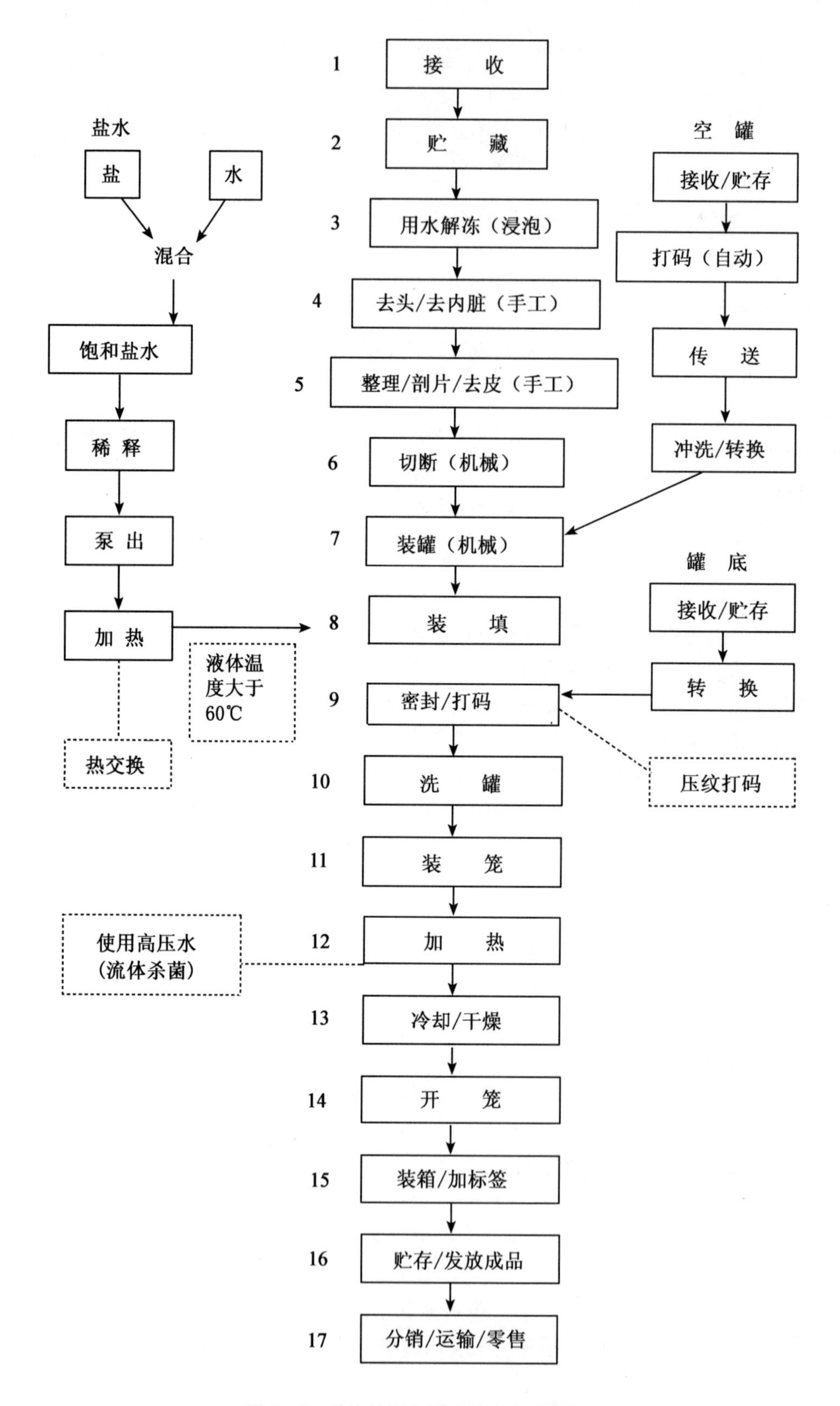

图 5－2 盐水沙丁鱼罐头的生产流程图

表 5-2　鱼类和贝类收获前和收获过程中的危害

生物的		化学的		物理的	
寄生虫 Parasites	对公共健康影响的寄生虫：收虫、线虫、绦虫	化学物质	杀虫剂、除莠剂、除海藻剂、杀真菌剂、抗氧化剂（加入饲料中）	异物	鱼钩
致病菌 Pathogenic bacteria	沙门氏菌、志贺氏菌、大肠埃希氏菌、霍乱孤菌、副溶血性孤菌、创伤孤菌	兽药残留	抗生素，促生长剂（激素）其他兽药和饲料添加剂		
肠病毒 Enteric Viruses	诺沃克病毒	重金属	从海洋沉淀物和土壤，从工业废物，从污水或动物肥料，富集的金属		
生物毒素 Biotoxins	生物毒素 鲭毒素				
		混杂物	石油		

表 5-3　鱼类和贝类在收获后和深加工中引入的危害*

生物的		化学的		物理的	
致病菌 Pathogenic bacteria	李斯特菌 肉毒梭菌 金黄色葡萄球菌	化学物质	消毒剂 卫生剂或润滑剂（误用）	异物	金属碎片，硬或锋利的物质
肠病毒 Enteric Viruses	甲型肝炎病毒 旋转病毒				
生物毒素 Biotoxins	鲭毒素 葡萄球菌肠毒素 肉毒毒素				
		配料和添加剂	误用或非许可		

注：* 由于生物危害、环境因素（如温度、氧气、pH 值和 AW）对鱼和贝类的活性和生长起着重要作用，于是鱼类或贝类在其加工及后续的贮存过程中会受这些因素的影响，因此，它们决定了产品对人类健康影响的风险程度以及食品安全管理计划的重要内容。另外，一些与水有关的危害会在两种操作水平间呈现一定交迭状态

对于特定产品的危害，参见有关加工部分。

对于本节以金枪鱼罐头为例，可以确定如表 5-4 所述基本的潜在危害。对于本节以金枪鱼罐头为例，可以确定如表 5-5 所述基本缺陷。

表 5-4　金枪鱼罐头的潜在危害

	生原料（冻金枪鱼）	加工或贮藏或运输过程中
生物的 Biological	存在肉毒梭菌、存在鲭毒素	肉毒梭菌的污染、生长，其孢子的存活，金黄色葡萄球菌的污染和生长；热加工后的微生物再污染，加工中组胺产生；肠毒素产生
化学的 Chemical	存在重金属	罐体金属的再污染，洗涤剂，盐水，机油的再污染
物理的 Physical	存在异物	加工中的再污染（如刀片，由罐体）

表 5－5 金枪鱼罐头的基本缺陷

	生原料（冻金枪鱼）	在加工或贮藏或运输过程中
生物的 Bilological	腐败	分解，腐败微生物存活，……
化学的 Chemical		贮藏中的氧化，……
物理的 Physical		令人厌恶的异物（如内脏、鱼鳞、皮……）鸟粪石结晶，容器缺陷（如渗漏……）
其他 Others	替代品种	异味，重量不准，标签不对，标识不对

5.3.3.1.1 危害

同样需要加以重视的是在鱼类或贝类的收获环境中自然发生的食品安全危害。一般来说，只要对从未污染海域捕捞的海产品在生产线上的处理是依据良好操作规范，那么其给消费者健康带来的风险不高。然而，对所有食品而言，某些产品的消费会给健康带来一定的风险，收获后误处理也会增加风险。某些海洋环境的鱼（例如，热带珊瑚礁鱼类）会因天然海洋毒素（例如，西加毒素）而引起消费风险。相对于海洋环境下捕捞的的鱼类和甲壳类，养殖的水产品所在的养殖环境会使风险上升。与养殖产品相关的食源性疾病的风险，与内陆和沿海生态有关，与捕捞鱼所在的环境相比，那里潜在的环境污染大得多。在世界某些地方，人们消费的是生的或半熟的鱼类或贝类，这样会带来很高的感染食源性寄生虫或细菌疾病的风险。为了完成作为 HACCP 计划一部分的危害分析，加工者必须拥有与原料和产品进一步加工有关的潜在危害的科学信息。

5.3.3.1.2 缺陷

潜在缺陷可以概括为基本质量、标签和成分等要求（见食品法典标准，附录 7）。没有法典标准的应当制定国家规范和（或）商业规格。

附录 2 至附录 6 简述的最终产品规格，描述了可选择性要求可以帮助买卖双方描述产品，以应用于商业交易或设计最终产品的规格。这些要求可用于商业合作伙伴的非官方申请及政府的非必须申请。

5.3.3.2 显著危害和缺陷

在食品加工过程中，认定某一项已确定的危害或缺陷是否是显著的是一项必须要进行的重要的行动，它也是食品安全管理体系中的一部分。决定危害或缺陷是否显著的两个基本要素是对健康造成不利影响的危害发生的可能性和其影响的严重程度。例如，肉毒梭菌毒素引起的死亡，影响严重，这样的危害也许发生的可能性非常低，但社会风险大，因此要保证应用 HACCP 控制显著危害—HACCP 的目的。所以，加工金枪鱼罐头中，肉毒梭菌应当考虑为显著危害，并通过应用热加工程序来加以控制。另外，例如，轻微肠胃炎，这种相对较低的严重效应的危害，可以不必用 HACCP 控制，因为它是可能性较低的风险，因此不是显著危害。并不严重的危害（例如，轻微肠胃炎）可以不必用 HACCP 控制，同样，发生的可能性很低的危害也不是 HACCP 计划中的显著危害。

在产品描述过程中收集的资料中（参见 5.3.1 节——描述产品）也能帮助确定显著危害，因为，危害或缺陷的存在受诸多与消费者有关的因素的影响，如食用方法（如生吃），消费者类型（如消费者可能是免疫缺陷的人、老人或儿童等），贮存和配送方式（如冷藏或冻结）。

一旦显著危害和缺陷被确定下来，应对加工过程的每一个步骤所引入或控制的危害进行评估。使用流程图（参考 5.3.2 节——流程图）有利于该项工作。控制措施应当用来消除每个步骤的显著危害或缺陷发生的可能性，或使其降低到可接受的水平。危害或缺陷可以由不止一个控制措施来控制。为了说明这个问题，表 5－6 和表 5－7 显示了热加工中的显著危害和缺陷及相应的控制措施。

表 5－6　金枪鱼罐头热加工步骤（显著危害：肉毒梭菌的存活）

加工步骤	潜在危害	潜在危害是否显著	判断理由	控制措施
12. 热加工	肉毒梭菌活性孢子	是	热加工不够，导致肉毒梭菌孢子存活而很可能产生毒素。产品必须经过商业杀菌	确保在杀菌锅中获得足够的热及相应的时间

表 5－7　冻金枪鱼罐头的贮藏（显著缺陷：酸败）

加工步骤	潜在缺陷	潜在缺陷是否显著	说明理由	控制措施
2. 冻金枪鱼的贮存	持久而明显的恶臭或出现酸败味	是	产品不能满足质量要求和顾客要求	控制贮存温度； 货架管理程序； 制冷系统维护程序； 人员培训和资格审定

5.3.4　确定关键控制点和缺陷作用点

一个过程的关键控制点和缺陷作用点的确定要全面简明，这对保证食品安全、确保其符合法典规范所规定的基本质量、成分和标签等相关内容是十分重要的。食品法典判断树（图 5－1，步骤 7）是一个工具，它能用来确定 CCP，同样也能用来确定 DAP。利用这个判断树，通过一系列逻辑问题，能够评估出各步骤的显著危害或缺陷。各个步骤中的 CCP 和 DAP 一旦确定，就必须在该点采取控制措施，从而防止、减少或消除可能发生的危害或缺陷，或使其降低到可接受的水平。为简要说明其用途，表 5－8 和表 5－9 列举了利用法典判断树确定金枪鱼罐头加工线中的危害和缺陷的示例。

表 5－8　以图 5－2 为例的加工过程步骤 12 用食品法典判断树确定的关键控制点和使用相应的控制措施进行危害分析的图解

加工步骤 12：热加工		食品法典判断树应用			
潜在危害	控制措施				
肉毒梭菌活性孢子	杀菌时，根据杀菌时间确保杀菌温度足够高	Q1：控制措施存在？ 是－Q2 否－考虑控制措施在加工过程中是否是有用的或必要的	Q2：该步骤能否消除或减少肉毒梭菌的产生至可接受水平	Q3：污染是否超过可接受水平或可能增加至不可接受水平	Q4：下步骤能否将危害消除或减少到可接受水平
			是－该步骤是 CCP	是－到 Q4	是－不是 CCP
		继续进行下一个已确定的危害	否－到 Q3	否－不是 CCP	否－CCP 前一步是怎么考虑的
		A：是，加热程序（杀菌规范、方法）是确定的	A：是，该步骤设计为专用于消除孢子		
		结论：加工步骤 12（热加工）是一个关键控制点			

5.3.5　建立关键限值

对于每个 CCP 和 DAP，必须确定控制危害或缺陷的关键限值。对任何一个给出的危害或缺陷而言，为每一个控制措施确定一个以上的关键限值可能是必要的。关键限值的确定应以科学证据和技术专家的意见为基础，以保证控制危害或缺陷的确定值有效。表5－10示例：金枪鱼罐头加工线CCP和DAP的关键限值。

表5－9 以图5－2为例的加工步骤2用食品法典判断树确定缺陷作用点和使用相应的控制措施进行缺陷分析的图解

加工步骤2 冻金枪鱼的冷藏		食品法典判断树应用			
潜在缺陷	控制措施				
持久而明显的恶臭，或出现酸败味	控制贮存间温度 货架管理程序	Q1：有无控制措施？ 是－到Q2 否－考虑控制措施在加工过程中是否是有用的或必要的 继续进行下一个已确定的危害	Q2：该步骤是否能消除恶臭产生的可能性或使其减少到可接受水平 是－该步骤是DAP 否－到Q3	Q3：恶臭是否超过可接受水平或可能增加至不可接受水平 是－到Q4 否－不是DAP	Q4：下步能否将恶臭消除或使其减少至可接受水平 是－不是DAP 否－DAP前一步是怎么考虑的
		A：是，贮存温度受控，有控制程序	A：否	A：是，是否贮存时间过长或贮存温度过高	A：否
		结果：加工步骤2（冻结金枪鱼的贮存）是一个缺陷作用点			

5.3.6 建立监控程序

监控系统必须由多学科小组设计，以检测出CCP和DAP的关键限值是否失控。对CCP和DAP的监控行动应简明记录，其内容包括观察或测量的负责人、采用的方法、参数以及检查的频率等细节。也应注意监控程序的复杂性。注意事项还包括充分运用各个检测标准的数字和所选用的适宜的方法，这些将会产生不同结果（例如，时间、温度、pH值）。对于CCP来说，监控程序的记录应由专人负责并记录日期，以便验证。

因为任何产品的每个生产过程都是独特的，所以仅为说明（也可能只是介绍），以金枪鱼生产线中对CCP和DAP的监控程序的用法的作为示例（表5－10）。

5.3.7 建立纠正措施

尽管我们预想建立一个有效的HACCP或DAP计划，但纠正措施总是必要的。应当建立一个文件化的纠正措施程序，以处理超过CCP或DAP的关键限值以及发生失控的情况。这个程序的目的在于保证采取适合的全面、有效的控制，并且防止受影响批次的产品不能到达消费者手里。例如，如果鱼类和贝类含有用常规的分拣或准备方法不能消除或使其降低至可接受水平的有毒物质和缺陷，就应该将其隔离或丢弃。同样重要的是要用工厂管理和合格人员来确定失控的根本原因。接下来，修改HACCP和DAP计划会很必要。当CCP和DAP发生失控情况时，其调查结果和所采取的行动都应由专人负责记录。记录应该证明已经重新建立控制过程，已经处理相应的产品，预防措施已经开始生效。金枪鱼罐头生产线的CCP或DAP的纠正措施的示例见表5－10。

表 5－10　金枪鱼罐头加工过程（表 5－8 和表 5－9）应用 HACCP 原则两个特定步骤（CCP 和 DAP）的结果

关键控制点（CCP）

加工过程步骤 12：热加工
危害：肉毒梭菌活性孢子

关键限值	监控程序	纠正措施	记录	验证
与热加工相关的特定参数	谁：具有资格的热加工人员 什么：所有参数 频率：每批 怎样：执行杀菌规范、并检查其他因素	谁：合格的人员 什么：人员再培训；重新热加工或成批报废；设备的正确维护；扣压产品到评估出其安全性为止	谁：受训过的合格人员监控记录、纠正措施记录、产品评价记录、校准记录、确认记录、审核记录、HACCP 计划评审记录	确认，成品的评价，内审记录的评审，机器校准（可以是前提条件）HACCP 计划评审、外审

缺陷作用点（DAP）

加工步骤 2：冻金枪鱼贮存
缺陷：持久而明显的恶臭或出现酸败味

关键限值	监控程序	纠正措施	记录	验证
总样品中出现油耗味样品数量不能超过建立采样程序的可接受数目	谁：受训过的专业人员 怎样：感官检验；化学测试；检测冷藏温度；检查堆垛形式 什么：依据食品法典的鱼的质量和可接受程度 频率：需要时	什么：加大监控力度；并根据其结果立即加工、分拣或丢弃超出关键限值的冻金枪鱼；调整贮存温度；个人培训 谁：受训过的合格人员	分析结果 堆垛形式 温度记录	现场审核 评审监控和纠正措施报告

5.3.8　建立验证程序

加工厂应当建立验证程序，并由专业人员来执行，以定期评估 HACCP 计划和 DAP 计划是否充分及其运行情况。这个步骤将帮助确定 CCP 和 DAP 是否受控。验证活动例子包括对 HACCP 计划各要素的确认：HACCP 体系的书面评审、它的程序和记录、对纠正措施和产品处理行动的评审（当未满足关键限值和已建立的关键限值的确认时）。当发生不明的系统失效时，加工过程、产品或包装发生显著变化时以及确定了新的危害或缺陷时，后者特别重要。使用加工设备的观察、测量和检验活动也应作为验证程序的一部分。验证活动应由合格、称职的人员来完成。HACCP 和 DAP 计划的验证频率应足以保证能够防止发生食品安全问题，同样保证防止与食品法典标准规定的基本质量、成分和标签有关的问题的发生，一旦发生能够探测出并及时处理。为了说明，金枪鱼罐头生产线的 CCP 或 DAP 的验证程序的示例见表 5－10。

5.3.9　建立文件和记录保持程序

文件包括危害分析、确定 CCP、确定关键限值、监控程序、纠正措施和验证。

及时、准确、简明的记录系统将大大增强 HACCP 计划有效性，并且使验证程序容易进行。需要文件化的 HACCP 计划的要素已在本节中简明展示。检验和纠正措施记录应该实效，并且收集所有适当的数据以表明 CCP 在“实时”控制或偏离控制。对于 DAP，除发生失控外，建议（但不是必须）记录。为了说明，金枪鱼罐头加工线的 CCP 和 DAP 记录保持方法的示例见表 5－10。

5.3.10　HACCP 和 DAP 计划的评审

如果图 5－1 概述的 HACCP 和 DAP 计划的所有步骤已完成，则要进行一个全面的评审。评审的目的在于验证计划能否满足目标。

5.4 结论

第五部分描述了 HACCP 原则以及在加工过程中应如何利用其来保证食品安全。同样的原则也可用来确定在某个加工过程中必须控制的缺陷点。由于每个工厂和加工生产线是不同的，本规范仅展示了必须加以考虑的潜在危害和缺陷的类型。另外，由于显著危害和缺陷的特性，如果没有准确评估加工的过程、加工的目标和它的环境以及预期结果，是不可能直接确定过程中的某步骤是否是 CCP 和 DAP 的。金枪鱼罐头加工线的例子只是为了说明如何应用原则，列举出商业杀菌产品的结果；为什么 HACCP 和 DAP 计划对于每个操作都是独特的。

本规范其余部分主要介绍在水产养殖品和软体动物类的生产以及对于鱼类、贝类及其产品的处理和加工，并展示在广义范围的加工过程的不同阶段中的潜在危害和缺陷。为了制定 HACCP 或 DAP 计划，在为得到特定建议参看具体过程的相应部分前，有必要先参看第三部分和第五部分。第八部分涉及鲜鱼、冻鱼和碎鱼肉的加工过程，还提供了许多很有用处的其他加工操作的指南。

第六部分 养殖水产品

前言

水产养殖企业的运作应以负责任的方式进行，例如，遵从《责任渔业行为规范》（FAO，罗马，1995 年）的建议，使任何对人类健康和环境（包括任何潜在生态改变）的不利影响减到最小。

水产养殖场应采取有效的鱼类健康和福利管理措施。应避免苗种和幼鱼感染病害，并且应遵循《OIE 操作规范》（国际水产动物健康规范，第六版，2003 年）。对生长期内的鱼进行病害监控。当在水产养殖场内使用化学品时，应采取专门措施，以保证这些物质不会被释放到周围环境中。

尽管在水产养殖活动中，鱼类的健康、环境以及生态等方面的问题是重要的，但本部分重点关注食品安全和质量方面的问题。

本部分适用于工业化和商业化的养殖模式，生产除哺乳动物、水生爬行动物和两栖动物以外的所有供人类直接消费的水生动物，但不包括本规范第七部分规定的双壳贝类，下文中的“鱼”指用于人类直接消费。这样的精养或半精养水产养殖体系采用高密度养殖，苗种来自育种场，主要采用配合饲料，并且可能使用药物治疗和疫苗。本规范不涵盖许多发展中国家中盛行的粗放型鱼类养殖系统或集家畜饲养与养鱼于一体的系统。本部分内容包括水产养殖生产过程中的饲喂、养殖、收获和运输环节。鱼类的进一步的处理和加工在本规范的其他部分加以规定。

在识别各操作步骤控制措施的内容中，本部分提供了潜在危害和缺陷的例子，并且描述了技术性指南，这些可以用于制定控制措施和纠偏行动。针对每一特定步骤，只列出了危害和缺陷，这些危害和缺陷很可能在本步骤被引入或受控。应注意在准备 HACCP 或 DAP 计划时，有必要参考第五部分，该部分提供了实施 HACCP 原则和 DAP 分析的应用指南。然而，限于本操作规范的范围，不可能提供对于每个步骤的关键限值、监控、记录保持和验证的细节，因为对于不同的危害和缺陷这些细节都不相同。

图 6－1 为水产养殖生产过程中的一些通用的步骤提供了指南。

本流程图只用于说明目的。为在工厂实施 HACCP 原则，应为每个加工过程绘制一个完整和综合的流程图。可参照本规范相关部分。

6.1 总则

除以下内容外，第三部分的一般原则适用于水产养殖生产。

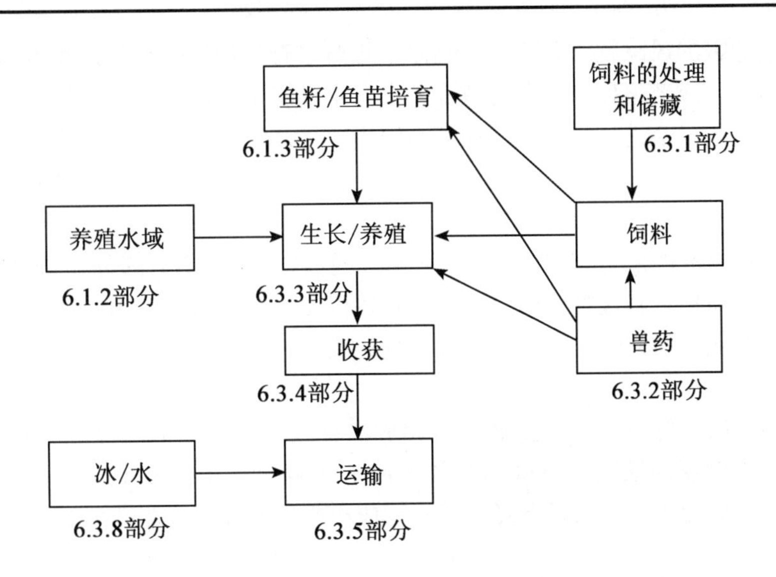

图 6－1 水产品养殖流程样图

6.1.1 场址选择

（1）水产养殖场的选址、设计和构筑物应遵守良好水产养殖操作原则，并适合其养殖品种；

（2）应检查该区域的温度、水流、盐度和水深等自然环境，这是由于不同的品种对环境的要求不同；封闭的循环系统应当能够调节环境，以使其满足养殖鱼类对环境的要求；

（3）水产养殖场应建于化学、物理或微生物危害污染风险最小的地区，也是污染源受控的地区；

（4）建造土池所用土壤所含化学和其他物质的浓度不应达到导致鱼受到污染的不可接受水平；

（5）池塘应具有分开的进水和排水渠，使供水和排水不会混合；

（6）排水道应该具有适当的处理装置，使得废水在排放到自然水体之前有足够的时间进行废物和有机物的沉淀；

（7）池塘的进水口和排水口应设滤网，预防不必要的物种进入；

（8）肥料、限制性投入品和其他化学、生物投入品的使用应符合良好水产养殖操作规范；

（9）所有渔场应以环境上可接受的方式的运作，以免因食用养殖鱼影响人类健康。

6.1.2 养殖水质

（1）养鱼用水应为适用于生产人类消费安全的产品；

（2）严格控制养殖用水的质量以保持养殖鱼的健康与卫生，从而保证消费者食用水产品的安全；

（3）水产养殖场不应建于养殖用水有被污染的风险的地点；

（4）水产养殖场应进行适当的设计和建造，以确保危害得到控制并防止水质受到污染。

6.1.3 苗种和幼鱼的来源

后期幼体、苗种和幼鱼的来源应避免给养殖产品带来潜在危害。

6.2 危害和缺陷的确定

有很多人类健康方面的危害与消费鱼类和水产加工品有关。养殖水产品中出现的危害与野生捕捞的相应品种（第 4.1 条）大体相同。在某些情况下，与野生捕捞的鱼相比，在养殖水产品中，受到某个特定危害源的风险可能会增加，例如，没有遵守兽药残留的休药期。相对于自然条件，高密度养殖可能导致因部分鱼感染病原菌而产生交叉感染的风险上升，同时也可能导致水质的下降；另一方面，养殖水产品也可能降低一些损害风险。在养殖系统中，鱼类接受人工合成饵料，这样由于鱼类摄食而传播危害的风险就可以减少。例如，相对于野生捕捞的鲑鱼，养殖鲑鱼中就没有感染线虫类寄生虫的情况，或者这种情况大大减少。在海水环境中，用网箱养鱼的危害少且风险低。在封闭的循环系统中，危害会进一步降低。在该系统中，水体经常更新和重复使用，并采用安全控制措施控制水质。

6.2.1　危害

在养殖水产品中出现的危害与野生捕捞的相应品种（第5.3.3.1部分）大体相同。养殖水产品所特有的潜在危害包括：超过推荐指南的兽药残留以及水产养殖中使用的其他化学品、由于养殖设施靠近人类居住区或动物饲养区而造成的排泄物污染。

6.2.2　缺陷

在养殖水产品中出现的缺陷与野生捕捞的相应品种（第5.3.3.1部分）大体相同。令人反感的口味或气味是可能出现的缺陷。在鲜活鱼类产品的运输过程中，减少（鱼类）紧张十分重要，这是因为鱼类紧张可能导致产品质量下降。另外，应小心操作以尽可能减少对鱼的机械损伤，因其会导致伤痕。

6.3　生产操作

6.3.1　投喂

水产养殖生产中使用的饵料应符合法典《动物饲养良好操作规范草案》（CAC/RCP-54(2004)）。

潜在危害：化学污染、霉菌毒素和微生物污染。

潜在缺陷：变质饵料、真菌造成的腐败。

技术指南：

（1）饵料和新鲜原料应在其保质期内购买、周转和使用；

（2）干制的鱼饲料应贮存于低温、干燥的环境下，以防止其变质、霉变或被污染。湿的饲料应该按照生产厂家的说明进行适宜的冷冻贮存；

（3）饵料配料内不应含有超过安全水平的农药、化学污染、微生物毒素或其他掺杂的异物；

（4）工业化生产的全价饲料和工业化生产的饲料原料应正确标识。其成分应满足卫生的要求并符合其标识内容；

（5）原料在病原体、霉菌毒素、除草剂、农药和其他可能引起人类健康危害方面的污染等方面应达到标准规定的可接受水平；

（6）饲料中只能添加经过批准的色素，且浓度符合要求；

（7）湿饲料或饲料配料应新鲜，且其化学和微生物指标应满足要求；

（8）新鲜或冷冻鱼类到达水产养殖场时应保证足够新鲜；

（9）青贮饲料和鱼的废弃物，如果要使用的话，应经过适当的煮制或精加工以减少其对人体健康存在的潜在危害；

（10）工业化生产和养殖场自行配制的饲料，应只含有官方主管机构准许应用于鱼类的添加剂、促生长物质、鱼类着色剂、抗氧化剂、胶凝剂以及兽药等；

（11）产品应经国家相关主管机构登记注册；

（12）贮藏和运输条件应符合标签说明；

（13）渔药和其他化学治疗手段应依据推荐的操作规范并符合国家法规要求；

（14）加药饲料应当在包装上有清楚的标注并单独存放，避免出错；

（15）养殖场应按照生产厂的说明使用加药饲料；

（16）应保持记录以确保对所有饲料成分的追溯。

6.3.2 渔药

潜在危害：渔药残留。

潜在缺陷：无。

技术指南：

（1）水产养殖场所使用的所有渔药应符合国家法规和国际指南［依据《国际推荐性操作规范—兽药应用的控制》（CAC/RCP 38－1993）和《建立食品中兽药残留控制管理程序的指导规范》（CAC/GL 16－1993）］；

（2）在使用渔药之前，应建立适当的体系以监控渔药的使用从而确定已用渔药的鱼类批次的停药时间；

（3）产品应在国家相关主管机构登记注册；

（4）渔药处方和分销应在国家法规规定下由持证人员进行；

（5）贮藏和运输条件应遵守标识说明；

（6）通过药物进行病害防治应在准确诊断的基础上进行；

（7）在水产养殖生产过程中，应保持渔药使用记录。屠宰前控制是一种控制鱼体内药物残留的方法。如果在测试鱼体内的平均药物聚集度超过 MRL（在某些国家，为工业强制的底线），对批次的收获必须推迟，直到其符合 MRL。对于所有未达到食品法典中渔药残留要求的鱼类产品在屠宰后控制中应拒收；

（8）体内的平均药物聚集度超过 MRL（在某些国家，为工业强制的底线）的鱼，对该批次的收获必须推迟，直到其符合 MRL。参考《良好水产养殖操作》对预收获措施进行评估后，应采取合适的措施对药物残留控制系统进行适当修改；

（9）对于所有未达到国家相关法规中渔药残留要求的鱼类产品在收获后的控制中应拒收。

6.3.3 养殖

潜在危害：微生物病原体和化学污染。

潜在缺陷：颜色不正常、泥腥味、机械损伤。

技术指南：

（1）后期幼苗、苗种和幼鱼的来源应受控以保证鱼群健康；

（2）放养密度应以养殖技术、鱼类品种、体长和年龄、水产养殖场容量、预期成活率以及预期的收获规格为基础；

（3）如果需要和可能的话，病鱼应隔离；而死鱼应立即以不会导致病害传播的卫生方式销毁，并调查其死亡原因；

（4）应通过确保养殖数量和投饲比率不超过养殖系统的养殖容量来维持良好水质；

（5）养殖水质应定期监控，以发现潜在危害和缺陷；

（6）水产养殖场应具有管理计划，其内容应包括卫生程序、监控措施和纠偏行动，农用化学品的正确使用，养殖操作的验收程序等，并保持系统的记录；

（7）水箱和网等设备的设计和制造应确保养殖期间的损伤达到最小化；

（8）所有的设备和支撑装置应便于清洗和消毒，并应该定期恰当的进行清洗和消毒。

6.3.4 收获

潜在危害：无。

潜在缺陷：机械损伤、由于活鱼受惊造成的身体或生理方面的改变。

技术指南：

（1）应采用尽量减少机械损伤和鱼类紧张的捕捞技术；

（2）不应使活鱼处于过冷或过热的环境，或遭受温度和盐度的突然改变；

（3）捕捞后，应尽快在用清洁的海水或淡水及适当的压力清洗鱼体，以使其不附着过多的泥和杂草；

（4）在需要的地方，鱼体应进行清洗，以减少在后续加工过程中鱼体内脏的杂物和污染；

（5）应依据本规范第四部分中的指南以卫生的方式处理鱼体；

（6）捕捞操作应迅速，以保证鱼体不会过度暴露于高温下；

（7）所有的设备和支撑装置应便于清洗和消毒，并应该定期恰当的进行清洗和消毒。

6.3.5 暂养和运输

潜在危害：微生物和化学污染。

潜在缺陷：机械损伤、由于活鱼受惊造成的身体或生理方面的改变。

技术指南：

（1）处理过程应尽量避免不必要的压力；

（2）鱼的运输不应过度延迟；

（3）活鱼运输设备的设计应使操作方便、快捷，并不造成机械损伤和鱼的紧张；

（4）所有的设备和支撑装置应便于清洗和消毒，并应定期恰当的进行清洗和消毒；

（5）应保持鱼的运输记录，以保证所有产品可追溯；

（6）鱼不应与可能导致污染的其他产品一起运输。

6.3.6 活鱼的贮存和运输

本节适用于养殖或捕捞活鱼的贮存和运输。

潜在危害：微生物病原体、生物毒素、化学污染（例如，油、清洁剂和消毒剂）。

潜在缺陷：死鱼、机械损伤、味道变差、由于鲜鱼受惊造成的身体或生化方面的改变。

技术指南：

（1）只挑选健康和未受损伤的鱼作为活鱼贮存和运输。在放入活鱼运输箱前，应清除受损伤的、患病的鱼以及死鱼；

（2）在贮存和运输期间，应定期检查暂养箱。受损伤的、患病的鱼以及死鱼一经发现应立即清除；

（3）用于注入运鱼箱、或在运鱼箱之间泵鱼类、或者用于调理鱼的清洁水在特性和成分上应与捕鱼处的水相似，以减少鱼的紧张；

（4）用水不应受人类生活污水和工业污染的影响。运鱼箱和运输系统应以卫生的方式设计和操作，以防止水和设备污染；

（5）在将鱼放入之前，活鱼运输箱中的水应很好充气；

（6）活鱼运输箱中使用海水时，对易被海藻毒素污染的品种，应避免使用细胞浓度高的海水，或应对其适当地过滤；

（7）活鱼在贮存和运输过程中不应进行投喂。投喂会迅速污染运输箱中的水，一般来说，运输前的24小时以内不再进行投喂；

（8）活鱼运输箱的材料、泵、过滤器、水管、温控系统、中间和最后的包装或容器不应对鱼有害或对人产生危害；

（9）所有设备和器具应定期和根据需要进行清洁和消毒。

6.3.6.1 活鱼的常温贮存和运输

潜在危害：微生物污染、生物毒素、化学污染（例如，油、清洁剂和消毒剂）。

潜在缺陷：死鱼、机械损伤、口味变坏、由于鲜鱼受惊造成的身体或生化方面的改变。

技术指南：

（1）根据水源、品种要求以及贮存或运输的时间，可能有必要循环用水，并通过机械或生物过滤器对水进行过滤；

（2）渔船甲板上运鱼箱进水口的位置，应避免被渔船污水、废弃物和排放的发动机冷却水所污染。当渔船进港或航行在靠近污水或工业排污水域时应避免抽水。在陆地上的进水也应采取相同的预防措施；

（3）鲜鱼的贮存和运输设备（运鱼箱）应能够：通过使水持续流动、直接充氧（使用氧气或气泡）或者定期和根据需要更换运鱼箱中的水，保持箱中水的溶解氧；对于对温度波动敏感的品种，保持贮存和运输温度。可能有必要对运鱼箱进行隔热处理并安装温控系统。准备备用水，万一运鱼箱必须排水时可能会需要。固定设施（仓库）的容量应至少与使用中的运鱼箱总容量相同。陆地运输设备的容量应至少能够补偿蒸发、泄漏、净化、过滤器清洗以及最终的为控制目的而混合水的水量；

（4）某些鱼种在受惊时，会出现诸如强烈领域性、同种相残或过度活跃等现象，这些鱼应分别放入单独的箱中或采取适当的保护措施（一种可以替代的方法是降低温度），以防损伤。

6.3.6.2　活鱼的低温贮存和运输

潜在危害：微生物污染、生物毒素、化学污染（例如：油、清洁剂和消毒剂）。

潜在缺陷：死鱼、机械损伤、口味变坏、由于鲜鱼受惊造成的身体/生化方面的改变。

技术指南：

（1）调理的目的是降低鱼体的新陈代谢速度，以减小鱼的（精神）压力。低温下鱼的调理应根据该品种的特性进行（最低温度、冷却速度、水分/湿度要求、包装条件）进行。调理是一种生物方法，用于减小对鱼的（精神）压力，降低其新陈代谢的速度；

（2）所要达到的温度应根据品种、运输和包装条件而定。在某一个温度范围内，鱼不会出现身体活力降低的情况。该温度限值即为当鱼的新陈代谢速度达到最小，且没有对鱼造成负面影响（基础新陈代谢速度）时的温度；

（3）在进行调理时，只能使用法规允许的麻醉剂和操作程序；

（4）经过调理的鱼类，应立即装入隔热容器；

（5）剩余的水或用于调理过的鱼的包装材料的水应清洁，其成分和pH值应与捕鱼处的水相类似，但温度应为贮存温度；

（6）用于包装经调理的鱼的吸水垫、碎木屑、刨花或锯屑以及栓系材料等应清洁，系首次使用，没有可能的危害，并在包装时弄湿；

（7）经过调理和包装的鱼，应在确保控制在适当温度的条件下进行贮存和运输。

第八部分　鲜鱼、冻鱼和碎鱼肉的加工

在意识到每个操作步骤的控制点之后，本部分给出潜在危害和缺陷的示例并且对技术指南做出了解释，这些有助于制定控制措施和纠正措施。在特定步骤中，只列出了可能引入的或控制住的危害和缺陷。应该认识到，在准备HACCP和（或）DAP计划时有必要参考第五部分，该部分提供了应用HACCP和DAP分析原则的指南。然而，在操作法典的范围内，由于存在不同的特定的危害和缺陷，不可能给出各个步骤的关键限值、监控、记录和验证等详细部分。

一般来讲，鲜鱼、冻鱼和碎鱼肉的加工很难分类。最简单的加工方式是可能将鲜鱼和冻鱼去皮、去内脏、切片，做成碎鱼肉，然后就在市场上销售或供加工厂使用。对于后者，鲜鱼、冻鱼和碎鱼肉的加工通常是生产高附加值产品（如第十二部分的熏鱼，第十六部分的鱼罐头，第十五部分的沾面包屑冻鱼或碎鱼）的中间步骤。在设计加工过程中，通常司空见惯的是使用传统方法。然而，现代科学食品工艺在延长产品的保质期和提高货架稳定性上具有不断上升的重要作用。不考虑特定加工过程的复杂性，目标产品的制造仍然依赖于各个步骤的连续成功操作。本规范特别强调的是，在这些步骤中必备程序（第三部分）和 HACCP 原则（第五部分）适当部分的应用，能为加工者提供合理的保证：遵守相应的食品法典标准规定的基本质量、成分和标签，从而控制食品安全问题。

流程图（图 8－1）提供鱼片生产线各个步骤的操作指南及三个最终产品类型的示例：气调贮藏（MAP）、碎鱼肉和冻鱼。由于气调贮藏鲜鱼、碎鱼肉或冻鱼的深加工，本部分在适当的部分使用了“鱼的准备”一词，并将其作为其他鱼类加工操作（见第九至第十六部分）的基础。

该流程图仅作说明使用。工厂实施 HACCP 的各个过程的流程图要完整全面的绘制出来。

参考本法典的相关部分。

8.1 鱼的准备

鱼的准备过程中各步骤的卫生条件和技术工艺相似，它们不会受预期用途（直接销售或再加工）影响。然而，鲜鱼肉的使用形式很多，这些形式可以包括（但不仅限于此）：去皮、去内脏、切片等。

8.1.1 生鱼、鲜鱼或冻鱼的接收（加工步骤 1）

潜在危害：微生物致病菌、存活的寄生虫、生物毒素、鲭鱼毒素、化学物质（包括渔药残留）及物理污染。

潜在缺陷：腐败、寄生虫、物理污染。

技术指南：

（1）对于生鱼原料，产品说明应包括下列特征：

感官特征：外形、气味、肌理等；

腐败和（或）污染的化学指数：如挥发性盐基氮（TVBN）、组胺、重金属、杀虫剂残留、硝酸盐等；

微生物指标，特别是中间原料，要防止其加工中含有微生物毒素；

外来异物；

形态特征，如鱼的大小；

品种一致；

（2）对鱼的处理者和相关人员要进行种类鉴别培训和传达产品规范的信息，以确保鱼的来源是安全的，并且应该有书面材料。要特别注意具有诸如热带和亚热带大型食肉鱼中的西加毒素、鲭鱼类中的鲭鱼毒素或寄生虫等生物毒素风险的鱼的接收和分拣；

（3）鱼的处理者和相关人员应学到鱼感官评价的技艺，以确保原料鱼符合相关法典标准规定的基本质量要求；

（4）在加工厂中需要除内脏的鱼，其加工过程应该高效进行，不得延误并且要小心以避免污染（见 8.1.5 部分－清洗或除内脏）；

（5）如果发现含有不能用常规分拣或准备程序消除或减少到可接受水平的有害物质、腐败和外来异物的鱼，应该拒收；

（6）获取原料收获地区的信息。

8.1.1.1 鱼类的感官评价

确定鱼新鲜或腐败的最好方法是感官评价技术[1]。我们推荐将适宜的感官评价指标用于评价鱼的可接受程度以消除相关法典标准规定的鱼的基本质量的损害。例如，当鲜白鱼品种出现下列特

[1] 鱼、贝类实验室感官评价准则（CAC/GL 31－1999）。

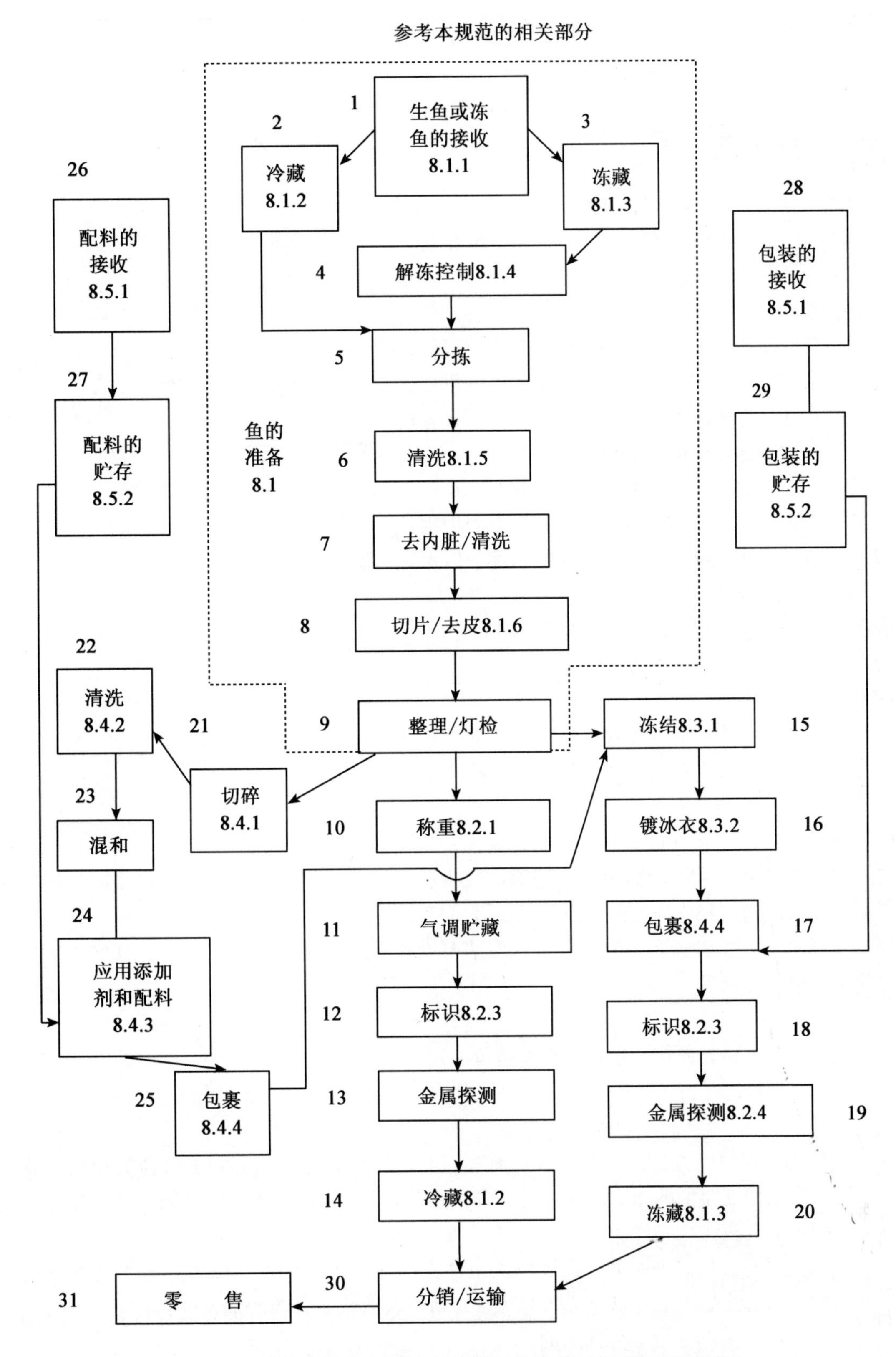

图 8－1　鱼片的生产流程图（包括气象调包装、切碎和冻结操作）

征，视为不能接受。

表皮/黏液	阴暗、沙砾色且具有棕黄色斑点状黏液
眼睛	凹陷、不透明、内陷处褪色
鳃	棕灰色或变白、黏液不透明呈黄色、黏稠或凝块状
气味	鱼肉呈胺味、氨味、酸味、硫化物味、粪便物味、腐烂味、恶臭味

8.1.2 冷却贮藏（加工步骤2和14）

潜在危害：微生物病原体、生物毒素和鲭鱼毒素。

潜在缺陷：腐败、物理损伤。

技术指南：

（1）应及时将鱼移至冷却贮藏设备中，不得延误；

（2）工厂应该有相应的设备保证将鱼的温度维持在0~4℃；

（3）应在冷却室中装有标准的温度计，强烈推荐用温度记录仪；

（4）存货周转计划应能保证鱼的适当利用；

（5）在加工前应将鱼贮存在薄冰层里并有充足的细碎冰或冰水混合物在其周围；

（6）保存鱼时要防止因装得太多太满带来的损害；

（7）要有适当冰源补充或改变房间的温度，以保证鱼温。

8.1.3 冻藏（加工步骤3和20）

潜在危害：微生物病原体、毒素、存活的寄生虫。

潜在缺陷：干耗、恶臭、营养质量损失。

技术指南：

（1）设备有能力维持鱼温度在-18℃或以下，并且只有最小的温度波动；

（2）贮存间应有标准的温度指示计，强烈推荐用温度记录仪；

（3）要制定且维持定期存货周转计划；

（4）产品应镀冰衣和（或）包裹以防止脱水；

（5）如果鱼含有在后续步骤中不能消除或减少到可接受水平的缺陷，就应该拒收。还应该进行适当的评估，以确定发生失控的原因并且修订DAP计划；

（6）为杀死对人类健康有害的寄生虫，冻结温度和对冻结时间的监控应当与良好存货控制结合，以确保足够的冷处理。

8.1.4 解冻（加工步骤4）

潜在危害：微生物病原体、生物毒素和鲭鱼毒素。

潜在缺陷：腐烂。

技术指南：

（1）应明确规定解冻方法，并且应该明确解冻的时间与温度，温度测量仪器的使用，放置测量仪器位置。应该仔细监测解冻时间表（时间和温度参数）。要根据解冻产品的尺寸大小、厚度仔细选择适合的解冻方法；

（2）应当选择适宜的解冻时间、温度以及鱼体温度作为关键限值，以防止微生物和某些高风险品种可能发生的组胺的产生，以及持久的特殊异味、腐败或恶臭的产生；

（3）作为解冻介质的水应是可饮用的；

（4）当水循环使用时，应小心避免微生物繁殖；

（5）当使用水时，水循环量应足以解冻；

(6) 解冻期间（根据所用的解冻方法）产品不要暴露在过高的温度下；

(7) 要特别注意控制冷凝水，并有效排水；

(8) 解冻后，鱼应该立即加工或冷藏并保存在适宜的温度当中（冰点温度）；

(9) 如有必要，解冻时间表应该及时评审和改进。

8.1.5 清洗和除内脏（加工步骤6和7）

潜在危害：微生物病原体、生物毒素和鲭鱼毒素。

潜在缺陷：内脏残存、擦伤、风味损失、错切。

技术指南：

(1) 当肠道和内部器官已除去时，认为内脏清除彻底；

(2) 应得到充足的干净海水或饮用水以清洗：清洗整鱼以去除外来残渣并减少除内脏前的微生物量；清洗除内脏后的鱼以除掉腹腔中的血和内脏；清洗鱼的表面以除掉松动的鳞；清洗除内脏设备以减少黏液、血和残渣的存留；

(3) 依靠船或加工厂的产品流动设备控制组胺或某缺陷而建立了贮存时间和温度关键限值时，除去内脏后的鱼应当沥水，并且在干净容器内冰冻或冷却，并且贮存在加工厂内适宜的指定区域；

(4) 鱼卵、鱼白和鱼肝，如果以后还要利用，则要提供单独的、适当的贮存设备。

8.1.6 切片、去皮、整理和灯检（加工步骤8和9）

潜在危害：存活的寄生虫，微生物病原体，生物毒素和鲭鱼毒素，骨刺。

潜在缺陷：寄生虫、骨刺、异物（皮、鳞等）、腐败。

技术指南：

(1) 为尽可能减少延误的时间，切片生产线和灯检生产线，应当连续不断且保证速度从而保证始终如一的速度，没有中断或放慢，并保证废物的处理；

(2) 应能得到充足的干净海水或饮用水以冲洗切片、切段前的鱼和去鳞后的鱼；冲洗切片前和切片时、除皮或整理时，以去除血迹、鳞和内脏等的痕迹；冲洗切片设备以最大限度地减少黏液、血和残渣的留存；为了销售的目的，鱼片中不能含有骨刺，鱼类加工者应该采用适合的检测技术和必要的工具，以去除不符合食品法典标准6、标准7及商业规范的骨刺。

(3) 由技术人员在照明效果最好的场所进行的对无皮鱼片的灯检，是控制（鲜鱼中）寄生虫的一项有效的技术。当使用可能含有寄生虫的鱼时，应采用上述操作；

(4) 应经常清洗灯检台，以最大限度地减少接触面的微生物活性和由于灯热烘干的鱼残渣；

(5) 如果为了控制组胺或某一缺陷建立了贮存温度和时间的关键限值时，鱼片应在干净的容器中冰冻或冷却，以防止干耗，并且贮存在加工区适当的区域。

8.2 真空或气调包装鱼类的加工

本部分是在鲜鱼加工过程部分的基础上，增加了气调包装鱼类加工过程的特殊的操作步骤（见附录I）。

8.2.1 称重（加工步骤10）

潜在危害：不可能。

潜在缺陷：净重不准确。

技术指南：

称量刻度要定期与标准物质进行校准以确保准确性。

8.2.2 真空或气调包装（加工步骤11）

潜在危害：后续的微生物病原体和生物毒素、物理污染（金属）。

潜在缺陷：后续腐烂。

技术指南：

真空或气调包装（MAP）的鱼类产品的货架寿命取决于种类、脂肪含量、初始细菌量、气体混合度、包装材料类型，贮存温度尤为重要。气调包装（MAP）的过程控制问题参见附件I。

（1）气调包装应通过以下方式严格控制：

—监测气体与产品比例；

—气体混合物类型和比例；

—薄膜的种类；

—密封的完整性和类型；

—产品贮存期间的温度控制。

（2）适宜的真空度和包装；

（3）鱼肉与接缝界面间应清洁；

（4）使用前应检验包装材料，保证其未受损或污染；

（5）最终成品包装的完整性，应该由培训过的人员定期进行检验，以验证密封的有效性以及包装机械是否正常运行；

（6）密封后，MAP或真空产品应当小心转运，并及时冷却贮存；

（7）确保合适的真空度，并且包装密封完好。

8.2.3 标签（加工步骤12和18）

潜在危害：不可能。

潜在缺陷：标签错误。

技术指南：

（1）使用标签前要验证标签，以确保所有内容满足《食品预包装标签通用标准》、《食品法典标准标签规定》以及其他相关的本国法律法规要求；

（2）有时，可能对标签不正确的产品重新标签。并进行适当的评估以确定标签不正确的原因并且必要时要修订DAP计划。

8.2.4 金属探测（加工步骤13和19）

潜在危害：金属污染。

潜在缺陷：不可能。

技术指南：

（1）生产线速度调整到与金属探测器性能相宜是很重要的；

（2）运行常规范序以保证金属探测器探测出应拒收的产品并找出拒收的原因；

（3）如果使用金属探测器，则要定期将其与公认标准进行校准以保证其正常操作。

8.3 冻鱼加工

本部分是在鲜鱼加工过程内容的基础上，增加了冻鱼加工过程的特殊操作步骤。

8.3.1 冻结过程（加工步骤15）

潜在危害：存活的寄生虫。

潜在缺陷：组织腐败、产生恶臭味、冻烧。

技术指南：

鱼产品应尽快冻结，因为冻结前不必要的延误，会引起鱼产品温度上升，由于微生物和不良化学作用而加快产品质量恶化、减少其货架寿命。

（1）应确定冻结时间和冻结温度，并且要考虑冻结设备和冻结能力；考虑鱼产品的特性包括热传导、厚度、形状、温度和生产量等，以确保尽可能快地通过最大冰结晶生成温度带；

(2) 进入冻结过程的产品厚度、形状和温度应尽可能均匀;

(3) 工厂生产量应适合于冻结能力;

(4) 为保证冻结彻底,应定期检查冻鱼的中心温度;

(5) 应经常检查以确保冻结过程操作正确;

(6) 应正确记录并保存所有冻结操作过程;

(7) 为了杀死所有对人体有害的寄生虫,冻结温度和冻结时间的监测应与良好存货控制相结合,以保证充足的冷处理。

8.3.2 镀冰衣(加工步骤16)

潜在危害:微生物病原体。

潜在缺陷:干耗、不正确净含量。

技术指南:

(1) 镀冰衣应彻底,冻鱼产品整个表面被冰层保护能够防止干耗(冻烧)发生;

(2) 如果镀冰使用添加剂,应小心保证其适合的比例并按产品说明使用;

(3) 当需要产品标签时,应保存产品或? 镀冰衣的重量或比例的信息,并用来确定除冰衣外的产品净重;

(4) 应监测喷嘴以确保其不会受堵;

(5) 用浸沾方法来镀冰衣时,重要的是要定期更换镀冰溶液以最大限度的减少细菌量和鱼蛋白的合成,从而减少其对冻结操作的影响。

8.4 碎鱼肉的加工

本部分在鲜鱼加工过程(鱼碎之前)和冻鱼加工过程(鱼碎之后)的基础上,额外增加了适合于碎鱼肉加工的特别操作步骤。

8.4.1 用机械分离法碎鱼(加工步骤21)

潜在危害:微生物病原体、生物毒素和鲭鱼毒素、物理污染(金属、骨头、分离机传送带上脱落的橡胶等)。

潜在缺陷:不正确的分离(例如,异物)、腐败、骨刺、寄生虫存在。

技术指南:

(1) 应连续不断的向分离机中供鱼,但供鱼量又不能过大;

(2) 对怀疑有寄生虫的鱼建议进行灯检;

(3) 应将打碎的鱼或鱼片放入分离机,以便于鱼的切割表面能够与分离机的有孔表面接触;

(4) 鱼的尺寸要与分选机的处理能力相匹配;

(5) 为避免机械调整时间的浪费和最终产品质量的变化,应分隔开不同种类和类型的原料鱼并且精心计划其分离加工批次;

(6) 应根据最终产品的特征要求调整分离机表面孔尺寸与施加在原料鱼上的压力;

(7) 分离的残料应当连续或半连续地转移到下一道工序;

(8) 温度监测应当保证避免产品温度上升过高。

8.4.2 鱼碎肉的清洗(加工步骤22)

潜在危害:微生物病原体和鲭鱼毒素。

潜在缺陷:色泽不正、肉质不良、水分过多。

技术指南:

(1) 如有必要鱼碎肉应该根据产品类型的需要进行充分的清洗;

(2) 应小心完成清洗过程的混搅,但应该尽可能温和以免鱼碎肉被过分打碎(这样会形成肉屑而减少产量）；

（3）清洗后的鱼碎肉可以用旋转滤网和离心设备脱水，通过压力使产品达到适当的湿度；

（4）如有必要，依据最终用途，可对脱水鱼碎肉进行过滤或将其乳化；

（5）特别应注意的是低温过滤；

（6）产生的废水应用适当方式处理掉。

8.4.3 鱼碎肉中添加剂和配料的混合及使用（加工步骤23和24）

潜在危害：物理污染、未被认可的添加剂和（或）配料。

潜在缺陷：物理污染、不正确的加入添加剂。

技术指南：

（1）如果在鱼中加入配料和（或）添加剂，应使混合比例恰当，以得到预期的感官质量；

（2）添加剂应符合法典食品添加剂通用标准的要求；

（3）制备好的鱼碎肉应包装并立即速冻，即使不立即速冻和使用也应将其冷冻处理。

8.4.4 包装（加工步骤17和25）

潜在危害：微生物病原体。

潜在缺陷：干耗、腐败。

技术指南：

（1）包装材料应干净、健康、耐用、可满足预期用途且是食品级材料；

（2）包装操作应尽可能减少交叉污染和腐败的风险；

（3）产品应满足标签和重量标准。

8.5 包装、标识和配料

8.5.1 接收——包装、标识和配料（加工步骤26和28）

潜在危害：微生物病原体、化学和物理的污染。

潜在缺陷：错误描述。

技术指南：

（1）只有配料、包装材料和标识与加工说明一致，才可接收。

（2）直接与鱼接触的标签应由防吸收材料制成，标签所用的墨水或染料应由法定官方机构许可。

（3）配料和包装材料如不是法定官方机构许可的，则应在接收时查明并拒收。

8.5.2 贮存——包装、标签和配料（加工步骤27和29）

潜在危害：微生物病原体，化学和物理污染。

潜在缺陷：包装材料或配料的质量下降。

技术指南：

（1）配料和包装应根据温度、湿度情况适当地贮存；

（2）应制定和维持一个系统的存货周转计划以免材料过期；

（3）应适当保护和隔离配料和包装以防止交叉污染；

（4）不得使用有缺陷的配料和包装。

第九部分 速冻鱼糜的加工

在意识到每个操作步骤的控制点之后，本部分给出潜在危害和缺陷的示例并且对技术指南做出了解释，这些有助于制定控制措施和纠正措施。在特定步骤中，只列出了可能引入的或控制住的危害和缺陷。应该认识到，在准备 HACCP 和（或）DAP 计划时有必要参考第五部分，该部分提供了应用 HACCP 和 DAP 分析原则的指南。然而，在操作法典的范围内，由于存在不同的特定的危害和缺陷，不可能给出各个步骤的关键限值、监控、记录和验证等详细部分。

速冻鱼糜是一种中间食品配料，它是通过对碎鱼肉不断冲洗和脱水而由鱼组织蛋白分离出来的肌原纤维鱼蛋白制成的。加入防冻剂以便鱼糜能冻结而且保持在解冻热加工后能形成凝胶的能力。冻鱼糜通常与其他成分混合并且经进一步加工而成为以鱼糜为基础的产品，例如，模拟蟹肉，就是利用了它可以形成凝胶的特性。

速冻鱼糜制作有多种方法，此流程图为最典型的过程。本流程图仅作说明，而对于工厂实施的 HACCP 计划，其每个加工过程都应有一个完整、全面的流程图（图 9－1）。

本部分主要给出利用海洋底层鱼类制造鱼糜时的指南，例如，用日本流行的机械加工方法加工阿拉斯加鳕鱼和太平洋白鱼，美国及其他一些国家也有加工者应用类似的机械加工方法。

大多数鱼糜由诸如阿拉斯加鳕鱼和太平洋白鱼等海洋底层鱼类加工而成。但是，技术的进步和鱼糜产品的原料鱼种类的变化，使定期修订操作法典本部分的内容变得十分必要。

9.1　速冻鱼糜危害和缺陷的一般说明

9.1.1　危害

速冻鱼糜是一种中间配料，经进一步加工可以制成以鱼糜为基础的产品（例如，模拟蟹肉）。一些潜在食品安全危害会在后续的加工过程中得到控制。例如，病原菌（例如，李斯特菌）、毒素形成菌（例如，肉毒梭菌）（因为成品的气调包装，它们成为危害），应该在最后的加工过程中的烹煮或巴氏消毒中加以控制；可能的金黄色葡萄球菌的污染（产生热稳定肠毒素）应当在预备程序以前得以充分控制；因为最终产品要煮熟或巴氏消毒，寄生虫不是危害。

金枪鱼、鲭鱼容易形成鲭鱼毒素，热带珊瑚鱼易聚集西加毒素，如果用它们来制作鱼糜，就要采取控制这些危害的适当措施。另外，由于鱼糜制造的高度机械化，要制定适当的控制措施，以保证在成品之中清除金属碎片（例如，轴承，螺钉、垫圈和螺母）。

一些国家用传统和非机械的方法，以当地的鱼为原料制作用于本地消费的鱼糜，就要重点考虑第三部分给出的预备程序。

9.1.2　缺陷

速冻鱼糜某些特性对于成功制作以鱼糜为基础的产品（例如，鱼肉丸或模拟蟹肉）（这可以满足消费者质量需求）是重要的。颜色、水分含量、pH 值和凝胶强度是这些重要因素中的一部分。这些因素和其他的要求在附录 X（冻鱼糜最终产品可选择性要求）中有更详细的描述。

黏孢子虫是一种海洋底层鱼（例如，太平洋白鱼）体内常见的寄生虫。这种生物含有蛋白酶，即使含量很少也可以分解化学蛋白影响鱼糜凝胶强度。如果在鱼肉丸或模拟蟹肉生产中使用含有这种寄生虫的鱼，需要加入蛋白酶抑制剂（例如，血浆蛋白或蛋白），以得到必要的凝胶强度。

腐败的鱼不能作为冻鱼糜生产的原料。其感官质量不能满足生产出可接受的模拟蟹肉产品要求。还有必须注意，腐败的鱼不能作为冻鱼糜生产的原料，是因为腐败细菌繁殖会使盐溶蛋白变性，从而对冻鱼糜凝胶的形成产生不利的影响。

清洗和脱水应充分以确保从肌原纤维蛋白中分离出水溶蛋白。如果水溶性蛋白存在于产品中，会对其凝胶形成能力及冻藏货架期造成不利的影响。

应当尽可能减少异物（例如，小骨刺、鳞及腹腔黑膜），因为它们会影响鱼糜在加工成成品时的使用性。

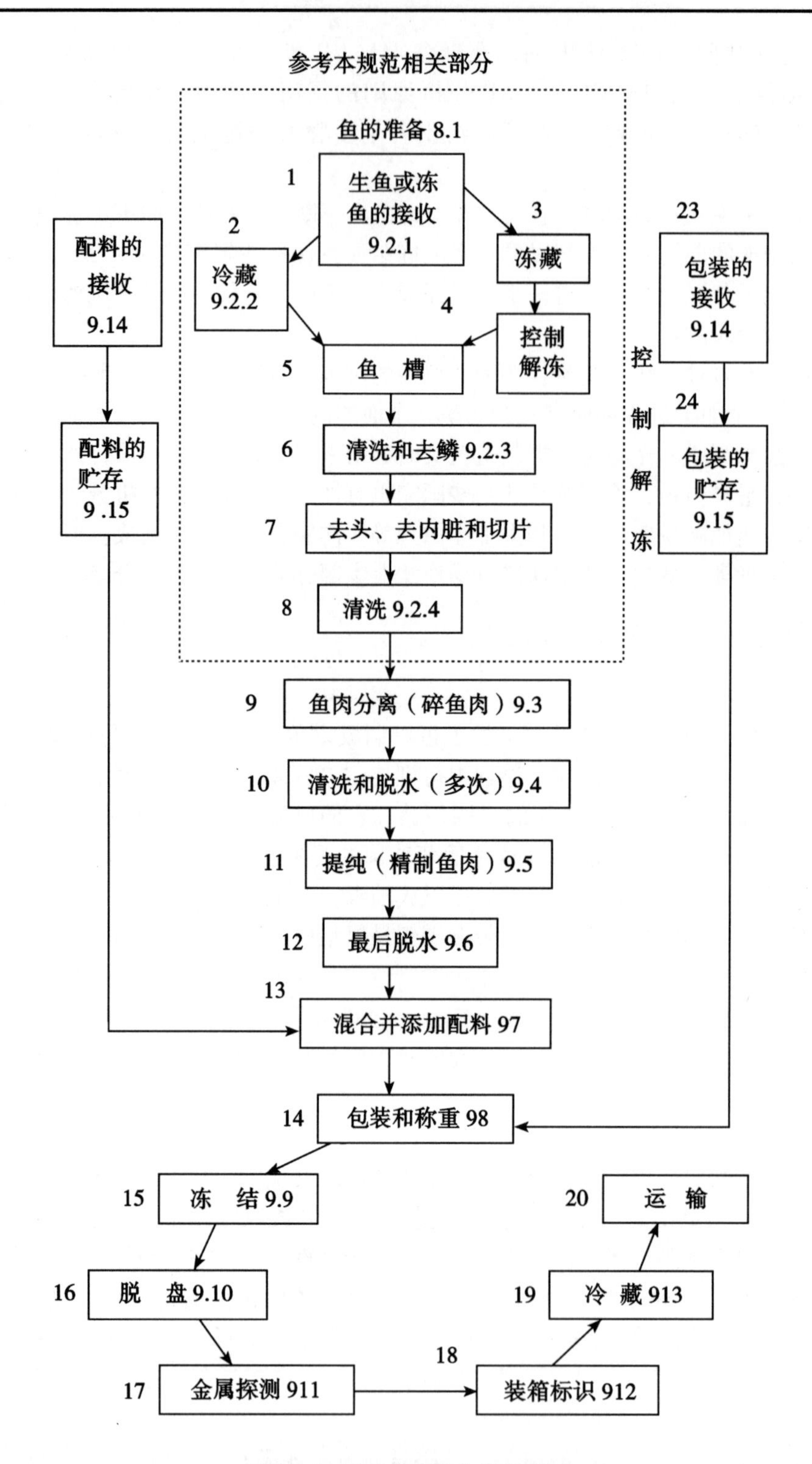

图 9－1 速冻鱼糜的生产流程图

由于原料鱼糜的分散特性，使用食品添加剂对于达到预期的质量水平是必要的。鱼糜中使用的添加剂应当符合相应法规和操作说明的规定，以免引起质量问题和纠正行动。

应当注意鱼蛋白的热稳定性。在正常的室温下，大多数蛋白会变性从而限制产品中凝胶形成的能力。对于阿拉斯加鳕鱼和其他冷水海洋鱼类来说，在加工中应保证温度在 10℃以下；温带鱼蛋白变性速率低，对温度敏感性差。

一些国家用传统的非机械化方法，以本地的鱼为原料，生产用于当地消费的速冻鱼糜，则要特别注意一些缺陷。因为腐败菌的生长会导致产品随温度的增加而腐败，并且导致其中的蛋白质发生变性，应仔细监测原料和加工过程中产品的状况。

9.2 鱼的准备（加工步骤1至8）

有关加工鱼的准备信息参见8.1部分步骤1至8。对于冻鱼糜的加工，应考虑下列每一步骤：

9.2.1 生鱼和冻鱼的接收（加工步骤1）

潜在危害：当用海洋底层鱼作为原料时不太可能。

潜在缺陷：腐败、蛋白质变性。

技术指南：

（1）原料鱼最好在捕捞后保存于在4℃或4℃以下的温度条件下；

（2）原料鱼的年龄和收获环境会影响产品的凝胶形成能力，因此应当注意。特别要注意生鱼收获后接收的时间。例如，收获后可接受期限如下（收获后尽快加工将能得到质量更好的鱼糜）：完整的；收获后14天内，贮存温度4℃或更低；去皮的；剥皮24小时后，贮存温度4℃或更低；

（3）收获日期、时间、产品来源、收获者或卖主应作好记录和鉴定工作；

（4）不允许原料鱼存在腐败，因为它对成品的凝胶强度有不利影响。恶劣条件下收获的鱼也不能得到外观良好的产品；

（5）原料鱼应当具有保证足够凝胶形成强度的新鲜程度。例如，阿拉斯加鳕鱼（狭鳕）pH值应在7.0±0.5；

（6）由于尺寸过大而被挤压、窒息而死以及收获后因保存时间过长而产生不良气味的鱼，应该从生产线上剔除，以避免对凝胶形成能力产生负面影响。

9.2.2 冷却贮存（加工步骤2）

潜在危害：不可能。

潜在缺陷：蛋白质变性。

技术指南：

（1）为了尽可能减少蛋白质变性及凝胶形成能力的损失，应缩短冷却贮藏过程并及时进行加工。

（2）生鱼应贮存在4℃或4℃以下，收获日期和接收的时间应能识别出用于加工的鱼的批次。

9.2.3 冲洗和去鳞（加工步骤6）

潜在危害：不可能。

潜在缺陷：蛋白质变性、色泽和异物问题。

技术指南：表皮（黏液层）、鱼鳞、色素应在去头和去内脏前除掉。这样可以减少杂质和外来异物，有利于凝胶强度和最终产品色泽形成。

9.2.4 清洗（加工步骤8）

潜在危害：不可能。

潜在缺陷：杂质、外来异物。

技术指南：去头和去内脏后的鱼应再次清洗。这样将减少杂质和外来异物，有利于凝胶强度和最终产品的色泽。

9.3 鱼肉分离加工（加工步骤9）

潜在危害：金属碎片。

潜在缺陷：杂质。

技术指南：

（1）鱼肉是用机械分解过程打碎的，因此应当将金属探测设备（用来探测产品从而检测出其大小可能伤害人类的金属碎片）安装在生产线上最适当的位置从而消除危害；

（2）应建立程序以保证不可能发生产品化学污染；

（3）分离后的鱼碎肉应立即放入水中并且转到清洗和脱水步骤，以防止凝结的血融化造成凝胶形成能力的损失。

9.4 清洗和再脱水（加工步骤10）

潜在危害：致病微生物生长。

潜在缺陷：腐败、蛋白质变性、水溶性蛋白残留。

技术指南：

（1）应得到有效控制旋转滤筛中的水和碎鱼肉以及冲洗水的温度，以防止致病微生物的生长；

（2）为使水溶蛋白分解，冲洗水的温度应在10℃以下；太平洋白鱼冲洗水温度应低于5℃，因为这种鱼通常具有高蛋白酶活性；某些温水鱼可以在15℃以上的温度下加工；

（3）产品应尽快加工以最大限度地减少可能的致病微生物生长；

（4）碎鱼肉应均匀地放置于水中，以确保稀释水溶性杂质并且彻底分离肌原纤维蛋白；

（5）清洗和脱水步骤的设计，应当注意根据所要求的产量、质量及鱼种；

（6）应能得到充分的清洗用的饮用水；

（7）清洗的水pH值应接近7.0。清洗水硬度最好在100mg/kg以下（根据改性$CaCO_3$）；

（8）在清洗的最后阶段可以加入盐（盐度小于0.3%）或其他脱水辅助剂以提高脱水效率；

（9）食品添加剂的使用要遵守国家法规和生产说明；

（10）废水应以适当的方式处理；

（11）清洗用水不得循环使用，除非有适当的微生物控制措施。

9.5 精制过程（加工步骤11）

潜在危害：致病微生物生长、金属碎片。

潜在缺陷：异物、蛋白质变性。

技术指南：

（1）应适当控制提纯过程中碎鱼肉温度以防止微生物的生长；

（2）为了防止蛋白质变性，碎鱼肉温度不能超过10℃；

（3）产品应尽快加工以尽可能减少致病微生物的生长；

（4）金属探测设备（用来探测产品从而检测出其大小可能伤害人类的金属碎片）安装在生产线上最适当的位置从而消除危害；

（5）在进入最后脱水之前，应当从精制设备冲洗后的鱼肉中去除异物（例如，小骨刺、黑膜、鱼鳞、血肉和连接组织）；

（6）设备要适当调整到与生产量匹配；

（7）精制的产品不允许在过滤筛子处，长时间聚集。

9.6 最后脱水过程（加工过程12）

潜在危害：致病微生物生长。

潜在缺陷：腐烂蛋白质变性。

技术指南：

（1）应适当控制最后脱水过程鱼肉温度，以防止致病菌生长；

（2）冷水鱼种如阿拉斯加鳕鱼的鱼肉温度不要超过10℃，对太平洋白鱼的鱼肉温度不要超过5℃，因为该种鱼常具有较高的蛋白质活性，某些热带鱼可以在15℃以上加工；

（3）产品应尽快加工以最大可能减少致病微生物生长；

（4）应使用适当的脱水设备（例如，离心分离，水平、螺旋压榨）将精制产品的湿度控制到一定水平；

（5）应当考虑到因鱼龄、原料鱼捕捞条件和形式状态而造成的不同的湿度水平。某些情况下精制前应脱水。

9.7 辅配料混合添加（加工步骤13）

潜在危害：致病微生物生长、金属碎片。

潜在缺陷：食品添加剂的不适当使用、蛋白质变性。

技术指南：

（1）混合过程的产品温度应适当控制以避免致病菌生长；

（2）对冷水鱼类如阿拉斯加鳕鱼来说，混合过程中脱水鱼肉的温度不应超过10℃；对于太平洋白鱼温度不能超过5℃，因为该种鱼常具有较高的蛋白酶活性；某些温水鱼的加工温度可以高于15℃；

（3）产品应尽快加工以尽可能减少可能的致病微生物的生长；

（4）金属探测设备（用来探测产品从而检测出其大小可能伤害人类的金属碎片）安装在生产线上最适当的位置从而消除危害；

（5）食品添加剂要符合《食品添加剂通用标准》；

（6）食品添加剂应混合均匀；

（7）防冻剂可以用于冻鱼糜中。糖和（或）酒精经常用来防止冻结状态下的蛋白质变性；

（8）食品级酶抑制剂（例如，蛋清、牛肉蛋白血浆）可以用于某些水解蛋白酶活性较高的鱼（会降低鱼肉丸或模拟蟹肉中的凝胶形成能力），例如，太平洋白鱼。应适当标识蛋白血清的使用。

9.8 包装和称量（加工步骤14）

潜在危害：致病微生物的生长、交叉感染。

潜在缺陷：外来异物（包装）、净含量不正确、包装不完整、蛋白质变性。

技术指南：

（1）应充分控制包装过程中产品温度以免致病菌生长；

（2）产品应尽快包装以便最大限度地减少可能的致病菌的生长；

（3）应建立包装操作程序避免交叉污染；

（4）产品应装于干净塑料袋或包装在干净的容器中，贮藏适当；

（5）产品形状应适宜；

（6）要进行快速包装以尽可能减少污染或腐败的风险；

（7）包装后的产品不得含有空气；

（8）产品净重应符合标准。

参见8.2.1部分“程序”和8.4.4部分“包装”。

9.9 冻结操作（加工步骤15）

水产品冻结的通用要求参见8.3.1部分。

潜在危害：不可能。
潜在缺陷：蛋白质变性、腐败。
技术指南：
（1）产品包装和称量后应迅速冻结以维持产品质量；
（2）应建立程序并详细说明从包装到冻结的最大时间值。

9.10 脱盘（加工步骤16）

潜在危害：不可能。
潜在缺陷：塑料袋、产品有损伤。
技术指南：应小心避免塑料袋和产品本身破损，以避免在长时间贮存中发生严重干耗。

9.11 金属探测（加工步骤17）

参见8.2.4部分“金属探测”的一般信息。
潜在危害：金属碎片。
潜在缺陷：不可能。
技术指南：金属探测设备（用来探测产品从而检测出其大小可能伤害人类的金属碎片）安装在生产线上最适当的位置从而消除危害。

9.12 装箱和标签（加工步骤18）

参见8.2.3部分“标识”和8.4.4“包装”。
潜在危害：不可能。
潜在缺陷：标识不正确、包装损坏。
技术指南：
（1）箱子要干净、耐用并且适合预期用途；
（2）进行装箱操作要避免包装材料损坏；
（3）箱子损坏的产品应重新装箱。

9.13 冻藏（步骤19）

参见8.1.3部分“冻藏”可以了解水产品冻藏的一般信息。
潜在危害：不可能。
潜在缺陷：腐败、蛋白质变性。
技术指南：
（1）冻鱼糜应贮存在-20℃或-20℃以下的温度下，以防止蛋白质发生变性，如果贮存在-25℃以下质量更好、货架寿命更长；
（2）冻结时，良好的空气循环以确保冻结彻底。这包括在贮藏时防止直接将产品放在冻结室的地板上。

9.14 原料接收——包装和配料（加工步骤21和步骤22）

参见8.5.1部分“原料接收——包装、标签和配料”。

9.15 原料贮存——包装和配料（加工步骤23和步骤24）

参见8.5.2部分“原料贮存——包装、标签和配料”。

第十部分　速冻裹面鱼制品的加工

在意识到每个操作步骤的控制点之后，本部分给出潜在危害和缺陷的示例并且对技术指南做出了解释，这些有助于制定控制措施和纠正措施。在特定步骤中，只列出了可能引入的或控制住的危害和缺陷。应该认识到，在准备 HACCP 和（或）DAP 计划时有必要参考第五部分，该部分提供了应用 HACCP 和 DAP 分析原则的指南。然而，在操作法典的范围内，由于存在不同的特定的危害和缺陷，不可能给出各个步骤的关键限值、监控、记录和验证等详细部分。

图 10－1 只用于说明目的。为在工厂实施 HACCP 原则，应为每个加工过程绘制一个完整和综合的流程图。可参照本规范相关部分。

10.1　必需加工步骤的一般原则

（1）设计和建造未裹面和裹面鱼制品的运输系统，以防止损伤或污染产品；

（2）用于加工成形鱼制品的薄片在锯切和褪火过程中，应注意控制温度以防止产品基本质量的下降；

（3）如果全部加工过程应连续不断地进行，则应具有数量充足的生产线，以避免出现加工中断和不连续现象。如果加工过程必须中断，中间产品必须在深度冻结条件下贮存直至进行下一步加工；

（4）预炸箱、用于再冻结的冷冻柜应设有永久的温度和带速控制装置；

（5）应使用合适的锯切设备使锯屑的比例降到最小；

（6）锯屑应与用于裹面产品的鱼块部分隔离开，应控制其温度，常温下放置时间不应过长，并且在进一步加工成合适的产品之前，最好应贮存在冷冻状态下。

10.2　危害和缺陷的确定

参考第 5.3.3 部分和附录 XI。

本部分描述了速冻裹面鱼和贝类产品特有的主要危害和缺陷。

10.2.1　危害

参考第 5.3.3.1 部分。

用于鱼块、鱼片等的面糊，其生产和贮存可能包括商业面糊混合物的复水处理或者用原配料制备。在制备和使用面糊的过程中，必须控制可能出现的金黄色葡萄球菌和芽孢杆菌的繁殖和产生毒素等潜在危害。

10.2.2　缺陷

潜在缺陷是根据相关的《裹面包屑或挂面糊的速冻鱼柳（鱼条）、鱼块和鱼片的法典标准》（CODEX STAN. 166－1989）中规定的基本质量、标签和成分等要求给出的。

附录 XI 中给出的最终产品说明描述了专门针对速冻裹面水产品的非强制性要求。

10.3　加工工艺

参考图 10－1，裹面鱼制品加工流程图示例。

10.3.1　接收

10.3.1.1　鱼

潜在危害：化学和生物化学污染，组胺。

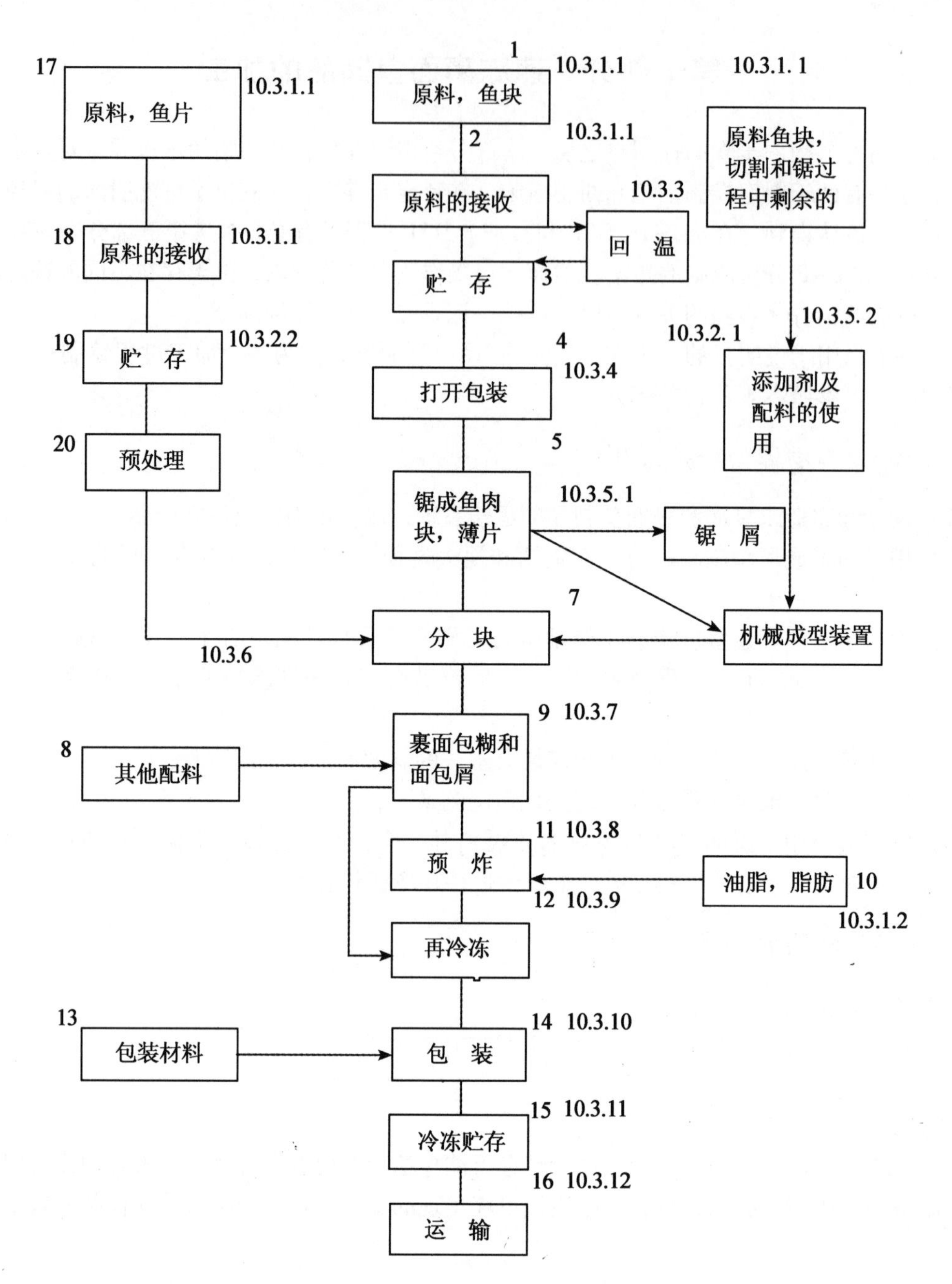

图 10－1 裹面鱼制品生产加工过程流程图示例

潜在缺陷：污渍、鱼块不规整、包装材料中进水或进气、外来异物、寄生虫、脱水、腐烂。

技术指南：

（1）应记录所有接收批次的温度；

（2）应对冷冻产品包装材料进行污垢、裂口以及是否化冻等进行检查；

（3）应检查运送冷冻鱼类产品的运输工具的清洁度和适用性；

（4）推荐在运输过程中使用温度记录设备；

（5）应抽取具有代表性的样品，以进一步检测可能的危害和缺陷。

10. 3. 1. 2 其他配料

潜在危害：化学、生物化学和微生物污染。

潜在缺陷：霉变、变色、不洁、沙子。

技术指南：

（1）应检查面包屑和面糊中有无包装材料碎片、啮齿动物啃咬和出现昆虫的痕迹，以及包装材料上的污垢和潮湿等损坏情况；

（2）应检查冷冻食品运输车的清洁度和适用性；

（3）应抽取具有代表性的配料样品进行检测以确保产品未被污染并达到应用于成品的要求；

（4）应使用适于处理食品和配料的运输车运送配料。先前运载过潜在不安全或有害物质的车不应用来运载食品和配料。

10.3.1.3　包装材料

潜在危害：外来异物。

潜在缺陷：污染产品。

技术指南：

（1）使用的包装材料应清洁、健康、耐用、满足其预期用途并且为食品级材料；

（2）预炸产品的包装材料应不应渗透脂肪和油脂；

（3）应检查食品包装材料运输车的清洁度和适用性；

（4）应检查预打印的标签和包装材料的准确性。

10.3.2　原料、其他配料和包装材料的贮存

10.3.2.1　鱼（冷冻贮存）

参见第8.1.3部分。

10.3.2.2　鱼（冷藏贮存）

用于非冷冻鱼类产品的贮存，参见8.1.2。

10.3.2.3　其他配料和包装材料

潜在危害：生物、物理和化学污染。

潜在缺陷：配料质量和特性丧失、腐臭。

技术指南：

（1）所有其他配料和包装材料应在在卫生条件下，贮存于干燥清洁处；

（2）所有其他配料和包装材料应在合适的温度和湿度下贮存；

（3）应制定并遵守系统化的库存周转计划，以避免物料过期；

（4）配料的贮存应防范昆虫、啮齿动物和其他害虫；

（5）不应使用有缺陷的配料和包装材料。

10.3.3　冷冻鱼块/片的回温

潜在危害：无。

潜在缺陷：由于锯切过软的鱼肉，造成尺寸不正确（对于鱼块）。

技术指南：

（1）根据鱼的用途，冷冻鱼块/鱼片的褪火应保证鱼体温度上升但不会融化。

（2）在冷藏库中对冻鱼块/片进行褪火是一个通常需要至少12个小时或更长时间的缓慢过程。

（3）外层过分软化是不合要求的（锯切过程不易操作），应避免出现。如果褪火设备的温度维持在0~4℃，并且鱼块/片分层堆放，则可避免。

（4）微波褪火是一种可供选择的方法，但同样应该控制，防止外层过分软化。

10.3.4　打开包装

潜在危害：微生物污染。

潜在缺陷：残存未发现的包装材料、污染。

技术指南：

（1）在打开鱼块包装的过程中，应小心处理以防污染鱼体；

（2）特别注意避免将纸板和（或）塑料材料的一部分或全部嵌入鱼块中；

（3）所有包装材料应快速和妥当地处置；

（4）在休息期间核换班期间对加工生产线进行清洁和消毒时，应对带包装的鱼块和已打开包装的鱼块加以保护。

10.3.5 鱼肉块的保护

10.3.5.1 锯切

潜在危害：外来异物（锯的金属或塑料部件）。

潜在缺陷：外型不规整的片或块。

技术指南：

（1）锯切工具应保存在清洁卫生的条件下；

（2）必须对锯刃进行定期检测，以防止撕裂产品或破损；

（3）如果要将锯屑用于进一步加工，不能将锯屑堆积在切锯台上，而必须使用专用的容器收集；

（4）当锯下的薄片用于机械压制做成外型不规则的鱼块时，则在进一步加工前，应保存在清洁、卫生的条件下。

10.3.5.2 添加剂和配料的使用

可参考第8.4.3条。

潜在危害：外来物质、微生物污染。

潜在缺陷：添加剂的不正确添加。

技术指南：应充分控制混合过程中产品的温度，以防止病原菌的生长。

10.3.5.3 成型

潜在危害：外来物质（来自机器的金属或塑料）和（或）微生物污染（专指鱼类混合物）。

潜在缺陷：成型不好的鱼肉块，鱼肉块受到过大压力（糊状、腐臭）。

技术指南：

（1）鱼肉块成型是一种高机械化方法生产用于挂糊和黏面包屑的鱼肉块。既可利用液压方法将薄片（从鱼块锯下的部分）压入模具中，并送到传送带上，或机械成型鱼肉混合物。

（2）成型机械应保持卫生；

（3）应严格检查成型后的鱼肉块是否具有合格的外型、重量和肉质。

10.3.6 鱼片的分离

潜在危害：无。

潜在缺陷：附着的片或小块。

技术指南：

（1）从鱼片冻块或其他不规则形状的速冻鱼肉原料上锯下的鱼肉块必须彼此分开，不应互相粘连；

（2）经过挂糊程序时，彼此接触的鱼肉块应挑出来，并重新放回传送带，以使其均匀地挂糊和黏面包屑；

（3）裹面前，应监控鱼肉块中是否有外来异物以及其他危害和缺陷；

（4）从生产中去除任何破损、外型不好以及不合规格的产品。

10.3.7 裹面

在工业生产中，裹面操作步骤的顺序和数量可能与本部分内容不同。

10.3.7.1 挂糊

潜在危害：微生物污染。

潜在缺陷：裹面不充分或过多。

技术指南：

（1）应使鱼块各面充分挂糊；

（2）应再次使用的剩余液体，再次运送的条件应干净卫生；

（3）鱼块上的多余液体应用清洁空气去除；

（4）应控制水合面糊混合物的黏性和温度在一定范围内，以使黏起的面包屑数量恰到好处；

（5）为避免水合面糊被微生物污染，应采用适当措施以确保不出现显著的微生物生长，如温度控制、控制加入的液体内容物，以及在换班期间定期或按计划清洁和/或卫生。

10.3.7.2 裹干面

潜在危害：微生物污染。

潜在缺陷：裹面不充分或过多。

技术指南：

（1）干面必须覆盖整个产品，并且与面糊粘连良好；

（2）应使用清洁空气吹风或（和）通过传送带的振动去除多余的面糊，如果面糊还要进一步使用，则去除方式应清洁卫生；

（3）加料斗中流出的面包屑应流动自由、均匀和连贯；

（4）应监控裹面过程的缺陷并符合《法典标准的冷冻鱼条、鱼块和鱼片－裹面包屑或挂面糊》（Codex Standard 166－1989）；

（5）面包屑和鱼肉块的比例应符合《冷冻鱼柳、鱼块和鱼片－粘面包屑或裹面糊法典标准》（Codex Standard 166－1989）。

10.3.8 预炸

目前，工业生产的速冻裹面产品经完全油炸（包括鱼肉块），然后再冷冻这一工艺有一些变化。因此，必须描述不同的危害和缺陷，而且，并非所有本部分的内容都适用。在一些地区，生产生的（未预炸）裹面鱼制品是常见的做法。

潜在危害：无。

潜在缺陷：过度氧化的油脂、煎炸不充分、裹面黏附松动、炸糊的块。

技术指南：

（1）煎炸用油的温度应在160℃和195℃之间；

（2）裹面鱼块应根据油温，煎炸足够的时间，以获得满意的色泽、风味，并使裹面与鱼肉块附着紧密。但在整个过程，鱼肉块应保持冷冻状态；

（3）当煎炸用油颜色变黑或脂肪降解物浓度超过某个限值时则必须加以更换；

（4）应定期去除在煎炸锅底部聚集的裹面残留物，以防止由于油的上涌而导致裹面产品部分变黑；

（5）预炸后，应使用适当设备去除裹面产品上过多的油。

10.3.9 再冷冻 最终冷冻

潜在危害：外来物。

潜在缺陷：冷冻不充分导致产品粘在一起或粘在冻结设备的壁上，容易造成面包屑或面糊的脱离。

技术指南：

（1）产品预炸后应立即进行再冷冻，并使整个产品温度全部降低到－18℃或更低；

（2）应将产品在冷冻室中放置足够长的时间，以保证其中心温度为－18℃或更低；

（3）低温冷冻设备应具有充足的压缩气流以对产品进行彻底的冷冻；

（4）操作人员在使用风冷冻结器进行冷冻加工前，可将产品装入零售包装中。

10.3.10 包装和标签

参考第8.2.3部分“标签”、第8.4.4部分“包装和装箱”以及第8.2.1部分“称重”。

潜在危害：微生物污染。

潜在缺陷：包装不足或过度、密封不好的容器、错误或误导的标签。

技术指南：

（1）再冷冻后，应立即在清洁卫生的条件下进行包装。如果不马上进行包装（如成批次包装）则应将再冷冻产品保存在低温冷冻条件下，直到进行包装时；

（2）应定期对包装进行重量监测，对成品使用金属探测器和（或）其他适用的检测方法进行检查；

（3）应在卫生条件下，立即将硬纸箱或塑料袋等包装装入大型集装箱；

（4）零售包装和集装箱上都应正确标注批号，以便于出现产品召回情况时，进行产品追溯。

10.3.11 成品的贮存

参见第8.1.3部分。

潜在危害：无。

潜在缺陷：由于温度波动而造成的变质和变味、深度冻烧、冷藏味、纸板味。

技术指南：

（1）所有成品应冷冻贮存于清洁、牢固、卫生的环境中；

（2）应避免严重的贮存温度波动（大于3℃）；

（3）避免贮存时间过长（根据所用鱼类品种的脂肪含量和裹面的类型）；

（4）应适当防止产品出现干耗、污垢以及其他形式的污染；

（5）所有成品应在空气循环良好的冷库中贮存。

10.3.12 成品的运输

详见第3.6部分“运输”和第17部分“运输”。

潜在危害：无。

潜在缺陷：冷冻产品化冻。

技术指南：

（1）在运输期间应维持深度冷冻条件，即维持温度在－18℃（最大温度波动±3℃），直到将产品运至目的地；

（2）应检查冷冻食品运输车的清洁度和适用性；

（3）推荐在运输过程中使用温度记录装置。

第十四部分 虾和对虾的加工处理

范围：整个的或去头的或生的无头的，去皮的，去皮和去内脏的或者在捕获的船只或加工船只或海岸上的加工厂煮熟的为进一步加工而冷冻的虾。

为了识别每一个加工步骤的控制方法，本节将举例说明潜在的危害和缺陷，并提供技术指南，以便采取控制措施和纠正措施。在每一个步骤中，只列举本步中可能引入或控制的危害和缺陷。应该认识到，在准备一个HACCP和（或）DAP计划时，有必要查阅第五部分，该部分对HACCP和（或）DAP分析的应用提供了指南。然而，在本操作规范的范围内，不可能列举每一个步骤的关键限制，监测，记录保持和验证等详细内容，因为这些都是针对特定危害和缺陷的具体内容。

14.1 冷冻虾和对虾——概述

（1）作为冷冻产品的虾来源广泛。有的来自深海、热带浅海、河流，也有的来自热带和亚热带养殖区。

（2）捕捞/收获和加工的方法同样多种多样。北方的虾可能会被现代化的大型冷冻船捕获、蒸煮并在船上速冻、包装。但更常见的是，这些虾在船上仍是生的单体速冻形式，送到岸上的工厂后再进一步加工，或上岸后才冷冻。这些品种都由岸上的工厂一律经过生产线先进行预蒸煮，再进行机械去壳、蒸煮、冷冻、镀冰衣及包装。在热带和亚热带地区，一个更大的产品线生产包括野生捕获的和养殖的对虾科虾类：整条的，去头的、去皮的、去皮和去内脏的生/熟产品，表现为不同的商品形式（易去皮的，带尾的，去尾的，清洗的，去肠的，寿司虾）。这些范围广泛的产品是在虾加工厂准备好的，这些工厂可能很小且使用手工技术，或者很大，且用大型的机械设备。蒸煮的虾产品通常在蒸煮以后去皮。

（3）温水虾也可以进一步加工增值，例如，进行浸泡调味、裹面包屑和加食品屑涂层。

（4）既然一些生虾产品以及烹调过的虾产品可能会在未进一步加工前被消费，那么安全问题是首要考虑的问题。

（5）上述加工过程已被描绘在两张流程图上，但一定要认识到，由于各种生产方式的不同，必须为每一种产品设计其特有的 HACCP/DAP 方案。

（6）除了上述对船上蒸煮的描述外，没有在其他在海上或在养殖场加工虾的参考资料。通常都认为产品按照操作规范中有关部分的规定进行正确加工处理，以及在产品送往工厂之前，进行适当的预处理，例如，去头。

14.2 虾的制备

14.2.1 生鲜虾和冻虾的接收（加工步骤）

潜在危害：植物毒素（如 PSP）、微生物污染、抗生素、亚硫酸盐、杀虫剂、燃料油（化学污染）。

潜在缺陷：货批质量不一、品种混杂、腐败变质、黑斑、头部酶作用而变软、腐烂。

技术指南：

（1）检查协议应包括商定的质量，HACCP 及 DAP 计划参数，并对承担这项工作的检查人员进行适当的培训；

（2）接收虾时应进行检验，确保冰冻或者深度冷冻状况良好而且有合适的记录以确保其可追溯性；

（3）从虾的来源及其过去已知的历史可以查出海捕虾的植物毒素或养殖虾的潜在抗生素等，尤其在供应商不提供保证书时。另外，可以应用其他化学指标来检查重金属和杀虫剂，用 TVBN 值来检查腐败程度；

（4）应当将虾保存在专用的合适的设备中，并及时进行加工以保证质量参数符合终产品要求；

（5）对大宗来料虾时，应该在其捕获时监测亚硫酸盐；

（6）对大宗来料虾要进行感观评价以确保产品具有可接收的质量且没有腐烂分解；

（7）接收到新鲜的虾后，有必要在一系列低速喷射的干净的冷水设备中用对其进行清洗。

图 14－1 仅用作解释问题的目的。对于在工厂里实施 HACCP 而言，必须为每一个加工过程制作全面的而详细的流程图。

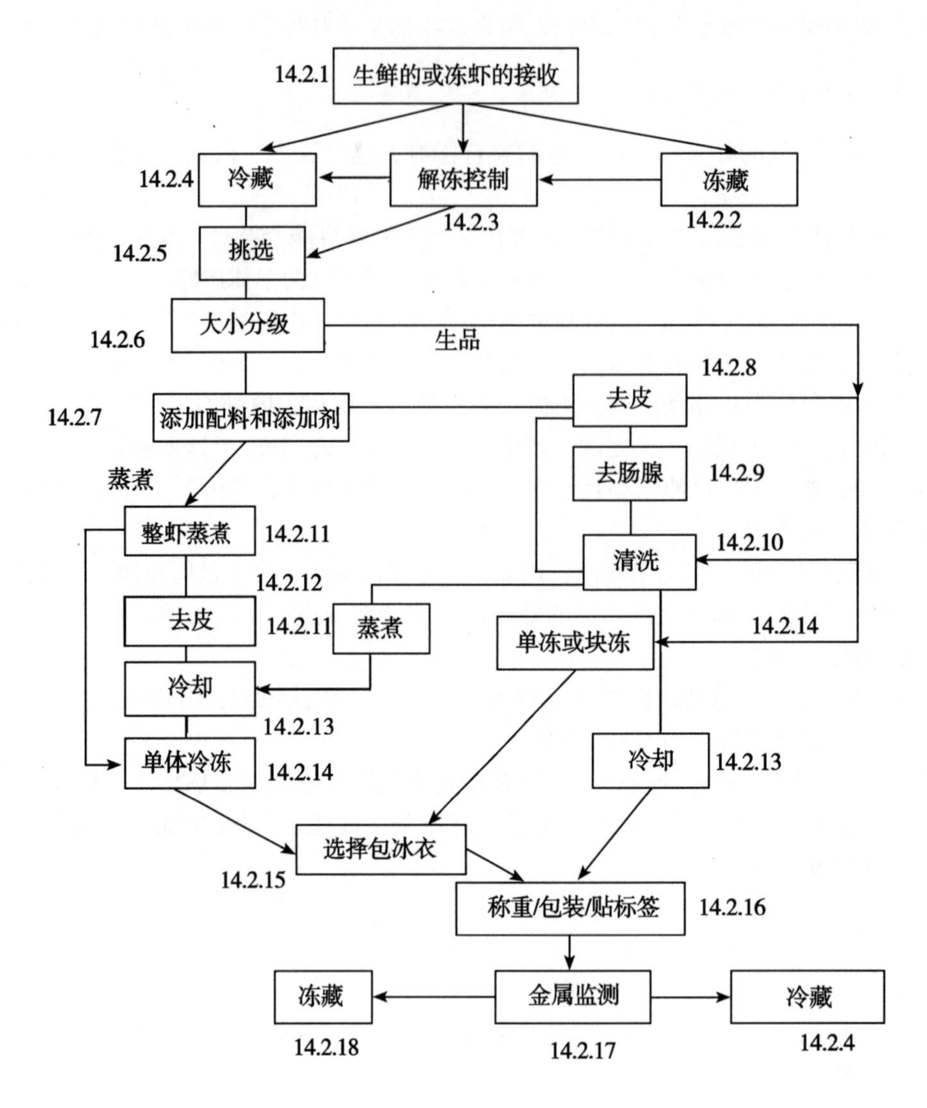

图 14－1 虾和对虾加工链流程图

14.2.2 冷冻贮藏

潜在危害：不太可能有。

潜在缺陷：蛋白质营养下降，脱水。

技术指南：

（1）保护性包装不能损坏，否则要重新包装以排除可能造成的污染和脱水；

（2）冷藏温度应适宜，波动应限制在最小范围之内；

（3）产品尽可能在包装前的最佳时间内进行加工或在接收时规定的时间之前进行加工；

（4）冷藏设备应该有温度监测装置，最好有连续的记录装置，用于合理监测和记录环境温度。

14.2.3 解冻控制

潜在危害：微生物污染、包装污染。

潜在缺陷：腐烂。

技术指南：

（1）解冻过程可以根据原料的不同分别进行，对块冻虾或单冻虾在解冻前去掉内外包装，以防止污染，对块冻对虾要格外注意，因为里面的蜡和聚乙烯包装可能会渗入虾体内；

（2）解冻池应专门设计，同时还要考虑通过逆向水流除霜，以保持解冻池必需的可能最低温度，但水不可再利用；

（3）解冻水应为清洁的海水或可饮用的淡水或冰水。加入适量冰使水温不高于20℃（68 ℉）；

（4）为了保证质量，解冻应尽可能快地完成；

（5）对连接到解冻池出口的输送器而言，较理想的做法是装备一系列低速喷水装置，用冷却的干净水清洗虾；

（6）解冻后，应立即将虾重新冰冻或放在冷藏环境中，以避免在进一步加工前引起温度变化。

14.2.4 冷藏

潜在危害：微生物污染。

潜在缺陷：腐烂。

技术指南：

（1）冷藏，最好在接收后放置在冷藏室的冰下，室温要低于4℃；

（2）冷藏设备应带有温度监测装置（最好是连续记录单位），用来更好监测和记录周围的温度。

14.2.5 挑选

潜在危害：不太可能有。

潜在缺陷：腐烂。

技术指南：根据规格要求，按照不同的质量等级对虾进行挑选。这个过程应尽量短，然后将虾重新进行冰冻。

14.2.6 大小分级

潜在危害：微生物污染。

潜在缺陷：腐烂。

技术指南：

（1）将虾按大小进行分级，典型的做法是用各种机械分级机。但虾有可能被分级机的栏杆拦截，因此，需要进行定期检查，以避免这些虾滞留到下一批，造成细菌污染；

（2）进一步加工前，应将虾重新冰冻和冷藏；

（3）分级应快速进行，以防止微生物滋生和产品的腐烂。

14.2.7 添加配料和使用添加剂

潜在危害：化学和微生物污染、亚硫酸盐。

潜在缺陷：腐烂、添加剂使用不当。

技术指南：

（1）根据规格和法规要求，对虾可以进行一定的处理，以提高感官质量、保持产量或为进一步加工提供保障；

（2）具体方法如下：用亚硫酸氢钠减少虾壳变黑，用苯甲酸钠延长在各加工环部分间的货架期，用多磷酸钠保持产品在加工过程中的多汁性以及防止在剥皮后产生黑斑，同时还要加入普通的盐水进行调味；

（3）这些配料可用在很多阶段，比如普通的盐和多磷酸钠可用在解冻阶段，冷却的盐水可用作蒸煮和冷冻阶段之间的水流传送带或镀作冰衣；

（4）配料无论在哪个阶段添加，都必需对加工步骤和产品进行监测，以确保没有超出任何法定标准、确保符合质量参数。另外，在用消毒水浸泡时，成分的改变要符合已制定方案所确立的常规标准；

（5）要始终保持冷却条件；

（6）用于防止黑斑形成的亚硫酸盐的使用应该符合厂商的说明和良好操作规范。

14.2.8 完全或部分去壳

潜在危害：微生物交叉污染。

潜在缺陷：腐烂、外壳碎片、外源性物质。

技术指南：

（1）这个过程主要应用于对温水对虾、经简单检查和制备的冻整对虾、降级处理的瑕疵对虾进行完全去壳；

（2）其他去壳阶段可能包括完全去壳或尾部不动的部分去壳；

（3）无论是哪个阶段，都必须有水冲洗，确保去壳工作台干净，不受污染的虾和虾皮碎片，并要对虾进行漂洗，确保无虾皮碎壳残留。

14.2.9 去肠线

潜在危害：微生物交叉污染、金属污染。

潜在缺陷：令人不快的物质、腐烂、外源性物质。

技术指南：

（1）肠线是长在对虾肉背脊上面的一条黑线。大型温水对虾的肠线不仅不雅观，而且里面含有大量泥沙，同时也是细菌污染源；

（2）去除肠线的方法为：用剃刀刀片沿纵向切开虾的背脊，抽掉肠线。部分虾是在去头带皮情况下完成的；

（3）这是一个机械过程，尽管是劳动密集型作业；

（4）应制定清洁和维护计划，同时要满足加工前、加工中和加工后技术工人进行清理的需要；

（5）此外，必须从生产线上除去腐败的和受污染的虾，而且不允许出现碎片堆积情况。

14.2.10 清洗

潜在危害：微生物污染。

潜在缺陷：腐败、外源性物质。

技术指南：

（1）应对去壳和去掉肠线的虾进行清洗，以确保将壳和肠线的去除完全；

（2）在进一步加工前，应尽快地将虾沥干和冷却。

14.2.11 蒸煮处理

潜在危害：夹生，微生物交叉污染。

潜在缺陷：过度烹煮。

技术指南：

（1）应按照对终产品规格的要求，制定严格的蒸煮程序，尤其是时间和温度。例如是否该产品供直接食用，是否要保持原始的天然的品质，以及是否要求同一等级的虾大小均匀；

（2）对每一批虾进行加工前都应当对蒸煮程序表进行复查，当使用连续的蒸煮锅时，应对过程参数进行连续的记录；

（3）不论是水煮，还是用蒸气蒸，只能使用饮用水；

（4）监测方法和频率应该严格符合预定的程序；

（5）应有蒸煮锅的清洁和维护程序，而且所有的操作只能由经过全面培训的工作人员进行；

（6）必须对不同设备蒸煮的虾隔离，以确保没有交叉污染。

14.2.12 已蒸煮的对虾去壳

潜在危害：微生物交叉污染。

潜在缺陷：存在虾壳。

技术指南：

（1）在冷却或冻结过程中在生产线上使用机械或人工对煮虾进行去壳处理；

（2）应该制定维护和清洁程序，而且要由经过完全培训的员工进行操作，以确保进行安全高效的加工。

14.2.13 冷却

潜在危害：微生物交叉污染和形成毒素。

潜在缺陷：不太可能有。

技术指南：

（1）煮过的虾应尽快冷却，使产品温度降到能够限制细菌繁殖和毒素产生的温度范围；

（2）冷却程序应符合时间和温度要求，同时要制定维护和清洁程序，培训过的技术工人要遵守程序操作；

（3）尽管对连续性操作而言，应当限定一个修饰性程序和最长的运转周期，但只能用冷/冰的饮用水进行冷却，而且已经用过的水不能再用于下一批；

（4）生/煮虾隔离很重要；

（5）在冷却和沥干水分后，应当尽快将虾冷冻，以避免受到任何环境污染。

14.2.14 冷冻处理

潜在危害：微生物污染。

潜在缺陷：慢速冷冻——影响虾的肉质和结块。

技术指南

（1）随着产品种类不同，冷冻操作方式变化非常大。最简单的方式是将生的整虾或去头虾放在专门设计的纸箱里，进行大块冻结或平板冻结，做法是用饮用水向纸箱里浇水，使虾和保护性冰结成一体；

（2）另一种做法是用流化床系统对煮过的和去壳（Pandalus）冷水虾进行冷冻，而许多温水虾产品则是放在风冷冻结器中的托盘上或在履带式冻结机中进行单体速冻；

（3）不论是哪一种冷冻过程，都必需保证符合指定的冷冻条件。对于单体速冻产品来说，不能结块，例如，虾不能冻在一起。在达到操作温度之前将产品放入风冷冻结器，可能会形成镀冰衣的慢冻产品，并导致污染；

（4）冷冻器是复杂的机器，要求有清洁和维护程序，并由有经验的员工操作。

14.2.15 镀冰衣

潜在危害：微生物交叉污染。

潜在缺陷：镀冰不充分、镀冰过多、黏接、不正确贴标。

技术指南：

（1）对冻虾镀冰衣是为了防止产品在贮藏和销售中脱水，保持虾的品质；

（2）冰块冻虾是最简单的镀冰衣方式，在冷却的饮用水中对冻虾进行浸泡后沥水。一种更为复杂的加工程序是，在冷水喷射下由振动带传送分级的冻虾，使虾匀速速通过，从而形成一层均匀的可计算的镀冰层；

（3）理想情况下，镀冰衣后的虾在包装前应进行第二次重新冷冻。如果不进行这个步骤，也应尽快地将虾包装并送到冷库。如果做不到的话，虾就可能冻结在一起或随着冰衣的硬化而粘连或结块；

（4）有法典方法可用于冰衣的测定。

14.2.16 所有产品的称重、包装和标识

潜在危害：亚硫酸盐。

潜在缺陷：不正确贴标、腐烂。

技术指南：

（1）所有产品包装材料和包装物，包括胶水和墨水在内，应为食品级、无味，不含可能会转移到所包装食品中的对身体有害的物质；

（2）所有的食品应当带包装称重，然后适当扣除皮重，以确保重量准确；

（3）对产品进行镀冰衣、裹面或其他加工时，应当进行检查，以保证成分标准正确，并符合法律或包装上的标称内容；

（4）包装上的成分表，应以含量递减的顺序写明食品所含成分，包括曾用过的仍残留在食品中的添加剂；

（5）所有包装操作应保证冷冻产品保持冷冻，并在送回冷库前的温升最小；

（6）亚硫酸盐的使用应该符合厂商的说明和良好操作规范；

（7）在使用亚硫酸盐的工序中应该注意贴上适当的标签加以说明。

14.2.17 金属检测

潜在危害：存在金属。

潜在缺陷：不太可能有。

技术指南：

（1）产品最后的包装应用机器进行金属检测，机器要设置到所能达到的最大灵敏度；

（2）检测大包装箱灵敏度要低于小包装箱，所以应考虑包装前检测。但可能仍是包装后再检测更好些，除非能避免包装前再次污染的可能性。

14.2.18 最终产品的冻藏

潜在危害：不可能。

潜在缺陷：由于温度的波动造成质地和气味的改变、深度冻灼、冷藏气味、纸板箱气味。

技术指南：

（1）冷冻产品应冷冻贮藏在干净、牢固和卫生的环境里；

（2）设备应能使产品保持在 -18℃或更低的温度下，而且温度波动最小（±3℃）；

（3）贮藏区应当配备刻度经过校准的温度计。强烈推荐使用温度记录仪；

（4）应制定和保持一个系统的存货周转计划；

（5）应适当保护产品，防止脱水、灰尘和其他形式的污染；

（6）所有的最终产品都应当贮藏在空气循环良好的冷冻器里。

第十五部分　头足类动物的加工

为了识别每一个加工步骤的控制方法，本部分将举例说明潜在的危害和缺陷，并提供技术指南，以便采取控制措施和纠正措施。在每一个步骤中，只列举本步中可能引入或控制的危害和缺陷。应该认识到，在准备一个 HACCP 和（或）DAP 计划时，有必要查阅第五部分，该部分对 HACCP 和（或）DAP 分析的应用提供了指南。然而，在本操作规范的范围内，不可能列举每一个步骤的关键限制，监测，记录保持和验证等详细内容，因为这些都是针对特定危害和缺陷的具体内容。

本部分适用于新鲜的或加工后的头足类动物，包括用于人们消费的乌贼（Sepia 和 Sepiella），鱿鱼（Alloteuthis，Berryteuthis，Dosidicus，Ilex，Lolliguncula，Loligo，Loliolus，Nototodarus，Ommastrephes，Onychoterthis，Rossia，Sepiola，Sepioteuthis，Symplectoteuthis 和 Todarodes）和章鱼（Octopus 和 Eledone）。

新鲜的头足类动物极容易腐败，不论在什么时候处理都应格外小心，避免污染，抑制微生物生长。头足类动物应避免日晒、风吹或其他有害条件，但应尽快地仔细清洗，并将温度降至0℃（32 ℉）。

本部分给出了一个头足类动物加工的例子。图15－1列出了有关新鲜鱿鱼接收和加工的工艺流程。应当注意，对头足类动物而言，有多种加工流程，该流程仅作说明之用。

图15－1仅为用于说明问题。对于在工厂里实施HACCP而言，必须为每一个加工过程制作全

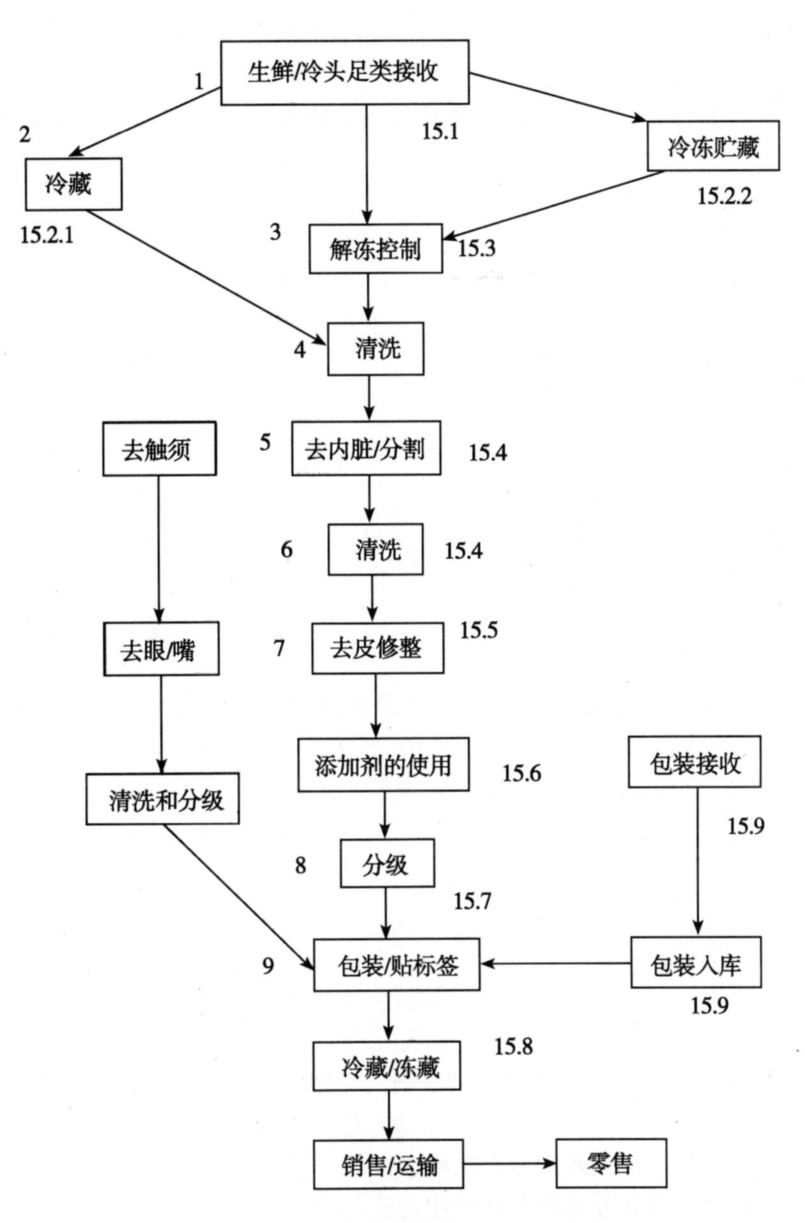

图15－1　可能的鱿鱼加工流程图示例

面的而详细的流程图。

15.1　头足类动物的接收（第一步）

潜在危害：微生物污染、化学污染、寄生虫。

潜在缺陷：产品受损、外来杂质。

技术指南：

（1）加工者应制定程序，对正在捕捞或到达工厂的产品进行检查。只能接收健康状况良好的产品用于加工。

（2）产品规格可以包括：感官特征，例如，外观、气味、质地等能作为消费健康的指标；腐败或污染的化学示剂，例如，TVBN、重金属（镉）；微生物学指标；寄生虫，例如，Anasakis 等杂质；存在破裂、破损，表皮变色或从内部的肝脏和消化器官扩散到外层的淡黄色（表明产品变质）。

（3）产品的检查人员应当接受培训，并有相关品种的工作经验，以识别任何缺陷和潜在的危害。

进一步信息参见第八部分“新鲜、冷冻和碎鱼肉的加工”和《实验室中鱼类和贝类动物感官评价法典指南》。

15.2 头足类动物的贮藏

15.2.1 冷藏（第 2 步和第 10 步）

潜在危害：微生物污染。

潜在缺陷：腐败、物理损伤。

技术指南：

参阅第 8.1.2 节“冷藏”。

15.2.2 冻藏（第 2 步和第 10 步）

潜在危害：重金属，例如，来自内脏的镉。

潜在缺陷：冻结烧。

技术指南：

参阅第 8.1.3 节“冻藏”。

（1）需要注意这样的事实，消化道内容物中镉含量较高时，此种重金属可能会转移到肌肉中去；

（2）产品应该通过充分的包装或镀冰来防止脱水。

15.3 解冻控制（第 3 步）

潜在危害：微生物污染。

潜在缺陷：腐败、变色。

技术指南：

（1）应当明确限定解冻参数，包括时间、温度，这对于防止发生浅粉红色变很重要；

（2）应当确定产品解冻时间和温度的关键限制。特别要注意解冻产品的体积，以控制变色的发生；

（3）如果用水进行解冻，所用水应当是饮用水；

（4）如果使用循环水，必须注意避免微生物繁殖。

若需进一步的指南，请参阅第 8.1.4 节“解冻控制”。

15.4 剖解，去内脏和清洗（第 4、5、6、11、12 及 13 步）

潜在危害：微生物污染。

潜在缺陷：消化道内容物存在、寄生虫、壳、黑墨褪色、嘴，腐败。

技术指南：

（1）去除内脏，将肠内物质和头足类的壳、头去除干净；

（2）如果加工的副产品供人类消费，那么该过程所有副产品，如触手和外皮都应当及时、卫生地进行处理；

（3）去除内脏后，应立即用干净的海水或饮用水进行清洗，去掉管腔内任何遗留的物质，以减少产品中微生物水平；

（4）在清洗整个头足类动物和其制品时，应有充足的干净海水或饮用水。

15.5 去皮，修整（第7步）

潜在危害：微生物污染。

潜在缺陷：异物、咬伤、皮伤、腐败。

技术指南：

（1）去皮方式不应导致产品污染，也不应造成微生物增长。酶法或热水技术去皮应设定时间/温度参数，以防止微生物的增长；

（2）应注意防止废料对产品的交叉污染；

（3）对产品进行清洗和去皮时，应有充足的干净海水或饮用水。

15.6 添加剂的使用

潜在危害：物理污染、未批准使用的添加剂、非鱼类过敏原。

潜在缺陷：物理污染、添加剂的超标使用。

技术指南：

（1）经过训练的操作者混合和合理使用添加剂；

（2）有必要对过程和产品进行监测以保证其不超过调整的标准并且符合质量参数要求；

（3）添加剂应该符合食品法典委员会《食品添加剂通用标准》的要求。

15.7 分级、包装、贴标签（第8，9步）

参阅第8.2.3节“标签”。

潜在危害：包装造成的化学或物理污染。

潜在缺陷：不正确贴标、重量不对、脱水。

技术指南：

（1）包装材料应干净，适合预定用途，并用食品级材料制成；

（2）分级和包装操作时间应尽可能短，以防止产品变质；

（3）使用亚硫酸盐的工序，应该保证它们贴上适当的标签。

15.8 冷冻（第十步）

潜在危害：寄生虫。

潜在缺陷：冻结烧、腐败、冷冻慢造成质量损失。

技术指南：

头足类动物应尽快冷冻，以防止产品变质、微生物增长和化学反应造成的保质期缩短。

（1）制定时间和温度参数时应确保能快速冷冻产品，另外要考虑冷冻设备的类型、容量和产品的尺寸、形状、产量。应当根据设备的冷冻容量来调整产量；

（2）如果将冷冻作为对寄生虫的控制点，那么需要设定时间和温度参数，以确保寄生虫无法存活；

（3）应对产品温度进行定时监测，以保证彻底冷冻，因其关系到产品的中心温度；

（4）对于所有的冷冻和冷冻贮藏过程应当保存足够的记录。

若需要进一步指南，请参阅第 8.3.1 节“冷冻过程”。

15.9 包装、标签和配料——接收和贮藏

应当考虑包装、标识和配料造成的潜在危害和缺陷。推荐本规范的使用者查阅第 8.5 节“包装、标识和配料”。

第十六部分 鱼类、贝类和其他无脊椎水产品罐头加工

本部分适用于鱼类、贝类和其他水产品。

在每个加工步骤公认的控制方面，本部分提供潜在危害和缺陷的例子以及技术指南，它们能帮助制定控制措施和纠正措施。在特定步骤中，只列出了可能引入的或控制住的危害和缺陷。应当认识到，HACCP 和 DAP 计划的准备过程应参见第 5 部分（危害分析和关键控制点 HACCP，缺陷作用点 DAP 分析），该部分提供了 HACCP 和 DAP 分析原则的应用方法。然而本操作规范不可能给出每个步骤关键限值、监测、记录保持和验证的细节，因为危害和缺陷都是具体的。

本部分涉及经过热加工杀菌的罐头鱼类和贝类产品的加工，这些产品是用密封容器包装的，供人类消费。

本规范强调的是，在这些步骤里恰当的应用前提条件（第三部分）和 HACCP 原则（第五部分）会为加工者提供以下保证：必要的质量、成分和标识规定会符合《食品法典标准》，食品安全问题会被控制。流程图（图 16－1）可以提供鱼类、贝类罐头生产流程中一些通用步骤的指导。

此流程图仅供说明。对于工厂实施的 HACCP 计划，其每个加工过程都应有一个完整、全面的流程图。

16.1 总则——必备程序的补充

第三部分（必备程序）提出了加工厂应用危害和缺陷分析之前，对良好卫生操作的基本要求。

对于鱼贝类罐头厂，除第三部分描述的准则外，还有一些补充要求，因为工艺是特定的。一些要求已在下面列出，但为进一步了解也应当去参考《推荐性国际法典－低酸和酸化低酸罐装食品卫生操作规范》（CAC/PRC 23－01979. Rev <1993>）。

（1）杀菌笼的设计、使用和维护以及其他杀菌设备的运用和卸载要适合罐的种类和材料。对这些装置的使用应防止装罐过量；

（2）应具备数量充足的、高效的密封设备以防止加工时间拖延；

（3）杀菌的锅应具足够的能量、蒸汽、水和（或）空气供应，以维持杀毒热处理过程中具有充足的压力。尺寸大小要与生产相适应，以免时间拖延；

（4）每个杀菌锅应装备可视温度计、压力计及时间温度记录仪；

（5）应在杀菌室里安装一个准确的可视钟表；

（6）使用蒸汽杀菌锅应当考虑安装自动蒸汽控制阀；

（7）用于热加工的控制和监测仪表应保持良好状态，应定期检验或校准。测量温度仪表的校准应与参照温度计对比。参照温度计应定期校准。应建立和保持仪表校准的记录。

16.2 危害和缺陷的识别

参见 4.1 节（与鱼类、贝类有关的潜在危害）。

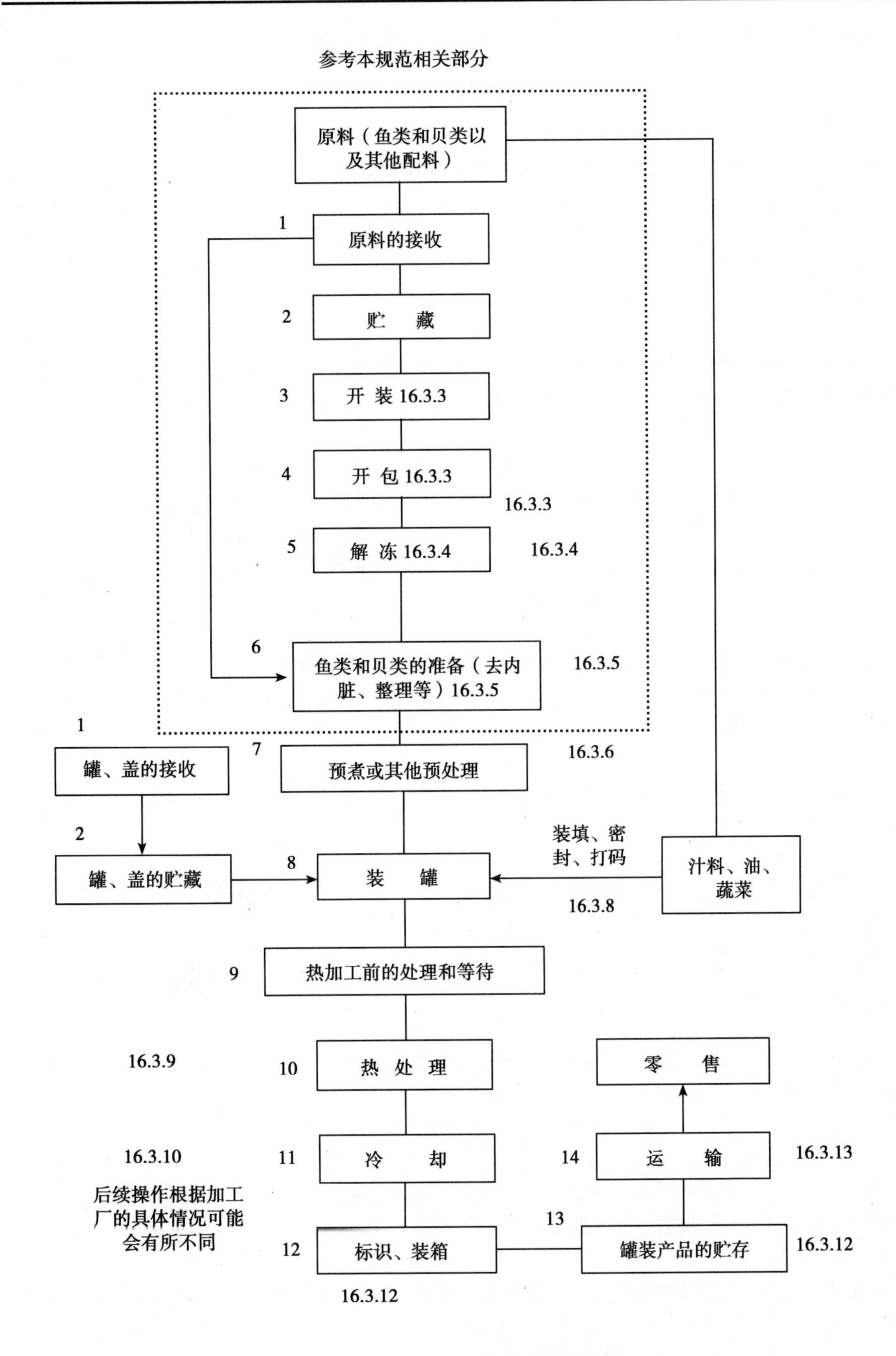

图 16－1　鱼类和贝类罐头的生产流程图

本节描述特定鱼类和贝类罐头的主要潜在危害。

16.2.1 危害

16.2.1.1 生物危害

（1）天然存在的海洋毒素

生物毒素（例如，河豚毒素或西加毒素）耐热性很强，因此了解用于加工的鱼类品种的一致性和来源是重要的。

藻类毒素（例如，DSP、PSP 或 ASP）耐热性也很强，因此，了解用于加工的软体动物类或其他品种来源地情况也很重要。

（2）鲭鱼毒素

组胺

组胺是耐热的，而且它的毒性能在容器内保持完好。从捕捞到热加工的保存和处理中的良好操作对于防止组胺产生是必要的。食品法典委员会制定了某些鱼品的组胺最大限量标准。

（3）微生物毒素

肉毒梭菌

肉毒中毒风险通常出现在热加工不充分和容器完整性不好时。该毒素具有热敏性，尤其是因为蛋白质分解作用，破坏肉毒菌孢子需要高杀毒值。热加工效果由操作时的污染水平决定。因此建议控制加工过程中的繁殖和污染风险。肉毒的高风险产生来自下列情况：热加工不充分、容器完整性不好、加工后冷却水不净和湿传送设备不洁。

金黄色葡萄球菌

金黄色葡萄球菌毒素会出现在严重污染的原料中或在加工中因细菌繁殖而产生。制罐后，如果热湿罐处理不卫生，也会存在金黄色葡萄球菌在加工后污染的潜在风险。这种毒素具有抗热性，因此应在危害分析中加以考虑。

16.2.1.2 化学危害

应当小心避免罐成分（例如，铅）和化学品（例如，润滑剂、卫生剂、清洁剂）引起的产品污染。

16.2.1.3 物理危害

罐装前罐可能含有诸如金属或玻璃碎片等物质。

16.2.2 缺陷：

食品法典标准中没有的，应该关注有关国家法规和/或商业规范。

16.3 加工操作

加工者要参考《推荐性国际法典－低酸和酸化低酸罐装食品卫生操作规范》（CAC/RCP23－1979，Rev.2 <1993>）可以得到罐头制作的详细建议。

16.3.1 原料、罐、盖子、包装材料和其他辅料的接收

16.3.1.1 鱼类贝类（加工步骤1）

潜在危害：化学和生物污染（例如，DSP、PSP、鲭鱼毒素、重金属）。

潜在缺陷：代用品、腐败、寄生虫。

技术指南：

参考8.1.1节（生鱼或冻鱼接收）和其他相关部分，同时，接收罐头加工用的活贝类（软体动物类）时，应进行检验以去除死的及严重破损的贝类。

16.3.1.2 罐、盖子和包装材料（加工步骤1）

潜在危害：后续的微生物污染。

潜在缺陷：残次品。

技术指南：

参考8.5.1节（原料接收——包装、标识和配料），并同时符合以下要求：

（1）罐、盖子和包装材料应适合于产品类型、贮存条件、罐装填充物、密封和包装设备以及运输条件；

（2）鱼和贝类罐头的罐应由适宜材料及结构制成，以便易于密封以防止污染物的进入；

（3）鱼类和贝类罐头的罐和盖子应符合下列要求：能保护内含物不会受到微生物或其他物质的污染；内表面不与内含物起反应以免对产品或罐产生不利影响；外表面在任何条件下贮存都不会腐蚀；耐用，能承受加工过程中的机械和热应力，可以防止销售过程中的物理损坏。

16.3.1.3 其他辅料（加工步骤1）

参考8.5.1节（原料接受——包装、标识和配料）。

16.3.2 原料、罐和包装材料的贮存

16.3.2.1 鱼类和贝类（加工步骤2）

参考8.1.2节（冷却贮存）、8.1.3节（冻结贮存）和7.6.2节（海水槽、池等中软体动物类的贮存条件）。

16.3.2.2 罐和包装（加工步骤2）

潜在危害：不可能。

潜在缺陷：外来异物。

技术指南：

参考8.5.2节（原料贮存——包装、标识和配料），同时符合以下要求：

（1）罐和包装的所有材料应贮存在令人满意的清洁卫生的条件下；

（2）贮存期间，空罐和盖子应防止灰尘、湿气和温度波动的影响，以免罐产生凝水，这样锡罐会被腐蚀；

（3）在装填、装载、运输和卸载空罐时，应避免任何冲击。罐不能踏踩。这项措施在罐放入袋子和货盘中时更有必要。冲击会使罐（罐身或边缘）变形，危及紧密性（冲击接缝，凸缘变形）或使外形不美观。

16.3.2.3 其他辅料（加工步骤2）

参考8.5.2节（原料贮存——包装，标识和配料）。

16.3.3 开装、开包（加工步骤3和4）

潜在危害：不可能。

潜在缺陷：外来异物。

技术指南：开装和开包操作时，应采取预防措施以保护产品不受污染并且防止异物进入产品中。为避免微生物繁殖，应尽量缩短深加工前的等待时间。

16.3.4 解冻（加工步骤5）

参考8.1.4节（控制解冻）。

16.3.5 鱼类和贝类准备过程（加工步骤6）

16.3.5.1 鱼类的准备过程（去内脏、整理等）

潜在危害：微生物污染、生物化学反应（组胺）。

潜在缺陷：异物（特定产品中的内脏、皮、鳞等）、异味、骨刺、寄生虫等。

技术指南：

参考8.1.5节（冲洗和去内脏）和8.1.6节（切片、去皮、整理和灯检），同时，当鱼皮浸泡在苏打溶液中时，应采取特殊措施以适当中和之。

16.3.5.2　软体水产品和软体动物类的准备过程

潜在危害：微生物污染、硬贝壳碎片。

潜在缺陷：异物。

技术指南：

参见7.7节（软体动物类的热处理/热冲击），同时，当使用鲜活的贝类时，应进行检验，以去除死的及严重受损的贝类，并采取特殊措施以保证去除贝肉中的贝壳碎片。

16.4　预烹制和其他加工方法

16.4.1　预烹制

潜在危害：化学污染（油氧化产物）、微生物或化学生物（鲭鱼毒素）生长。

潜在缺陷：最终产品出水（产品用油灌制）、异味。

技术指南：

16.4.1.1　通则

（1）应该设计预烹制罐装鱼类或贝类的方法以获得缩短延误时间、缩小处理量的效果；方法的选择通常受处理原料的影响。对于用油罐制的产品（例如，沙丁或金枪鱼）预烹制应该充分，以避免热加工时放出过多的水；

（2）应找到实用方法减少预烹制的后续过程中的处理量；

（3）如果用已去内脏的鱼进行加工，则应在蒸煮前将鱼以腹部位置向下排列，从而方便排除鱼体内可能积聚并影响热加工中产品质量的鱼油和汁液；

（4）适当时，软体贝类、龙虾和蟹、对虾、大虾和头足类的预烹制可以根据其他部分的技术指南来操作，例如，第七部分（软体贝类的加工）、第十三部分（龙虾和螃蟹的加工）、第十四部分（对虾、大虾的加工）和第十五部分（头足类的加工）；

（5）预烹制前应注意消除鲭鱼毒素危害的温度。

16.4.1.1.1　预烹制进度表

（1）预烹制方法（特别是时间和温度）应明确地决定下来。应检查预烹制进度表；

（2）同时进行预烹制的鱼，其尺寸应相当。送入蒸煮机时，它们的温度应一样。

16.4.1.1.2　预烹制用油和其他液体的质量控制

（1）只有质量良好的蔬菜油才能用于罐装鱼类和贝类预烹制中（参见相关蔬菜油食品法典标准）；

（2）烹制用油应经常更换以免形成极性化合物。用于预烹制的水也应常更换以免污染；

（3）必须注意油或其他所用液体（例如，蒸汽、水等）不能给产品带来异味。

16.4.1.1.3　冷却

（1）除非产品是在仍然很热时就进行了包装，预烹制的鱼或贝类的冷却应尽快进行，以使产品的温度可以限制细菌繁殖、毒素产生，避免产品受污染；

（2）用于冷却将立即去壳的甲壳类水产品的水，应是饮用水或干净的海水并且只能冷却一批产品。

16.4.1.2　熏制

参考第12部分（熏鱼的加工）。

16.4.1.3　盐水和其他溶液的使用

潜在危害：微生物和化学污染（因浸沾溶液）。

潜在缺陷：掺假（添加剂）、异味。

技术指南：

（1）用盐水或其他添加剂、风味剂（为达到其他目的或得到其他风味）浸渍罐装用鱼类或贝类时，溶液浓度和浸渍时间都应小心控制以带来最好效果；

（2）应当经常性地并定期地更换浸渍溶液，并且仔细清洗浸渍容器及其他浸渍用的器具；

（3）浸渍用的配料或添加剂是否被允许用于罐装鱼类和贝类的加工，要根据相关食品法典标准和出售该产品的国家的相关法律法规要求。

16.4.2 装罐（装填、密封和编码）（加工步骤8）

16.4.2.1 装填

潜在危害：微生物生长（等待期间）、微生物存活生长和热加工后再污染（由于不正确的装填或使用缺损的罐及外来异物）。

潜在缺陷：重量不准、异物。

技术指南：

（1）应在将罐和盖运送到装填机或包装台之前，抽检其中的一部分，以确保其干净、无损并且没有可以看到的裂纹等；

（2）如有必要，空罐应被清洗。在使用罐前，倒置所有的待用罐以确保罐中不含任何外来异物，这也是一种明智的预防措施；

（3）也应当注意去除有缺损的罐，因为它会堵塞罐装机或密封机，也可能在热加工中带来麻烦（消毒不好、渗漏）；

（4）在清洗包装台或传送带时，空罐不应留在上述设备上以避免污染或污迹的产生；

（5）在预处理结束后，为防止微生物繁殖，可以适当地将罐内充满热的鱼类或贝类（>63℃，如鱼汤），或尽快装罐（最短等待时间）；

（6）如果鱼和贝类在装罐包装之前必须放置很长时间，则应冷却；

（7）用于罐装鱼类和贝类的罐应当按计划的过程直接装罐；

（8）机械或手工装罐应当对装罐的速率和留出的顶部空间进行检查，以确保符合进度表的规定。有规则的装罐是重要的，这不仅是因为经济因素，而且因为热穿透和罐的完整性会受装入量过多的影响；

（9）顶部空间必要量部分地依赖于内含物的特性。装罐量也应考虑热加工的方法。顶部空间量由罐的制造者决定；

（10）另外，罐内的装填量可根据对成品的法规要求及关于内含物重量的公认的标准；

（11）如果罐装鱼类和贝类用手工包装，则应有稳定的鱼类、贝类和其他配料的供应。同时，也应避免在包装台上进行装填时，原料鱼类和贝类的堵塞；

（12）应特别注意装罐机的操作、维护、定期检验、校准和调整。应注意遵守机械制造者的产品说明；

（13）其他配料（例如，油、汁、醋……）的质量及用量应小心控制以达到所期望的最佳效果；

（14）如果鱼用盐水冻结或贮存在制冷盐水里，那么当其会影响到产品风味时，应对产品吸收的盐量加以考虑；

（15）装填罐检验应确保装填正确，并且满足对于内容物重量的公认标准；并应在封罐之前验证产品的质量和工艺；

（16）手工装罐产品（例如，小的远洋鱼）应由操作者仔细检查验证罐的边缘及密封面不应有任何产品残留，以免残留影响产品的密封性。对于自动装罐产品，应执行抽样计划。

16.4.2.2 密封

罐和盖的密封是罐头生产最基本的过程之一。

潜在危害：因密封不好而造成的后续的污染。

潜在缺陷：不可能。

技术指南：

（1）应特别注意密封机的操作、维护、定期检验和调校。应该调整密封机以适应不同的罐的类型及使用的封口方法。无论使用那种密封设备都应该谨慎遵守制造者或设备供应商的说明；

（2）接缝和其他的封口处的尺寸要符合特定罐的可接受的公差范围；

（3）应由合格人员进行此操作；

（4）使用真空方法包装时，应有效地防止罐在任何条件下（温度高或气压低），尤其是在产品销售过程中发生膨胀。对于罐体比较长的深罐或玻璃罐采用真空包装比较有效，但对于盖子宽大且易变形的较短的浅罐，很难也没有必要采用真空包装；

（5）过分的真空包装会导致罐变形，特别是当顶部空间较大时，则可能导致在接缝处有细微缝隙的罐吸入污染物；

（6）为了找到最好的方法建立真空，应当向权威的技术专家咨询；

（7）应在生产期间进行定期检验以检测出罐的潜在的、外在的缺陷。依据说明保证封口操作者、封口监督者或其他合格的人员定期检查所使用的其他类型的罐的接缝或封口系统。应考虑采用诸如真空测量和接缝拆除等方法进行检验。应在检查中使用抽样计划；

（8）特别是在每个生产线的开始和每次罐尺寸的变化时，堵塞后、新的调教后以及密封机长时间停止的重新启动后，要进行检查；

（9）应记录所有适当的观察资料。

16.4.2.3 打码

潜在危害：由于损坏的罐而造成的后续的污染。

潜在缺陷：因不正确的打印失去可追溯性。

技术指南：

（1）每一个鱼类和贝类罐头应具有不能拭去的码印，其上显示所有可能被确定的重要的制造细节（产品类型、罐头生产厂、生产日期等）；

（2）打码设备必须小心调整以使容器不被损坏并且码印持久易读；

（3）有时是在冷却步骤后进行打码操作。

16.4.3 封罐后续操作——加热前的分段运输（加工步骤9）

潜在危害：微生物生长（等待时期），由于罐的破坏而导致的污染。

潜在缺陷：不可能。

技术指南：

（1）罐子被密封以后应当轻拿轻放，防止可能引起过失和微生物污染的任何破坏；

（2）如有必要，装满的和密封的金属容器在加热步骤前应当经过彻底的清洗，洗掉壁上油脂，污垢和鱼贝器官等；

（3）为防止微生物繁殖，等待时间应当尽量的短；

（4）如果在加热操作进行前，已装满和封口的罐子需要放置较长时间的话，应当把产品置于能最抑制微生物生长的温度环境中；

（5）任何一个罐头加工厂都应当发展一个系统，来防止未经加热处理的罐装鱼和贝在贮藏过程中被偶然的污染或侵染。

16.4.4 热处理（操作步骤10）

在罐头生产中，加热处理是至关重要的一个步骤。

关于加热操作的细节操作步骤罐头制造商可查阅“国际推荐的低酸和酸化低酸罐头食品的卫

生操作代码”（CAC/RCP 23－1979，Rev. 2 in 1993）。这里仅涉及一些重要的原理。

潜在危害：存活的肉毒梭菌的孢子。

潜在缺陷：具有分解作用的微生物体的存活体。

技术指南

16.4.4.1　消毒时间表

（1）确定消毒时间表，首先应当确定加热步骤操作后达到商业无菌，这需要考虑一些因素（微生物群，维数和容器的属性，产品描述，等等）。一个消毒时间表是为在特定容积大小的容器的特定产品而设计的；

（2）应当进行正确的加热，温度分布要合理。标准的热加工步骤和试验方法确立时，要经过专家的审核和确证，确证方法和安排适合每一个产品和蒸器；

（3）在改良或变化操作步骤前（最初装罐的温度，产品组成，容器大小，蒸器的充满的状态），应当向技术专家质询，质询是否需要进一步的评估。

16.4.4.2　加热操作步骤

（1）只有有资格的和经过正确培训的人才能操作蒸器。因此必须蒸器操作人员来控制加工操作，并确保消毒进程跟上，包括在时间把握、温度和压力的监控和记录保持等方面；

（2）为避免处理不当，服从时间表上规定的温度要求是相当重要的。如果满装罐在加热处理前由于长时间的等待而导致了温度降低，消毒时间表就应当考虑这些温度问题；

（3）为了使加热操作高效、处理温度能得到很好的控制，应当通过一个清空步骤把罐中的空气清除，而且这个步骤要经过专家认可。应当考虑容器的大小和类型，蒸器装置和装载设备、步骤等因素；

（4）只有等到达到一定的温度时，加热操作的时间才能开始算，而且蒸器中应当始终保持一致的温度，最少安全排风时间过去以后才算完成；

（5）其他类型的蒸器（水，蒸汽/空气，火等）涉及到“国际推荐的低酸和酸化低酸罐头食品的卫生操作代码”（CAC/RCP 23－1979，Rev. 2 in 1993）；

（6）如果不同容积大小的鱼和贝罐放在一起加热的话，应当把握好加工时间，以便保证所有型号的罐子都能保持商业无菌；

（7）当对玻璃罐进行操作时，必须确保蒸器里水的最初温度要比往罐里装的时候要低一些。水温升起来之前应当利用大气压力。

16.4.4.3　热处理操作的监控

（1）加热处理中，应确保消毒步骤和各个因素如罐的填充，封口时内部压力最小化，蒸器装载，最初产品温度等，都与消毒进度保持一致；

（2）蒸器的温度应当由指示温度计确定，而不是通过温度记录；

（3）对每个蒸器都应当保持对时间、温度及其他相关的记录；

（4）为确保准确无误，应当有规律的用温度计来测量温度，检测记录应当保持下来；识读读数时要认真，不能超出指示的读数；

（5）应当周期性的检测蒸器，以保证操作合理，能正常高效的完成加热处理步骤。检验每个蒸器装备完全、填充和使用合理，来保证蒸器中的每个填充物都能快速的达到加工温度和在整个加工周期中能保持这个温度；

（6）检测应当在专家指导下进行。

16.4.5　*冷却（加工步骤 11）*

潜在危害：由于缝合和水污染的问题导致的再污染。

潜在缺陷：鸟粪石结晶的组成、扭曲的罐子、烧焦。

技术指南：

（1）加热处理后，罐装的鱼和贝实际上应当在适当的压力下用水进行冷却来防止变形，这样会导致紧密性降低。如果循环利用水的话，水应当经常用氯消毒（或者其他合理的处理）。冷却水中氯的残留水平和冷却时间检测以使快速加工的污染最小化。用非氯化方法处理的应当对处理方法的效率进行监控和核实；

（2）为避免鱼贝类罐头产生感官缺陷，例如，烧焦或烹调过度等，应当尽快把罐内温度降低；

（3）对于玻璃罐，冷却剂的温度应该在开始的时候缓慢下降，防止由于温度突然下降带来的危害；

（4）加热之后，没有用水冷却鱼贝类产品之前，其堆积方式应当能使之快速降温；

（5）加热处理的罐装鱼和贝在没有被冷却和彻底干燥前，没有必要的话不能用手或衣物触摸。处理不能太野蛮，也不能把使其表面和接缝暴露在污染之中；

（6）应当快速的冷却，避免鸟粪石结晶的形成；

（7）每个罐头厂都应当发展一个系统来防止已加工的罐头与未加工的混合。

16.4.5.1 热处理和冷却后的监控

（1）罐装鱼贝类产品在生产过程结束后、贴标签前应当进行缺点检查和质量评估；

（2）从每个代码批次中抽取样品进行检查，以确保罐体没有外部缺陷，确保产品符合重量标准、真空要求、作工和健康要求。质地、颜色、气味、滋味和包装材料的情况等等也要评估；

（3）如有要求，可以做稳定性试验来检验热处理步骤是否符合要求；

（4）鱼贝罐头生产出来以后应当尽可能符合实际的来检验，如果发现由于操作工人设备的问题导致产品缺陷的话，检查出来可以尽快地得到解决而不拖延。不适于人类消费的缺陷批次产品应当与合格产品分开放置。

16.4.6 标签，包装和成品的贮存（加工步骤 12 和 13）

参见 8.2.3——标签部分。

潜在危害：由于容器破坏或被置于极端环境所引起的污染。

潜在缺陷：标识不当。

技术指南：

（1）罐装鱼贝类产品的标签和包装材料不应当利于引起容器罐的腐蚀。包装容器应当大小适宜，不会因为箱内的震动而引起损伤。包装容器和箱子大小和质量达标，在鱼贝罐头的运输过程中能起到很好的保护作用；

（2）鱼贝罐头的罐上有的代码符号包装时也应当在容器上体现出来；

（3）鱼贝罐头在储藏时应尽量避免罐体受损。一般情况下，成品的货盘堆积的不宜太高，贮藏用的铲车使用方法应得当；

（4）罐装鱼贝罐头应当保持干燥，不应当被置于极端的温度下。

16.4.7 成品的运输（加工步骤 14）

潜在危害：由于容器破坏或被置于极端环境所引起的污染。

潜在缺陷：不可能。

技术指南：

参见第十七部分（运输），同时应符合以下要求：

（1）鱼贝罐头在运输过程中应当尽力防止罐体受损，装载和卸载中使用铲车的方法应当得当；

（2）容器和箱子应彻底干燥。实际上，湿度会对箱子的物理特性产生影响，如果过于湿润就会使罐体容易受损；

（3）金属容器在运输过程中应当保持干燥，以避免被腐蚀和锈化。

第十七部分 运输

参照《国际推荐操作规范 - 食品卫生总则》，第Ⅷ部分 - 运输，CAC/RCP1969，Rev.4（2003）和《法典卫生操作规范 - 运输》散装和半包装食品（CAC/RCP 47 - 2001）。

运输适用于所有部分，是流程图的一个步骤，需要专门的技术。应考虑给予与其他加工步骤相同的关注。本部分将举例说明潜在的危害和缺陷，并提供技术指南，以便采取控制措施和纠正措施。在每一个步骤中，只列举本步中可能引入或控制的危害和缺陷。应该认识到，在准备一个HACCP和（或）DAP计划时，有必要查阅第五部分，该部分对HACCP和（或）DAP分析的应用提供了指南。然而，在本操作规范的范围内，不可能列举每一个步骤的关键限制、监测、记录保持和验证等详细内容，因为这些都是针对特定危害和缺陷的具体内容。

在运输新鲜冷冻鱼或冷藏鱼、贝类及其产品的整个过程中，要特别注意将产品的温升控制到最小，适用时，应对冷藏或冷冻温度进行控制。而且，应采取措施尽量减少产品及其包装的损坏。

17.1 新鲜、冷藏和冷冻的产品

参照3.6运输。

潜在危害：生化产物（组胺），微生物的增长和污染。

潜在缺陷：腐败、物理损伤，化学污染（燃料）。

技术指南：

（1）装货前对产品温度进行检查；

（2）鱼类、贝类及其产品的装卸货过程中，应尽量避免暴露在升高的温度下；

（3）码放应保证产品与墙壁、地板、顶板之间空气流通良好；建议安装稳定装置；

（4）运输中监测货舱中的空气温度；建议使用温度记录装置；

（5）运输过程中，冷冻的产品应当保持在 -18℃或更低的温度下（最大波动 +3℃）；新鲜的鱼类、贝类及其产品温度应尽可能保持在接近0℃。新鲜整鱼应当放在浅盘里，裹以正在融化的碎冰；要有适当的排水装置，确保冰融化后产生的水不会滞留与产品接触，也不会从一个容器流向另一个容器引起产品交叉污染；鲜鱼放于带有干冷冻袋的容器中运输，适用时，可不考虑用冰；鱼放在冰碴、冷冻的海水或冷藏的海水里运输（例如，适用时，可考虑用于远洋鱼）；冷冻的海水和冷藏的海水应该达到可接收的条件；冷藏加工的产品应维持在加工者指定的温度（但是通常不应超过4℃）；对鱼、贝类及其产品进行应适当保护，防止灰尘污染，高温以及太阳或风的干化。

17.2 活鱼和贝类

参阅本规范有关部分的具体条款。

17.3 罐装鱼和贝类

（1）参阅第十六部分的具体条款；

（2）装货前应核实运输车的货舱是否干净、适合和卫生；

（3）装货和运输过程中应防止产品损坏和污染，确保包装的完整性；

（4）卸载以后，应该避免垃圾的堆积，而是应该以合理的方式对它们进行处理。

第十八部分 零售

为了识别每一个加工步骤的控制方法，本部分将举例说明潜在的危害和缺陷，并提供技术指南，以便采取控制措施和纠正措施。在每一个步骤中，只列举本步中可能引入或控制的危害和缺陷。应该认识到，在准备一个 HACCP 和（或）DAP 计划时，有必要查阅第五部分，该部分对 HACCP 和（或）DAP 分析的应用提供了指南。然而，在本操作规范的范围内，不可能列举每一个步骤的关键限制、监测、记录保持和验证等详细内容，因为这些都是针对特定危害和缺陷的具体内容。

零售的鱼、贝类及其制品的接收、操作、贮藏及对消费者的展示都应最大限度地减少食品安全方面的潜在危害和缺陷，同时应保持必需的产品质量。与用于食品安全和质量的 HACCP 和 DAP 方法一样，应从了解的或经过批准的并受健康主管机构控制的来源购买产品，该主管机构可以验证 HACCP 控制情况。零售商应制定和使用书面的购买说明，用于保证食品安全和期望的质量水平。零售商应该对产品的质量和安全负责。

要保持产品的安全和必需的质量，接收后保持合适的贮藏温度非常关键。冷藏产品应当以卫生的方式贮藏在 4℃（40 ℉）或更低的温度下，MAP 产品应当贮藏在 3℃（28 ℉）或更低的温度下，而冷冻产品则应当贮藏在 -18℃（0 ℉）或更低的温度下。

预处理和包装方式应符合第三部分必备程序和《法典标签标准》中的规范和建议。完全打开展示的产品，应进行保护不受环境的影响，例如，使用展示罩（防喷嚏）。无论何时，都应当为展示的海产品提供适宜的温度和条件，除了避免失去必需的品质外，还应防止细菌生长、毒素和其他危害的产生。

在购买处为消费者提供的资料对于保证产品的安全和质量很重要，例如，海报或小册子，告诉消费者关于贮藏、制备方法，以及如不正确处理或制备可能发生的危险等信息。

应建立鱼、贝类及其产品的来源和代码追溯系统，便于在预防性的健康保护过程和措施失败事件中，进行产品召回或公共健康调查。有些国家对软体贝类设置了该系统，做法是要求对软体贝类加标签。

18.1 零售鱼、贝类及其产品的接收——总则

潜在危害：见第 7.1、8.1 节“接收”。

潜在缺陷：见第 7.1、8.1 节“接收”。

技术指南：

（1）应对运输工具进行全面的卫生检查。受到污染的产品应拒收；

（2）应检查运输车是否可能引起生鱼、鱼产品和即食的鱼、渔产品的交叉污染。确定即食产品没有接触生的产品、汁液或活的软体贝类，以及生的软体贝类没有接触其他的生鱼或贝类；

（3）应对海产品进行定期检查，看其是否符合采购要求；

（4）所有产品在接收时都应检查是否腐烂或损坏，有腐败迹象的产品应拒收；

（5）当记录运输工具的温度的时候，应该检查记录确保它符合温度的要求。

18.1.1 零售冷藏产品的接收

潜在危害：病原体增长，微生物病原体，化学和物理污染，产生鲭鱼毒素，产生肉毒梭菌。

潜在缺陷：腐烂（腐败）、污染物、污物。

技术指南：货运时，应记录不同位置的产品温度。冷藏鱼、贝类及其产品的温度应当保持在 4℃（40 ℉）或更低的温度下，MAP 产品如果不是冷冻的则应当保持在 3℃（38 ℉）或更低的温度下。

18.1.2 零售冷冻产品的接收

潜在危害：不可能。

潜在缺陷：解冻、污染物、污物。

技术指南：

（1）接收冷冻海产品时，应检查是否有解冻的迹象及存在污物或污染，可疑的货物应拒收；

（2）应当检查购进的海产品的内部温度，记录货物中不同位置的温度。冷冻的鱼、贝类及其产品应当保持在－18℃（0 ℉）或更低的温度下。

18.1.3 零售产品的冷藏

潜在危害：产生鲭鱼毒素，微生物病原体，病原体增长，化学污染，产生肉毒杆菌。

潜在缺陷：腐烂，污染物，污物。

技术指南：

（1）冷藏产品应当保持4℃（40 ℉），MAP产品应当保持3℃（38 ℉）或更低；

（2）海产品应包装良好并与地板隔离，以保护其不受污物或其他污染物的影响；

（3）建议对海产品的冷藏装置使用连续的温度记录图表；

（4）冷藏室应当有适当的排水装置，以防止产品受到污染；

（5）即食产品和软体贝类以及冷库中其他生的食品应彼此隔离。在货架上，生的产品应当放在熟产品的下面，以防止滴水造成交叉污染；

（6）应当建立一个合理的产品周转体系，这一体系以先入先出，生产日期或者标签上标明的最佳日期，感官质量等为基础。

18.1.4 零售产品的冷冻贮藏

潜在危害：不太可能有。

潜在缺陷：化学分解（酸败）、脱水。

技术指南：

（1）产品应当保持在－18℃（0 ℉）或更低的温度下，应定期进行温度监测。建议使用温度记录装置；

（2）不应将海产品直接放在地板上。码放产品时应保证适当的空气流通。

18.1.5 零售冷藏产品的制备和包装

参阅第8.2.3节“标签”。

潜在危害：微生物病原体、鲭鱼毒素的产生、病原体增长、物理和化学污染、过敏原。

潜在缺陷：腐烂、贴标不正确。

技术指南：

（1）应注意保证产品的处理和包装按照第三部分程序前提中的指南进行；

（2）应注意保证产品的标识按照第三部分必备程序和《法典标签标准》中的指南进行，特别是对已知的过敏原；

（3）应注意确保产品在操作和包装过程中，不受温度变化的影响；

（4）应注意防止即食的和生的贝类、贝类及其产品，通过工作区、器具或人造成交叉污染。

18.1.6 零售冷冻海产品的制备和包装

参阅第8.2.3节“标签”。

潜在危害：微生物病原体、物理和化学污染、过敏原。

潜在缺陷：解冻、不正确贴标。

技术指南：

（1）应注意要确保按照第三部分必备程序和《法典标签标准》来识别过敏原；

（2）应注意防止即食产品和生的产品间的交叉污染；

（3）在较长时期内，冷冻海产品不应受室内环境温度的影响。

18.1.7 冷藏海产品的零售展示

潜在危害：产生鲭鱼毒素、微生物增长、微生物污染、产生肉毒杆菌毒素。

潜在缺陷：腐烂、脱水。

技术指南：

（1）冷藏展示的产品应保持在4℃（40 ℉）或更低的温度下，并定期记录产品温度；

（2）在柜台服务冷藏展示柜中，应将即食产品、软体贝类与生的食品彼此分开。建议使用展示框图，以确保不发生交叉污染；

（3）如果使用冰，则应安装冰水的排放装置，零售展示柜应能自己排水，要每天换冰，并确保即食产品不会放在生的产品曾用过的冰上；

（4）在柜台销售展示柜中的每一件商品都应有单独的容器和销售用具，以防止交叉污染；

（5）应注意产品摆放的量（厚度）不应过大，以致不能维持正常的冷藏，从而影响产品质量；

（6）在柜台销售展示柜中，应注意防止未进行保护的产品干化。建议在保证卫生的条件下使用喷雾器；

（7）以自助式销售带包装的产品时，码放的产品不应当超过“装载线”，此时无法保持冷藏状态；

（8）当供应/现货销售时，产品不应长时间暴露在室温下；

（9）在柜台销售展示柜中，海产品应该使用标签或布告的形式进行合理的表示以告知消费者该产品的通常被接受的名称。

18.1.8 冷冻海产品的零售展示

潜在危害：不太可能有。

潜在缺陷：解冻、脱水（冻解烧）。

技术指南：

（1）产品应保持在-18℃（0 ℉）或更低的温度下。应定期进行温度监测。建议使用温度记录装置；

（2）自助式销售时，产品不应超过货柜的“装载线”。立式冷冻自助售货柜应有能自动关闭的门或气幕来保持冷冻状态；

（3）在定单或现货销售时，产品不应当长时间暴露在室温下；

（4）应建立一个冷冻海产品周转系统，确保先入先出；

（5）应定期检查零售展示柜中的冷冻海产品，从而估计包装的完整性和脱水或冻结烧的程度。

附录 1

与鲜鱼、贝和其他无脊椎水产品相关的潜在危害

1 生物危害的例子

1.1.1 寄生虫

寄生虫可以引起人类疾病，并可以通过鱼或甲壳类动物传播，它们被划分为寄生虫和蠕虫。常见的有线虫、绦虫、吸虫。鱼能被原生动物寄生，但没有鱼的原生动物疾病传播到人身上的记录。寄生虫有完整的生命周期，会涉及一个或多个中间寄主，向人类传播的主要途径是通过伤口、粗略的加工或不充足的烹饪所引起的食源疾病。在 -20℃或更低的温度下冷冻七天或 -35℃下放置 20 小时就会杀死这些寄生虫。用盐或酸浸泡会减少寄生虫的危害，但是不会彻底清除。对着光进行镜检，整修罐体和寄生虫包囊的物理移动等方法不可能清楚但可以减轻危害。

线虫

已知有许多种类的线虫会广泛的侵入鱼贝等体内，而且事实上有很多种类的海洋鱼类是作为第二寄主。这些线虫中备受关注的是以下几种，海兽胃线虫，毛细线虫，棘口线虫和 Pseudoteranova spp.，它们会寄生在肝脏，腹腔和海洋鱼的肉中。线虫引起人类疾病的一个例子是异尖线虫；处于传染期的寄生虫在 60℃下加热一分钟和 -20℃下冷冻 24 小时的情况下被杀死。

绦虫

与消费相关的绦虫主要是阔节裂头虫。绦虫寄主范围广，海洋和淡水鱼类都会成为它的中间寄主。与其他寄生虫传染方法相似，主要是通过生食或加工不当引起。类似的冷冻和加热温度也适用于绦虫来杀死传染期的寄生虫。

吸虫

由于鱼引起的吸虫（扁虫）病传染是 20 多个国家共同面临的主要的地方性的公共健康问题。根据受感染的人数来看，最主要的吸虫是以下两个属的种类：无性繁殖和 *Ophisthorchis*（肝脏寄生），*Paragonimus*（肠道寄生），仅次之的是异形吸虫 *and Echinochasmus*（*intestinal flukes*）。这些吸虫的最终寄主主要是人类或其他哺乳动物。淡水鱼是 *Clonorchis and Ophistorchis* 生命周期的第二寄主，*Paragonimius.* 的第二寄主是淡水甲壳类。食物传染通过消费生品，烹饪不彻底或其他加工不当引起的带有感染性的寄生虫的食品。-20℃冷冻七天或 -35℃冷却 24 小时会杀死这些感染性的寄生虫。

1.1.2 细菌

鱼捕获时的污染水平取决于捕获水域的环境和细菌学质量。长须鲸的微生态组成受很多因素的影响：水温、盐度、捕获区域离人生活居住区的距离、鱼饵料的质量及来源和捕获方法。长须鲸可食用的肌肉组织在捕获时经过一般消毒，细菌常附在表皮、腮和肠道内。

在捕获时主要有两种类型的细菌对公共健康会起到重要的作用，一种是水域环境中通常或偶尔会出现的细菌，涉及当地环境的微环境（microflora）；另外一种是由于鱼体内部或工业废物引起的微生物污染。可能引起健康危害的本土环境中的细菌的例子是气单菌，肉毒梭菌，溶血弧菌，霍乱弧菌，创伤弧菌和李斯特菌。对公共健康有重要作用的非本土细菌包括肠杆菌，例如，沙门氏菌属、志贺菌属和埃希氏菌属。除此之外，引起食源疾病的，并在鱼肉中偶尔会分离到的其他细菌有迟钝爱德华式菌，*Pleisomonas shigeloides* 和耶耳辛氏肠炎杆菌。金黄葡萄球菌也有可能出现，并且会产生抗热毒素。

新鲜鱼体上的本土致病菌一般数目不多，通过充分的烹饪可以安全食用，食品安全危害是可以忽略的。在贮藏过程中，本土损坏细菌生长会超过本土致病菌，因此，鱼体在有毒前会先腐烂，消费者会拒绝食用。这些致病菌引起的危害能通过充分加热、把鱼放置在较低的温度下或避免加工过程中的交叉污染来控制。

弧菌在海岸和河口环境中常见，污染程度取决于水深和潮汐水平。弧菌多在暖热水域出现，温水域在夏季的时候会出现。弧菌是暖热环境下味道不好闻的水分的自然污染种类，常出现在这些区域的养殖鱼中。弧菌给长须鲸带来的危害可以通过彻底烹饪和防止烹饪产品的交叉污染的方法来避免，也可以通过捕获后快速冷却产品以减少这些生物体繁殖的方法来减少健康危害。副溶血性弧菌的特定的变形可能引起致病性。

1.1.3 滤过性病毒污染

在近海岸捕获的被人类或动物粪便中滤过性病毒污染的甲壳类软体动物对人类是具有致病性的。水产品疾病相关的肠病毒有A型肝炎病毒，环状病毒，星状病毒和诺瓦克病毒。后三种都是小的圆形病毒。所有这些水产品食源病毒都是通过粪便-口的途径传播的，多数滤过性病毒胃肠炎的爆发都与食用污染的贝、特定的生牡蛎相关。

一般情况下病毒有特定的种类，而且在食品或主体细胞的任何地方都不会生长或繁殖。没有可靠的标记可以指示甲壳类动物丰收水域的病毒的有无状况。水产品食源病毒很难检测，必须是非常精准的分子学方法才能确定病毒是否存在。

胃肠炎病毒的发生可以通过控制贝类养殖区域的下水道污染，贝类丰收和水域情况监控和控制加工过程中其他污染源的方法来使的污染最小化。要么会传播疾病，要么就要净化，而贝类对病毒的自净作用要比净化细菌的时间长。热处理（85~90℃温度下1.5min）会破坏贝中的病毒。

1.1.4 生物毒素

有很多重要的生物毒素要考虑。大约有四百多种有毒鱼类，通过定义可以知道主要是其中的毒素在起作用。毒素常限定在某些器官中，或者在一年之中的某个时期。

对于一些鱼来说，毒素存在于血液中，这些是ichtyohaemotoxin。相关的种类有亚得里亚海的鳗鲡、欧洲海鳗和七腮鳗。其他种类中，毒素遍布所有器官（肉，内脏，皮肤），这些是ichtyohaemotoxin。河豚毒素是属于这个范畴的，很多中毒是由河豚毒素引起的，而且常常是致命的。

一般情况下，这些毒素是热稳定性的，唯一可能的控制方法是明确鱼的种类。

藻毒

珊瑚礁鱼毒

另外一个重要的毒素是珊瑚礁鱼毒，它广泛存在于热带和亚热带珊瑚礁附近的食肉鱼栖息的浅水区域，已经发现很多种类。这种毒素的来源是腰鞭毛虫，超过400个种类的热带鱼与之相关。已知这种毒素具有热稳定性。对毒素进一步的情况尚没有了解，唯一的控制方法是避免有毒素记录的鱼的上市。

PSP/DSP/NSP/ASP

麻痹性贝毒、腹泻性贝毒、神经性病毒和记忆性病毒是有浮游植物产生的。双壳贝类通过滤食水中的浮游植物富集了这些毒素，一些鱼和甲壳类动物也对毒素有富集作用。

一般情况下，这种毒素热处理后还会有毒性，因此对贝和鱼的来源种类、身份的了解对其加工处理人员来说是很重要的。

河豚毒素

河豚鱼（充气鱼）可以积聚这种毒素，很多中毒事件都是这种毒素引起的，而且常常是致命的。这种毒素一般分布在鱼类肝脏、卵巢和内脏中，鱼肉中也有一些。这种毒素与其他鱼或贝聚集的毒素不同点在于，这种毒素不是有藻类产生的，具体的产生机制还不是很明确，但是明显的可以推测这与共生细菌有关。

1.1.5 鲭组蛋白毒素

鲭亚目鱼类青睐者有时会发生组胺中毒，那是由于鱼在捕获后没有正确的冷却引起的。鲭组蛋白中毒主要归咎于肠杆菌，捕捞后未能立即冷却的话，肠杆菌能在鱼体肌肉中产生高水平的组胺和其他生物胺。可能引起中毒的疾病主要是亚鲭鱼类，例如，金枪鱼，鲭鱼和鲣鱼，但也可以是如鲱科鱼类。这种毒素很少引起致命，多数是很温和的。捕获后快速冷却和加工过程中高标准的处理可以防止这种毒素的生成。正常的加热处理后毒素的毒性会尚失。此外，鱼体可能会有较高的组胺毒素而不存在任何的感官参数特征的损坏。

1.2 化学危害

捕获自沿海区域和内陆栖息地的鱼可能会暴露在各种各样的环境污染下。捕自沿海和河口的鱼类比在公海上捕获的鱼受污染的可能性要更高。化学无，有机金属化学物和重金属都可能污染产品而引起安全问题。如果没有遵循正确的休药期或这些化合物的销售和应用为很好的被控制，兽药残留可能在养殖产品中产生健康危害。当被不正当处理和去毒剂或消毒剂而未能很好的冲洗时，鱼也能被化学物质（例如，柴油）污染。

1.3 物理危害

物理危害包括物质的危害，例如，金属或玻璃碎片、贝壳、骨骼的危害等。

CODE OF PRACTICE FOR FISH AND FISHERY PRODUCTS

(*CAC/RCP 52 - 2003, REV. 2 - 2005*)

TABLE OF CONTENTS

	Introduction
	How To Use This Code
SECTION 1	Scope
SECTION 2	Definitions
2. 1	General definitions
2. 2	Aquaculture
2. 4	Fresh, Frozen and Minced Fish
2. 5	Frozen Surimi
2. 12	Canned Fish and Shellfish
SECTION 3	Pre-Requisite Programme
3. 1	Fishing and Harvesting Vessel Design and Construction
3. 2	Processing Facility Design and Construction
3. 3	Design and Construction of Equipment and Utensils
3. 4	Hygiene Control Programme
3. 5	Personal Hygiene and Health
3. 6	Transportation
3. 7	Product Tracing and Recall Procedures
3. 8	Training
SECTION 4	General Considerations for the Handling of Fresh Fish and Shellfish
4. 1	Potential Hazards Associated with Fresh Fish and Shellfish
4. 2	Time and Temperature Control
4. 3	Minimise the Deterioration of Fish and Shellfish - Handling
SECTION 5	Hazard Analysis Critical Control Point (HACCP) and Defect Action Point (DAP) Analysis
5. 1	HACCP Principles
5. 2	Defect Action Point Analysis
5. 3	Application
5. 4	Conclusion
SECTION 6	Aquaculture Production
6. 1	General
6. 2	Identification of hazards and defects
6. 3	Production operations

SECTION 8 Processing of Fresh, Frozen and Minced Fish
8. 1 Finfish Preparation
8. 2 Processing of Modified Atmosphere Packed Fish
8. 3 Processing of Frozen Fish
8. 4 Processing of Minced Fish
8. 5 Packaging, Labels and Ingredients
SECTION 9 Processing of Frozen Surimi
9. 1 General Considerations of Hazards and Defects
9. 2 Fish Preparation
9. 3 Meat Separation Process
9. 4 Washing and De-Watering Process
9. 5 Refining Process
9. 6 Final De-watering Process
9. 7 Mixing and Addition of Adjuvant Ingredients Process
9. 8 Packaging and Weighing
9. 9 Freezing Operation
9. 10 Dismantling Freezing Pan
9. 11 Metal Detection
9. 12 Boxing and Labelling
9. 13 Frozen Storage
9. 14 Raw Material Reception - Packaging and Ingredients
9. 15 Raw Material Storage - Packaging and Ingredients
SECTION 10 Processing of QF Coated Fish Products
10. 1 General Addition to Pre-requisite Programme
10. 2 Identification of Hazards and Defects
10. 3 Processing Operations
SECTION 14 Processing of Shrimps and Prawns
14. 1 Frozen Shrimps and Prawns - General
14. 2 Shrimp Preparation
SECTION 15 Processing of Cephalopods
15. 1 Reception of Cephalopods
15. 2 Storage of Cephalopods
15. 3 Controlled Thawing
15. 4 Splitting, Gutting and Washing
15. 5 Skinning, Trimming
15. 6 Application of Additives
15. 7 Grading / Packing/Labelling
15. 8 Freezing
15. 9 Packaging, Labels and Ingredients - Reception and Storage
SECTION 16 Processing of Canned Fish and Shellfish
16. 1 General - Addition to Pre-requisite Programme

16. 2	Identification of Hazards and Defects
16. 3	Processing Operations
16. 4	Pre-cooking and Other Treatments
SECTION 17	Transport
17. 1	Fresh, Refrigerated and Frozen Products
17. 2	Live Fish and Shellfish
17. 3	Canned Fish and Shellfish
17. 4	All Products
SECTION 18	Retail
18. 1	Reception of Fish, Shellfish and their Products at Retail – General Considerations

INTRODUCTION

This Code of Practice for Fish and Fishery Products has been developed by the Codex Committee on Fish and Fishery Products from the merging of the individual codes listed in Appendix XII * plus a section on aquaculture and a section on frozen surimi. These codes were primarily of a technological nature offering general advice on the production, storage and handling of fish fishery products on board fishing vessels and on shore. It also deals with the distribution and retail display of fish and fishery products.

This combined Code of practice has been further modified to incorporate the Hazard Analysis Critical Control Point (HACCP) approach described in the *Recommended International Code of Practice – General Principles of Food Hygiene* (CAC/RCP 1-1969, Rev. 3 1997), Annex: *HACCP System and Guidelines for its Application.* A pre-requisite programme is described in the Code covering technological guidelines and the essential requirements of hygiene in the production of fish, shellfish and their products, which are safe for human consumption, and otherwise meets the requirements of the appropriate Codex product standards. The Code also contains guidance on the use of HACCP, which is recommended to ensure the hygienic production of fish and fishery products to meet health and safety requirements.

Within this Code a similar systematic approach has been applied to essential quality, composition and labelling provisions of the appropriate Codex product standards. Throughout the code this is referred to as "Defect Action Point (DAP) Analysis". However DAP analysis is optional.

The Codex Committee on Fish and Fishery Products recommended at its Twentieth Session that defects of a commercial nature, i. e. workmanship defects, which had been removed from Codex fish product standards, be transferred to the appropriate Codex Code of practice for optional use between buyers and sellers during commercial transactions. The Committee further recommended that this detail should be described in a section on Final Product Specifications, which now appear as Appendices II-XI * of this document. A similar approach to HACCP has been incorporated into the Code as guidelines for the control of defects (DAP Analysis).

This Code will assist all those who are engaged in the handling and production of fish and fishery products, or are concerned with their storage, distribution, export, import and sale in attaining safe and wholesome products which can be sold on national or international markets and meet the requirements of the Codex Standards (see Appendix XII *).

HOW TO USE THIS CODE

The aim of this Code is to provide a user-friendly document as background information and guidance for the elaboration of fish and shellfish process management systems which would incorporate Good Management Practice (GMP) as well as the application of HACCP in countries where these, as yet, have not been developed. In addition, it could be used for training of fishermen and employees of the fish and shellfish processing industries.

The practical application of this *international* Code, with regard to *national* fisheries, would therefore require some modifications and amendments, taking into account local conditions and specific consumer requirements. This Code, therefore, is not intended to replace the advice or guidance of trained and experienced technologists regarding the complex technological and hygienic problems which might be unique to a specific geographical area or specific fishery and, in fact, is intended to be used as a supplement in such instances.

This Code is divided into separate, though interrelated, Sections. It is intended that in order to set up a HACCP or DAP programme these should be consulted as appropriate:

(a) *Section* 2 – *Definitions* – Being acquainted with the definitions is important and will aid the overall understanding of the Code.

(b) *Section* 3 – *Pre-requisite Programme* - Before HACCP or a similar approach can properly be applied to a process it is important that a solid foundation of good hygienic practice exists. This Section covers the groundwork which should be regarded as the minimum requirements for a facility prior to the application of hazard and defect analyses.

(c) *Section* 4 – *General Considerations for the Handling of Fresh Fish, Shellfish and Other Aquatic Invertebrates* – This Section provides an overall view of the potential hazards and defects which may have to be considered when building up a HACCP or DAP plan. This is not intended to be an exhaustive list but is designed to help a HACCP or DAP team to think about what hazards or defects should be considered in the fresh fish, shellfish and other aquatic invertebrates, and then it is up to the team to determine the significance of the hazard or defect in relation to the process.

(d) *Section* 5 – *Hazard Analysis Critical Control Point (HACCP) and Defect Action Point (DAP) Analysis* -Only when the groundwork in Section 3 has been satisfactorily achieved should the application of the principles outlined in Section 5 be considered. This Section uses an example of the processing of a canned tuna product to help illustrate how the principles of HACCP should be applied to a process.

(e) *Sections* 6 *and* 7 – *Aquaculture Production* and *Molluscan Shellfish production* deal with pre-harvest and primary production of fish, crustaceans and molluscan shellfish not caught in the wild*.

Although potential hazards and potential defects are listed for most steps in Sections 6 to 18, it should be noted that this is only for guidance and the consideration of other hazards and/or defects may be appropriate. Also, the format in these Sections has been designed for maximum 'ease of use' and therefore the 'potential hazards' or 'potential defects' are listed only where they may be introduced into a product or where they are controlled, rather than repeating them at all the intervening processing steps.

Additionally, it must be stressed that hazards and defects, and their subsequent control or action points, are product and line specific and therefore a full critical analysis based on Section 5 must be completed for each individual operation.

(f) *Section* 8 - *Processing of Fresh, Frozen and Minced Fish* – This Section forms the foundation for most of the subsequent processing Sections. It deals with the major process steps in the handling of raw fish through to cold storage and gives guidance and examples on the sort of hazards and defects to expect at the various steps. This Section should be used as the basis for all the other processing operations (Sections 9-16) which give additional guidance specific to the appropriate product sector*.

(g) *Sections* 9 *to* 16 – *Processing of Specific Fish and Shellfish Products* – Processors operating in particular sectors will need to consult the appropriate Section to find additional information specific to that sector*.

(h) *Sections* 17 *to* 18 - *Transportation and Retail* cover general transportation and retail issues. Transportation and retail apply to most if not all sections for processing of specific products. They should be considered with the same care as the other processing steps*.

(i) Additional information will be found in the *Appendices**.

SECTION 1 SCOPE

This Code of practice applies to the growing, harvesting, handling, production, processing, storage transportation and retail of fish, shellfish and aquatic invertebrates and products thereof from marine and freshwater sources, which are intended for human consumption.

SECTION 2 DEFINITIONS

For the purpose of this Code:

2.1 GENERAL DEFINITIONS

Biotoxins	means poisonous substances naturally present in fish and fishery products or accumulated by the animals feeding on toxin producing algae, or in water containing toxins produced by such organisms.
Chilling	is the process of cooling fish and shellfish to a temperature approaching that of melting ice.
Clean Water	means water from any source where harmful microbiological contamination, substances and/or toxic plankton are not present in such quantities as may affect the health quality of fish, shellfish and their products.
Cleaning	means the removal of soil, food residues, dirt, grease or other objectionable matter.
Contaminant	means any biological or chemical agent, foreign matter, or other substances not intentionally added to food which may compromise food safety or suitability.
Contamination	the introduction or occurrence of a contaminant in fish, shellfish and their products.
Control Measure	means any action and activity that can be used to prevent or eliminate a food safety hazard or reduce it to an acceptable level. For the purposes of this Code a control measure is also applied to a defect.
Corrective Action	means any action to be taken when the results of monitoring at the CCP indicate a loss of control. For the purposes of this Code this also applies to a DAP.
Critical Control Point (CCP)	a step at which control can be applied and is essential to prevent or eliminate a food safety hazard or reduce it to an acceptable level.
Critical Limit	is a criterion, which separates acceptability from unacceptability. For the purpose of this Code this also applies to a DAP.
Decision Tree	a sequence of questions applied to each process step with an identified hazard to identify which process steps are CCPs. For the purpose of this Code this also applies to a DAP.
Decomposition	is the deterioration of fish, shellfish and their products including texture breakdown and causing a persistent and distinct objectionable odour or flavour.
Defect	means a condition found in a product which fails to meet essential quality, composition and/or labelling provisions of the appropriate Codex product standards.

Defect Action Point (DAP)	a step at which control can be applied and a quality (non-safety) defect can be prevented, eliminated or reduced to acceptable level, or a fraud risk eliminated.
Disinfection	means the reduction, by means of chemical agents and/or physical methods, the number of micro-organisms in the environment, to a level that does not compromise food safety or suitability.
Dressed	means that portion of fish remaining after heading and gutting.
Facility	means any premises where fish and fishery products are prepared, processed, chilled, frozen, packaged or stored. For the purposes of this Code, premises also includes vessels.
Fish	means any of the cold-blooded (ectothermic) aquatic vertebrates. Amphibians and aquatic reptiles are not included.
Hazard	a biological, chemical or physical agent in, or condition of, food with the potential to cause an adverse health effect.
Hazard Analysis	the process of collecting and evaluating information on hazards and conditions leading to their presence to decide which are significant for food safety and therefore should be addressed in the HACCP plan.
Hazard Analysis Critical Control Point (HACCP)	a system which identifies, evaluates, and controls hazards which are significant for food safety.
Monitor	the act of conducting a planned sequence of observations or measurements of control parameters to assess whether a CCP is under control. For the purpose of this Code this also applies to a DAP.
Potable Water	is fresh water fit for human consumption. Standards of potability should not be lower than those contained in the latest edition of the "International Standards for Drinking Water", World Health Organisation.
Pre-Requisite Programme	is a programme that is required prior to the application of the HACCP system to ensure that a fish and shellfish processing facility is operating according to the Codex Principles of Food Hygiene, the appropriate Code of Practice and appropriate food safety legislation.
Raw Material	are fresh and frozen fish, shellfish and/or their parts which may be utilised to produce fish and shellfish products intended for human consumption.
Refrigerated Water	is clean water cooled by a suitable refrigeration system.
Shelf-Life	the period during which the product maintains its microbiological and chemical safety and sensory qualities at a specific storage temperature. It is based on identified hazards for the product, heat or other preservation treatments, packaging method and other hurdles or inhibiting factors that may be used.
Shellfish	means those species of aquatic molluscs and crustaceans that are commonly used for food.
Step	is a point, procedure, operation or stage in the food chain including raw materials, from primary production to final consumption.
Validation	means obtaining evidence that the elements of the HACCP plan are effective.

Verification	the application of methods, procedures, tests and other evaluations, in addition to monitoring to determine compliance with the HACCP plan. For the purposes of this Code this also applies to a DAP.
Whole Fish (or Round Fish)	are fish as captured, ungutted.

2.2 AQUACULTURE

Aquaculture	means the farming during part or the whole of their life cycle of all aquatic animals, except mammalian species, aquatic reptiles and amphibians intended for human consumption, but excluding species covered in Section 7 of this code. These aquatic animals are hereafter referred to as "fish" for ease of reference in Section 2.2 and Section 6.
Aquaculture Establishment	is any premises for the production of fish intended for human consumption, including the supporting inner infrastructure and surroundings under the control of the same management.
Chemicals	includes any substance either natural or synthetic which can affect the live fish, its pathogens, the water, equipment used for production or the land within the aquaculture establishment.
Colouring	means obtaining specifically coloured feature (e.g. flesh/shell/gonad) of a targeted organism by incorporating into the fish food a natural or artificial substance or additive approved for this purpose by the agency having jurisdiction.
Diseased Fish	means a fish on or in which pathological changes or other abnormalities that affect safety and quality are apparent.
Extensive farming	means raising fish under conditions of little or incomplete control over the growing process and production conditions where their growth is dependent upon endogenously supplied nutrient inputs.
Feed Additives	means chemicals other than nutrients for fish which are approved for addition to their feed.
Fish farm	is an aquaculture production unit (either land-or water based); usually consisting of holding facilities (tanks, ponds, raceways, cages), plant (buildings, storage, processing), service equipment and stock.
Fish Feed	means fodder intended for fish in aquaculture establishments, in any form and of any composition.
Good Aquaculture (or Good Fish Farming) Practices	are defined as those practices of the aquaculture sector that are necessary to produce quality and safe food products conforming to food laws and regulations
Harvesting	Operations involving taking the fish from the water.
Intensive farming	means raising fish under controlled growing process and production conditions where their growth is completely dependent on externally supplied fish feed.

Official Agency Having Jurisdiction	means the official authority or authorities charged by the government with the control of food hygiene (sometimes referred to as the competent authority) as well as/or with sanitation in aquaculture.
Pesticide	means any substance intended for preventing, destroying, attracting, repelling or controlling any pest including unwanted species of plants or animals during the production, storage, transport, distribution and processing of food, agricultural commodities, or animal feeds or which may be administered to animals for the control of ectoparasites. The term normally excludes fertilisers, plant and animal nutrients, food additives, and veterinary drugs.
Pesticide Residue	means any specified substance in food, agricultural commodities, or animal feed resulting from the use of a pesticide. The term includes any derivatives of a pesticide, such as conversion products, metabolites, reaction products, and impurities considered to be of toxicological significance.
Residues	means any foreign substances including their metabolites, which remain in fish prior to harvesting as a result of either application or accidental exposure.
Semi-intensive farming	means raising fish under conditions of partial control over the growing process and production conditions where their growth is dependent upon endogenously supplied nutrient inputs and externally supplied fish feed.
Stocking density	is the amount of fish stocked per unit of area or volume.
Veterinary Drug	means any substance applied or administered to any food-producing animal, such as meat or milk-producing animals, poultry, fish or bees, whether used for therapeutic, prophylactic or diagnostic purposes or for modification of physiological functions or behaviour.
Withdrawal Time	is the period of time necessary between the last administration of a veterinary drug to fish, or exposure of these animals to a veterinary drug, and harvesting of them to ensure that the concentration of the veterinary drug in their edible flesh intended for human consumption, complies with the maximum permitted residue limits.

2.4 FRESH, FROZEN AND MINCED FISH

Candling	is passing fillets of fish over a translucent table illuminated from below to detect parasites and other defects.
Dehydration	is the loss of moisture from frozen products through evaporation. This may occur if the products are not properly glazed, packaged or stored. Deep dehydration adversely affects the appearance and surface texture of the product and is commonly known as "freezer burn"
Fillet	is a slice of fish of irregular size and shape removed from the carcase by cuts made parallel to the backbone.

Freezer is equipment designed for freezing fish and other food products, by quickly lowering the temperature so that after thermal stabilisation the temperature in the thermal centre of the product is the same as the storage temperature.

Freezing Process is a process which is carried out in appropriate equipment in such a way that the range of temperature of maximum crystallisation is passed quickly. The quick freezing process shall not be regarded as complete unless and until the product temperature has reached −18℃ (0°F) or lower at the thermal centre after thermal stabilisation.

Frozen Storage Facility a facility that is capable of maintaining the temperature of fish at -18℃. Fresh Fish are fish or fishery products which have received no preserving treatment other than chilling.

Frozen Fish are fish which have been subjected to a freezing process sufficient to reduce the temperature of the whole product to a level low enough to preserve the inherent quality of the fish and which have been maintained at this low temperature, as specified in the Standard for Quick Frozen Finfish, Eviscerated and Uneviscerated during transportation, storage and distribution up to and including the time of final sale. For the purpose of this Code the terms "frozen", "deep frozen", "quick frozen", unless otherwise stated, shall be regarded as synonymous.

Glazing The application of a protective layer of ice formed at the surface of a frozen product by spraying it with, or dipping it into, clean sea water, potable water, or potable water with approved additives, as appropriate.

Minced Fish is comminuted flesh produced by separation from skin and bones.

Modified Atmosphere Packaging (MAP) means packaging in which the atmosphere surrounding the fish is different from the normal composition of air.

Separation is a mechanical process for producing minced fish whereby the skin and bone is substantially removed from the flesh.

Separator is a mechanical device used for separation.

Steak is a section of fish, removed by cutting approximately at right angle to the backbone.

2.5 FROZEN SURIMI

De-Watering means removal of excessive wash water from the minced fish flesh.

Frozen Surimi means the fish protein product for further processing, which has been processed by heading, gutting, cleaning fresh fish, and mechanically separating the edible muscle from the skin and bone. The minced fish muscle is then washed, refined, de-watered, mixed with cryoprotective food ingredients and frozen.

Gel Forming Ability means the ability of surimi to form an elastic gel when fish meat is comminuted with the addition of salt and then formed and heated. This elasticity is a function possessed by myosin as the primary component of myofibrillar protein.

Myofibrillar Protein	is a generic term of skeletal muscle proteins such as myosin and actin.
Refining	means a process of removing from washed meat by used of a strainer small bones, sinews, scales and bloody flesh of such sizes as may not be mixed in a final product, thereby concentrating myofibrillar protein.
Surimi-Based Products	means a variety of products produced from surimi with addition of ingredients and flavour such as "surimi gel" and shellfish analogues.
Water-Soluble Components	means any water-soluble proteins, organic substances and inorganic salts contained in fish meat.
Washing	means a process of washing away blood and water soluble components from minced fish with cold water by the use of a rotary filter, thus increasing the level of myofibrillar proteins thereof.
Washed	meat means fish meat that is washed and then drained of water.

2.6 QUICK-FROZEN COATED FISH PRODUCTS

Batter	liquid preparation from ground cereals, spices, salt, sugar and other ingredients and/or additives for coating. Typical batter types are: non-leavened batter and leavened batter.
Breading	dry breadcrumbs or other dry preparations mainly from cereals with colorants and other ingredients used for the final coating of fishery products. Typical breading types are: freeflowing breading, coarse breading, flour-type breading.
Coating	covering the surface of a fishery product with batter and/or breading.
Pre-frying	frying of breaded and battered fishery products in an oil bath in a way so that the core remains frozen.
Sawing	cutting (by hand or fully mechanised) of regular shapes QF fish blocks into pieces suitable for later coating.

2.10 SHRIMPS AND PRAWNS

Dehead	means to remove the head from the entire shrimp or prawn.
De-veined shrimp	means all the shrimp which have been peeled, the back of the peeled segments of the shrimp have been open out and the gut ("vein") removed.
Fresh shrimp	are freshly caught shrimp which have received no preserving treatment or which have been preserved only by chilling. It does not include freshly cooked shrimp.
Peeled shrimp	are shrimps with heads and all shell removed.
Raw headless shrimp	are raw shrimps with heads removed and the shell on.
Shrimp	The term shrimp (which includes frequently used term "prawn") refers to the species covered by the most recent edition of the FAO listing of shrimps, FAO Fisheries Synopsis No. 125, Volume 1, Shrimps and Prawns of the World.

2.11 CEPHALOPODS

Splitting	is the process of cutting cephalopods along the mantle to produce a single fillet.

2.12 CANNED FISH AND SHELLFISH

For the purpose of this Code, only the definitions of the main terms related to canning industry and used in section 13 *are given. For an overall set of definitions; please refer to the Recommended International Code of Hygienic Practice for Low-Acid and Acidified Low-Acid Canned Food* (*CAC/PRC* 23-1979, *Rev.* 2 (1993)).

Canned Food	means commercially sterile food in hermetically sealed containers.
Commercial sterility of thermally processed food	means the condition achieved by application of heat, sufficient, alone or in combination with other appropriate treatments, to render the food free from micro-organisms capable of growing in the food at normal non-refrigerated conditions at which the food is likely to be held during distribution and storage.
Hermetically Sealed Containers	are containers which are sealed to protect the content against the entry of microorganisms during and after heat processing.
Retort	means a pressure vessel designed for thermal processing of food packed in hermetically sealed containers.
Scheduled Process (or Sterilisation schedule)	means the thermal process chosen by the processor for a given product and container size to achieve at least commercial sterility.
Sterilisation Temperature	means the temperature maintained throughout the thermal process as specified in the scheduled process.
Sterilisation time	means the time between the moment sterilisation temperature is achieved and the moment cooling started.
Thermal Process	means the heat treatment to achieve commercial sterility and is quantified in terms of time and temperature.
Venting	means thorough removal of the air from steam retorts by steam prior to a scheduled process.

2.13 TRANSPORT

2.14 RETAIL

Retail	means an operation that stores, prepares, packages, serves, or otherwise provides fish, shellfish and their products directly to the consumer for preparation by the consumer for human consumption. This may be free standing seafood markets, seafood sections in grocery or department stores, packaged chilled or frozen and/ or full service.
Packaged	means packaged in advance and displayed chilled or frozen for direct consumer pick up.
Full Service Display	means a display of chilled fish, shellfish and their products to be weighed and wrapped by establishment personnel at the request of the consumer.

SECTION 3 PRE-REQUISITE PROGRAMME

Prior to the application of HACCP to any segment of the product processing chain, that segment must be supported by pre-requisite programmes based on good hygienic practice or as required by the competent authority.

The establishment of pre-requisite programmes will allow the HACCP team to focus on the HACCP application to food safety hazards which are directly applicable to the product and the process selected, without undue consideration and repetition of hazards from the surrounding environment. The pre-requisite programmes would be specific within an individual establishment or for an individual vessel and will require monitoring and evaluation to ensure their continued effectiveness.

Reference should be made to the *International Recommended Code of Practice – General Principles of Food Hygiene* (CAC/RCP 1-1969, Rev. 3 1997), Annex: HACCP *System and Guidelines for its Application* for further information to assist with the design of the pre-requisite programmes for a processing facility or vessel.

It should be noted that some of the issues listed below, e. g. those related to damage, are designed to maintain quality rather than food safety and are not always essential to a pre-requisite programme for a food safety oriented HACCP system.

HACCP principles can also be applied to defect action points.

3. 1 FISHING AND HARVESTING VESSEL DESIGN AND CONSTRUCTION

There are many different types of fishing vessel used throughout the world which have evolved in particular regions to take account of the prevailing economics, environment and types of fish and shellfish caught or harvested. This Section attempts to highlight the basic requirements for cleanability, minimising damage, contamination and decomposition to which all vessels should have regard to the extent possible in order to ensure hygienic, high quality handling of fresh fish and shellfish intended for further processing and freezing.

The design and construction of a fishing vessel and vessels used to harvest farmed fish and shellfish should take into consideration the following:

3. 1. 1 For Ease of Cleaning and Disinfection

- vessels should be designed and constructed to minimise sharp inside corners and projections to avoid dirt traps;
- construction should facilitate ample drainage;
- a good supply of clean water or potable water[1] at adequate pressure.

3. 1. 2 To Minimise Contamination

- all surfaces in handling areas should be non-toxic, smooth impervious and in sound condition, to minimise the build-up of fish slime, blood, scales and guts and to reduce the risk of physical and microbial contamination;
- where appropriate, adequate facilities should be provided for the handling and washing of fish and shellfish and should have an adequate supply of cold potable water or clean water for that purpose;

---

[1] WHO Guidelines for Drinking Water Quality, 2nd edition, Geneva, 1993.

· adequate facilities should be provided for washing and disinfecting equipment, where appropriate;

· the intake for clean water should be located to avoid contamination;

· all plumbing and waste lines should be capable of coping with peak demand;

· non-potable water lines should be clearly identified and separated from potable water to avoid contamination;

· objectionable substances, which could include bilge water, smoke, fuel oil, grease, drainage and other solid or semi-solid wastes should not contaminate the fish and shellfish;

· where appropriate, containers for offal and waste material should be clearly identified, suitably constructed with a fitted lid and made of impervious material;

· separate and adequate facilities should be provided to prevent the contamination of fish and shellfish and dry materials, such as packaging, by:

—poisonous or harmful substances;

—dry storage of materials, packaging etc. ;

—offal and waste materials;

· adequate hand washing and toilet facilities, isolated from the fish and shellfish handling areas, should be available where appropriate;

· prevent the entry of birds, insects, or other pests, animals and vermin, where appropriate.

3. 1. 3 To Minimise Damage to the Fish, Shellfish and Other Aquatic Invertebrates

· in handling areas, surfaces should have a minimum of sharp corners and projections;

· in boxing and shelving storage areas, the design should preclude excessive pressure being exerted on the fish and shellfish;

· chutes and conveyors should be designed to prevent physical damage caused by long drops or crushing;

· the fishing gear and its usage should minimise damage and deterioration to the fish and shellfish.

3. 1. 4 To Minimise Damage during Harvesting of Aquacultured and Molluscan Shellfish

When aquacultured products and molluscan shellfish are harvested using seines or nets or other means and are transported live to facilities:

· seines, nets and traps should be carefully selected to ensure minimum damage during harvesting;

· harvesting areas and all equipment for harvesting, catching, sorting, grading, conveying and transporting of live products should be designed for their rapid and efficient handling without causing mechanical damage; These should be easy cleanable and free from contamination;

· conveying equipment for live and slaughtered products should be constructed of suitable corrosion-resistant material which does not transmit toxic substances and should not cause mechanical injuries to them;

· where fish is transported live, care should be taken to avoid overcrowding and to minimise bruising;

· where fish are held or transported live, care should be taken to maintain factors that affect fish health (e. g. CO_2, O_2, temperature, nitrogenous wastes, etc) .

3. 2 FACILITY DESIGN AND CONSTRUCTION

The facility should include a product flow-through pattern that is designed to prevent potential sources of contamination, minimise process delays which could result in further reduction in essential quality, and prevent cross-contamination of finished product from raw materials. Fish, shellfish and other aquatic invertebrates are highly perishable foods and should be handled carefully and chilled without undue delay. The facility, therefore, should be designed to facilitate rapid processing and subsequent storage.

The design and construction of a facility should take into consideration the following:

3.2.1 For Ease of Cleaning and Disinfection

· the surfaces of walls, partitions and floors should be made of impervious, non-toxic materials;

· all surfaces with which fish, shellfish and their products might come in contact should be of corrosion resistant, impervious material which is light-coloured, smooth and easily cleanable;

· walls and partitions should have a smooth surface up to a height appropriate to the operation;

· floors should be constructed to allow adequate drainage;

· ceilings and overhead fixtures should be constructed and finished to minimise the build-up of dirt and condensation, and the shedding of particles;

· windows should be constructed to minimise the build-up of dirt and, where necessary, be fitted with removable and cleanable insect-proof screens. Where necessary, windows should be fixed;

· doors should have smooth, non-absorbent surfaces;

· joints between floors and walls should be constructed for ease of cleaning (round joints).

3.2.2 To Minimise Contamination

· facility layout should be designed to minimise cross-contamination and may be accomplished by physical or time separation;

· all surfaces in handling areas should be non-toxic, smooth impervious and in sound condition, to minimise the build-up of fish slime, blood, scales and guts and to reduce the risk of physical contamination;

· working surfaces that come into direct contact with fish, shellfish and their products should be in sound condition, durable and easy to maintain. They should be made of smooth, nonabsorbent and non-toxic materials, and inert to fish, shellfish and their products, detergents and disinfectants under normal operating conditions;

· adequate facilities should be provided for the handling and washing of products and should have an adequate supply of cold potable water for that purpose;

· suitable and adequate facilities should be provided for storage and/or production of ice;

· ceiling lights should be covered or otherwise suitably protected to prevent contamination by glass or other materials;

· ventilation should be sufficient to remove excess steam, smoke and objectionable odours and cross contamination through aerosols should be avoided;

· adequate facilities should be provided for washing and disinfecting equipment, where appropriate;

· non-potable water lines should be clearly identified and separated from potable water to avoid contamination;

all plumbing and waste lines should be capable of coping with peak demands;

· accumulation of solid, semi-solid or liquid wastes should be minimised to prevent contamination;

· where appropriate, containers for offal and waste material should be clearly identified, suitably constructed with a fitted lid and made of impervious material;

· separate and adequate facilities should be provided to prevent the contamination by:

—poisonous or harmful substances;

—dry storage of materials, packaging etc.;

—offal and waste materials;

- adequate hand washing and toilet facilities, isolated from handling area, should be available;
- prevent the entry of birds, insects, or other pests and animals;
- water supply lines should be fitted with back flow devices, where appropriate.

3.2.3 To Provide Adequate Lighting

- to all work surfaces.

3.3 DESIGN AND CONSTRUCTION OF EQUIPMENT AND UTENSILS

The equipment and utensils used for the handling of fishery products on a vessel or in a facility will vary greatly depending on the nature and type of operation involved. During use, they are constantly in contact with fish, shellfish and their products. The condition of the equipment and utensils should be such that it minimises the build-up of residues and prevents them becoming a source of contamination.

The design and construction equipment and utensils should take into consideration the following:

3.3.1 For Ease of Cleaning and Disinfection

- equipment should be durable and movable and/or capable of being disassembled to allow for maintenance, cleaning, disinfection and monitoring;
- equipment, containers and utensils coming into contact with fish, shellfish and their products should be designed to provide for adequate drainage and constructed to ensure that they can be adequately cleaned, disinfected and maintained to avoid contamination;
- equipment and utensils should be designed and constructed to minimise sharp inside corners and projections and tiny crevices or gaps to avoid dirt traps;
- a suitable and adequate supply of cleaning utensils and cleaning agents, approved by the official agency having jurisdiction, should be provided.

3.3.2 To Minimise Contamination

- all surfaces of equipment in handling areas should be non-toxic, smooth, impervious and in sound condition, to minimise the build-up of fish slime, blood, scales and guts and to reduce the risk of physical contamination;
- accumulation of solid, semi-solid or liquid wastes should be minimised to prevent contamination of fish;
- adequate drainage should be provided in storage containers and equipment;
- drainage should not be permitted to contaminate products.

3.3.3 To Minimise Damage

- surfaces should have a minimum of sharp corners and projections;
- chutes and conveyors should be designed to prevent physical damage caused by long drops or crushing;
- storage equipment should be fit for the purpose and not lead to crushing of the product.

3.4 HYGIENE CONTROL PROGRAMME

The potential effects of harvesting and handling of products, on-board vessel handling or in-plant production activities on the safety and suitability of fish, shellfish and their products should be considered at all times. In particular this includes all points where contamination may exist and taking specific measures to ensure the production of a safe and wholesome product. The type of control and supervision needed will depend on the size of the operation and the nature of its activities.

Schedules should be implemented to:

— prevent the build up of waste and debris;

— protect the fish, shellfish and their products from contamination;

— dispose of any rejected material in a hygienic manner;

— monitor personal hygiene and health standards;

— monitor the pest control programme;

— monitor cleaning and disinfecting programmes;

— monitor the quality and safety of water and ice supplies.

The hygiene control programme should take into consideration the following:

3.4.1 A Permanent Cleaning and Disinfection Schedule

A permanent cleaning and disinfection schedule should be drawn up to ensure that all parts of the vessel, processing facility and equipment therein are cleaned appropriately and regularly. The schedule should be reassessed whenever changes occur to the vessel, processing facility and/or equipment. Part of this schedule should include a 'clean as you go' policy.

A typical cleaning and disinfecting process may involve as many as seven separate steps:

Pre-cleaning	Preparation of area and equipment for cleaning. Involves steps such as removal of all fish, shellfish and their products from area, protection of sensitive components and packaging materials from water, removal by hand or squeegee of fish scraps, etc.
Pre-rinse	A rinsing with water to remove remaining large pieces of loose soil.
Cleaning	means the removal of soil, food residues, dirt, grease or other objectionable matter.
Rinse	A rinsing with potable water or clean water, as appropriate, to remove all soil and detergent residues.
Disinfection	Application of chemicals, approved by the official agency having jurisdiction and/or heat to destroy most microorganisms on surface.
Post-rinse	As appropriate a final rinse with potable water or clean water to remove all disinfectant residues.
Storage	Cleaned and disinfected equipment, container and utensils should be stored in a fashion which would prevent its contamination.
Check of the efficiency of the cleaning	The efficiency of the cleaning should be controlled as appropriate

Handlers or cleaning personnel as appropriate should be well trained in the use of special cleaning tools and chemicals, methods of dismantling equipment for cleaning and should be knowledgeable in the significance of contamination and the hazards involved.

3.4.2 Designation of Personnel for Cleaning

· In each processing plant or vessel a trained individual should be designated to be responsible for the sanitation of the processing facility or vessel and the equipment within.

3.4.3 Maintenance of Premises, Equipment and Utensils

· buildings, materials, utensils and all equipment in the establishment - including drainage systems - should be maintained in a good state and order;

· equipment, utensils and other physical facilities of the plant or vessel should be kept clean and in good repair;

• procedures for the maintenance, repair, adjustment and calibration, as appropriate, of apparatus should be established. These procedures should specify for each equipment, the methods used, the persons in charge of their application, and their frequency.

3.4.4 Pest Control Systems

• good hygienic practices should be employed to avoid creating an environment conducive to pests;

• pest control programmes could include preventing access, eliminating harbourage and infestations, and establishing monitoring detection and eradication systems;

• physical, chemical and biological agents should be properly applied by appropriately qualified personnel.

3.4.5 Supply of Water, Ice and Steam

3.4.5.1 Water

• an ample supply of cold and hot potable water2 and/or clean water under adequate pressure should be provided where appropriate;

• potable water[1] should be used wherever necessary to avoid contamination.

3.4.5.2 Ice

• ice should be manufactured using potable water2 or clean water;

• ice should be protected from contamination.

3.4.5.3 Steam

• for operations which require steam, an adequate supply at sufficient pressure should be maintained;

• steam used in direct contact with fish or shellfish or food contact surfaces should not constitute a threat to the safety or suitability of the food.

3.4.6 Waste Management

• offal and other waste materials should be removed from the premises of a processing facility or vessel on a regular basis;

• facilities for the containment of offal and waste material should be properly maintained;

• vessel waste discharge should not contaminate vessel water intake system or incoming product.

3.5 PERSONAL HYGIENE AND HEALTH

Personal hygiene and facilities should be such to ensure that an appropriate degree of personal hygiene can be maintained to avoid contamination.

3.5.1 Facilities and Equipment:

Facilities and equipment should include:

• adequate means of hygienically washing and drying hands;

• adequate toilet and changing facilities for personnel should be suitably located and designated.

3.5.2 Personnel Hygiene

• no person who is known to be suffering from, or who is a carrier of any communicable disease or has an infected wound or open lesion should be engaged in the preparation, handling or transportation;

• where necessary, adequate and appropriate protective clothing, headcovering and footwear should be worn;

---

[1] WHO Guidelines for Drinking Water Quality, 2nd edition, Geneva, 1993.

all persons working in a facility should maintain a high degree of personal cleanliness and should take all necessary precautions to prevent the contamination;

· hand-washing should be carried out by all personnel working in a processing area:

— at the start of fish or shellfish handling activities and upon re-entering a processing area;

— immediately after using the toilet;

· the following should not be permitted in handling and processing areas:

— smoking;

— spitting;

— chewing or eating;

— sneezing or coughing over unprotected food;

— the adornment of personal effects such as jewellery, watches, pins or other items that, if dislodged, may pose a threat to the safety and suitability of the products.

3.6 TRANSPORTATION

Vehicles should be designed and constructed:

· such that walls, floors and ceilings, where appropriate, are made of a suitable corrosion-resistant material with smooth non-absorbent surfaces. Floors should be adequately drained;

· where appropriate with chilling equipment to maintain chilled fish or shellfish during transport to a temperature as close as possible to 0℃ or, for frozen fish, shellfish and their products, to maintain a temperature of -18℃ or colder (except for brine frozen fish intended for canning which may be transported at -9℃ or colder);

· live fish and shellfish are to be transported at temperature tolerant to species;

· to provide the fish or shellfish with protection against contamination, exposure to extreme temperatures and the drying effects of the sun or wind;

· to permit the free flow of chilled air around the load when fitted with mechanical refrigeration means.

3.7 PRODUCT TRACING AND RECALL PROCEDURES

Experience has demonstrated that a system for recall of product is a necessary component of a pre-requisite programme because no process is fail-safe. Product tracing, which includes lot identification, is essential to an effective recall procedure.

· managers should ensure effective procedures are in place to effect the complete product tracing and rapid recall of any lot of fishery product from the market;

· appropriate records of processing, production and distribution should be kept and retained for a period that exceeds the shelf-life of the product;

· each container of fish, shellfish and their products intended for the final consumer or for further processing should be clearly marked to ensure the identification of the producer and of the lot;

· where there is an health hazard, products produced under similar conditions, and likely to present a similar hazard to public health, may be withdrawn. The need for public warnings should be considered;

· recalled products should be held under supervision until they are destroyed, used for purposes other than human consumption, or reprocessed in a manner to ensure their safety.

3.8 TRAINING

Fish or shellfish hygiene training is fundamentally important. All personnel should be aware of their role and responsibility in protecting fish or shellfish from contamination and deterioration. Handlers should have the necessary knowledge and skill to enable them to handle fish or shellfish hygienically. Those who handle strong cleaning chemicals or other potentially hazardous chemicals should be instructed in safe handling techniques.

Each fish and shellfish facility should ensure that individuals have received adequate and appropriate training in the design and proper application of a HACCP system and process control. Training of personnel in the use of HACCP is fundamental to the successful implementation and delivery of the programme in fish or shellfish processing establishments. The practical application of such systems will be enhanced when the individual responsible for HACCP has successfully completed a course. Managers should also arrange for adequate and periodic training of relevant employee in the facility so that they understand the principles involved in HACCP.

SECTION 4 GENERAL CONSIDERATIONS FOR THE HANDLING OF FRESH FISH, SHELLFISH AND OTHER AQUATIC INVERTEBRATES

Unless they can be reduced to an acceptable level by normal sorting and/or processing, no fish, shellfish and other aquatic invertebrates should be accepted if it is known to contain parasites, undesirable microorganisms, pesticides, veterinary drugs or toxic, decomposed or extraneous substances known to be harmful to human health. When fish and shellfish determined as unfit for human consumption are found they should be removed and stored separately from the catch and either reworked and/or disposed of in a proper manner. All fish and shellfish deemed fit for human consumption should be handled properly with particular attention being paid to time and temperature control.

4.1 TIME AND TEMPERATURE CONTROL

Temperature is the single most important factor affecting the rate of fish and shellfish deterioration and multiplication of micro-organisms. For species prone to scombrotoxin production, time and temperature control may be the most effective method in controlling food safety. It is therefore essential that fresh fish, fillets, shellfish and their products which are to be chilled should be held at a temperature as close as possible to 0℃.

4.1.1 Minimise the Deterioration - Time

To minimise the deterioration, it is important that:

- chilling should commence as soon as possible;
- fresh fish, shellfish and other aquatic invertebrates should be kept chilled, processed and distributed with care and minimum delay.

4.1.2 Minimise the Deterioration - Temperature Control

Where temperature control is concerned:

- sufficient and adequate icing, or chilled or refrigerated water systems where appropriate, should be employed to ensure that fish, shellfish and other aquatic invertebrates are kept chilled at a temperature as close as possible to 0℃;

· fish, shellfish and other aquatic invertebrates should be stored in shallow layers and surrounded by finely divided melting ice;

· live fish and shellfish are to be transported at temperature tolerant to species;

· chilled or refrigerated water systems and/or cold storage systems should be designed and maintained to provide adequate cooling and/or freezing capacities during peak loads;

· fish should not be stored in refrigerated water systems to a density which impairs its working efficiency;

· monitoring and controlling the time and temperature and homogeneity of chilling should be performed regularly.

4.2 MINIMISE THE DETERIORATION - HANDLING

Poor handling practices can lead to damage of fresh fish, shellfish and other aquatic invertebrates which can accelerate the rate of decomposition and increase unnecessary post-harvest losses. Handling damage can be minimised by:

· fish and shellfish should be handled and conveyed with care particularly during transfer and sorting in order to avoid physical damage such as puncture, mutilation, etc.;

· where fish and shellfish are held or transported live, care should be taken to maintain factors that can influence fish health (e.g. CO_2, O_2, temperature, nitrogenous wastes, etc.);

· fish and shellfish should not be trampled or stood upon;

· where boxes are used for storage of fish and shellfish they should not be overfilled or stacked too deeply;

· while fish and shellfish are on deck, exposure to the adverse effects of the elements should be kept to a minimum in order to prevent unnecessary dehydration;

· finely divided ice should be used where possible, which can help minimise damage to fish and shellfish and maximise cooling capacity;

· in refrigerated water storage areas, the density of the fish should be controlled to prevent damage.

SECTION 5 HAZARD ANALYSIS CRITICAL CONTROL POINT (HACCP) AND DEFECT ACTION POINT (DAP) ANALYSIS

The Hazard Analysis Critical Control Point (HACCP) is a science-based system which is aimed to prevent food safety problems from occurring rather than reacting to non-compliance of the finished product. The HACCP system accomplishes this by the identification of specific hazards and the implementation of control measures. An effective HACCP system should reduce the reliance on traditional end-product testing.

Section 5 explains the principles of HACCP as it applies aquaculture and molluscan shellfish production and to the handling and processing, but the Code can only provide guidance on how to use these principles and offer suggestions as to the type of hazards which may occur in the various fishery products. The HACCP plan, which should be incorporated into the food management plan should be well documented and be as simple as possible. This section will demonstrate one format, which may be considered in the development of the HACCP plan.

Section 5 also explains how a similar approach involving many of the principles can apply to the broader application covering the essential quality, composition and labelling provisions of Codex standards or other non-safety requirements which in this case are referred to as Defect Action Point Analysis. This approach for defect analysis is optional and other techniques, which achieve the same objective, may be considered.

Figure 5. 1 summarises how to develop a HACCP and Defect Analysis system.

5. 1 HACCP PRINCIPLES

The HACCP System consists of seven principles[1], which are

PRINCIPLE 1 - Conduct a hazard analysis.

PRINCIPLE 2 - Determine the Critical Control Points (CCPs) .

PRINCIPLE 3 - Establish critical limit (s) .

PRINCIPLE 4 - Establish a system to monitor control of the CCP.

PRINCIPLE 5 - Establish the corrective action to be taken when monitoring indicates that a particular CCP is not under control.

PRINCIPLE 6 - Establish procedures for verification to confirm that the HACCP system is working effectively.

PRINCIPLE 7 - Establish documentation concerning all procedures and records appropriate to these principles and their application.

These principles have to be followed in any consideration of HACCP.

HACCP is an important management tool, which can be used by operators for ensuring safe, efficient processing. It must also be recognised that personnel training is essential in order that HACCP will be effective. In following HACCP principles, users are requested to list all of the hazards that may be reasonably expected to occur for each product type at each step or procedure in the process from point of harvest, during unloading, transport, storage or during processing, as appropriate to the process defined. It is important that HACCP principles be considered on a specific basis to reflect the risks of the operation.

5. 2 DEFECT ACTION POINT ANALYSIS

Since the Code is intended to cover not only those hazards associated with safety but to include other aspects of production including the essential product quality, composition and labelling provisions as described in product standards developed by the Codex Alimentarius Commission, not only are critical control points (CCP) described but also defect action points (DAP) are included in the Code. The HACCP principles may be applied to the determination of a DAP, with quality instead of safety parameters being considered at the various steps.

5. 3 APPLICATION

Each aquaculture, molluscan shellfish, shellfish and fish facility should ensure that the provisions of the appropriate Codex standards are met. To accomplish this, each facility should implement a food safety management system based on HACCP principles and should at least consider a similar approach to defects, both of which are described in this code. Prior to the application of HACCP to any segment of the growing,

---

[1] *International Recommended Code of Practice - General Principles of Food Hygiene* (*CAC/RCP* 1 - 1969, *Rev.* 3 - 1997), *Annex*: *HACCP System and Guidelines for its Application.*

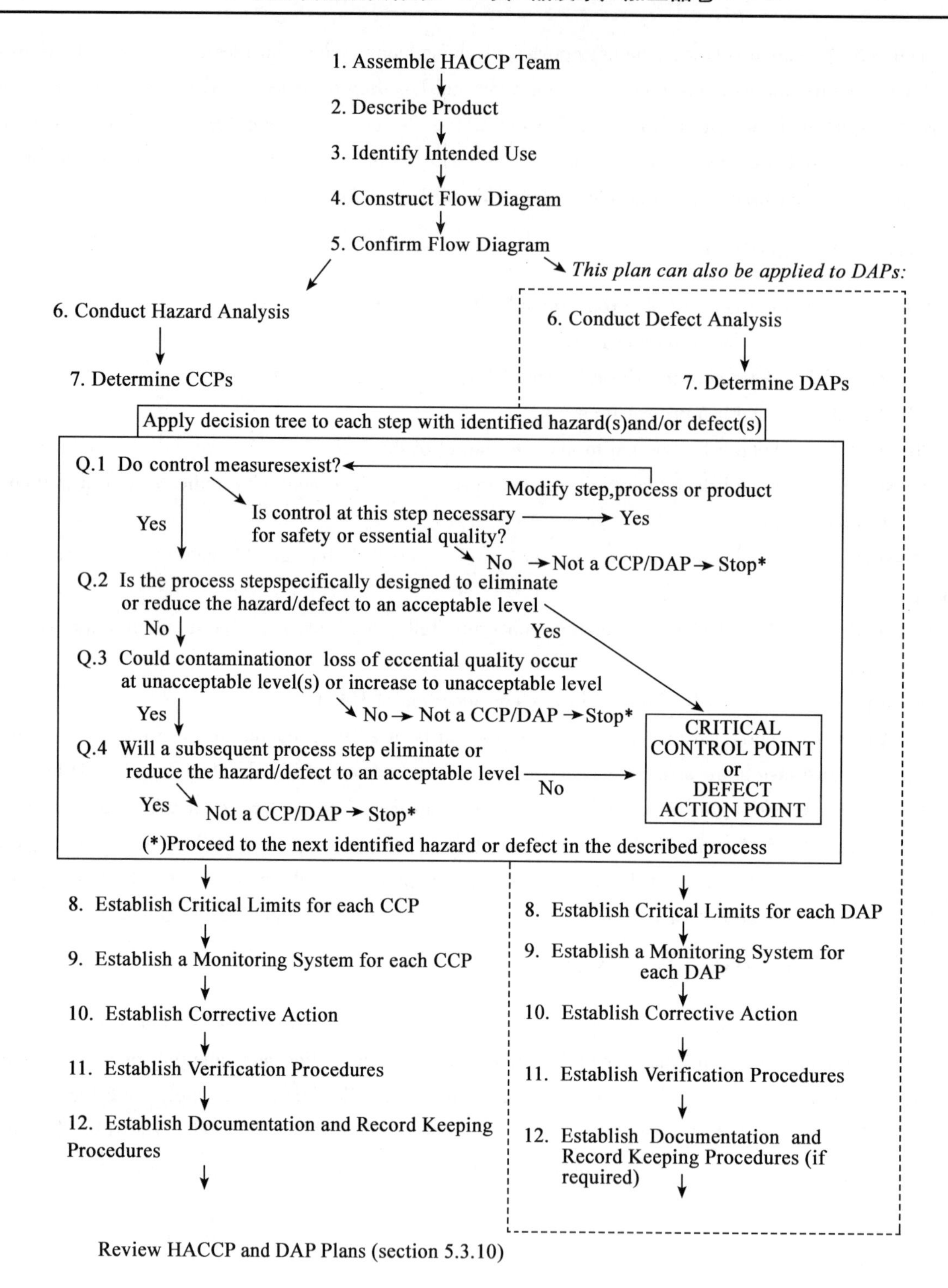

Figure 5.1 Summary of how to implement a HACCP and Defect Analysis

handling and processing chain, that segment must be supported by a pre-requisite programme based on good hygienic practice (see Section 3). It should be noted that parts of the pre-requisite programme may be classified as a CCP or DAP within a particular process.

The food management system developed should indicate responsibility, authority and the interrelationships of all personnel who manage, perform and verify work affecting the performance of such systems. It is important that the collection, collation and evaluation of scientific and technical data should be carried out by a multi-disciplinary team. Ideally, a team should consist of people with the appropriate level of expertise together with those having a detailed knowledge of the process and product under review. Examples of the type of personnel to include on the team are the processing facility manager, a microbiologist, a quality assurance/quality control specialist, and others such as buyers, operators, etc., as necessary. For small-scale operations, it may not be possible to establish such a team and therefore external advice should be sought.

The scope of the HACCP plan should be identified and should describe which segments of the food chain is involved and the general classes of hazards to be addressed.

The design of this programme should identify critical control points in the operation where the processing facility or product will be controlled, the specification or standard to be met, the monitoring frequency and sampling plan used at the critical control point, the monitoring system used to record the results of these inspections and any corrective action when required. A record for each critical control point that demonstrates that the monitoring procedures and corrective actions are being followed should be provided. The records should be maintained as verification and evidence of the plant's quality assurance programme. Similar records and procedures may be applied to DAPs with the necessary degree of record keeping. A method to identify, describe, and locate the records associated with HACCP programmes should be established as part of the HACCP programme.

Verification activities include the application of methods; procedures (review/audit) and tests in addition to those used in monitoring to determine:

- the effectiveness of the HACCP or DAP plan in delivering expected outcomes i. e. validation;
- compliance with the HACCP or DAP plan, e. g. audit/review;
- whether the HACCP or DAP plan or its method of application need modification or revalidation.

Table 5.1 A product description for Canned Tuna in Salted Water

Product name (s)	Objective	Example
	Identify the species and method of processing.	Canned tuna in salted water
Source of raw material	Describe the origin of the fish	Yellowfin tuna caught by purse seine in the Gulf of Guinea Whole brine frozen
Important final product characteristics	List characteristics that affect product safety and essential quality, especially those that influence microbial flora	Compliance with Codex Standard Canned Tuna and Bonito; 'low-acid' food; can seal integrity
Ingredients	List every substance added during processing. Only ingredients approved by the official agency having jurisdiction may be used	water, salt
Packaging	List all packaging materials. Only materials approved by the official agency having jurisdiction may be used	Container in coated chromium steel, capacity: 212 ml, total net weight: 185 g, fish weight: 150 g Traditional opening
How the end product is to be used	State how the final product is to be prepared for serving, especially whether it is ready to eat.	Ready to eat
Shelf life (if applicable)	State the date when the product can be expected to begin to deteriorate if stored according to instructions	3 years
Where the product will be sold	Indicate the intended market. This information will facilitate compliance with target market regulations and standards	Domestic retail market
Special labelling instructions	List all instructions for safe storage and preparation	"Best before the date shown on label."
Special distribution control	List all instructions for safe product distribution	None

The implementation of HACCP principles is better identified in the Logic Sequence for implementation of HACCP (Figure 5.1).

5.3.1 Describe Product

In order to gain a greater understanding and knowledge of the product under review, a thorough product description evaluation should be carried out. This exercise will facilitate in the identification of potential hazards or defects. An example of the type of information used in describing a product is given in Table 5.1.

5.3.2 Flow Diagram

For Hazard and Defect Analysis, it is necessary to carefully examine both the product and the process and produce a flow diagram (s). Any flow diagram should be as simple as possible. Each step in the process, including process delays from the selection of raw materials through to the processing, distribution, sale and customer handling, should be clearly outlined in sequence with sufficient technical data to avoid ambiguity. If a process is too complex to be easily represented by a single flow diagram, then it can be sub-divided into constituent parts, provided the relationship between each of the parts is clearly defined. It is helpful to number and label each processing step for ease of reference. An accurate and properly constructed flow diagram will provide the multi-disciplinary team with a clear vision of the process sequence. Once CCPs and DAPs have been identified they can be incorporated into the flow diagram specific for each processing facility. Figure 5.2 represents an example of a flow diagram for a canned tuna fish processing

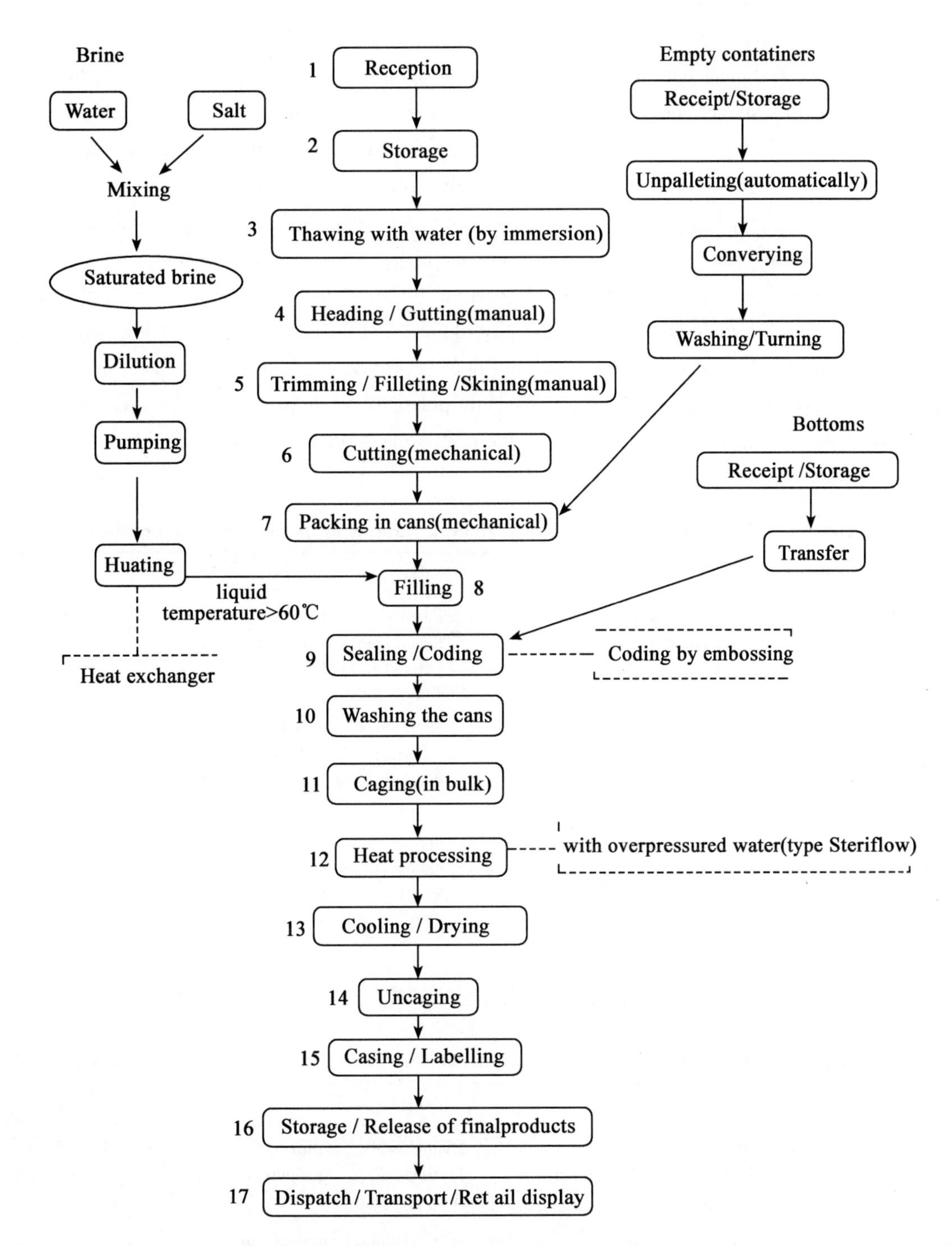

Figure 5. 2 Example of a flow diagram for a processing line of canned tuna fish in

line. For examples of different processes see Figures 8. 1 to 10. 1 in the individual processing sections of the code.

5. 3. 3 Conduct Hazard and Defect Analysis

The purposes of hazard analysis are to identify all such food safety hazards at each Step, to determine their significance and to assess whether control measures for those hazards are available at each Step. Defect analysis serves the same purpose for potential quality defects.

5.3.3.1 Identification of Hazards and Defects

It cannot be stressed enough that where practical and feasible each individual facility should gather sound scientific and technical data relevant to the businesses for each step, from primary production, processing, manufacture, storage and distribution until the point of consumption. The assembly and nature of this information should be such to ensure that the multi-disciplinary team is able to identify and list, at each step of the process, all of the hazards that may reasonably likely to occur and defects that, in the absence of control measure (s), may likely result in the production of an unacceptable food. Potential hazards, which have been known to be associated with fresh fish and shellfish, are described in Annex 1. Table 5.2 summarises possible pre-harvest and harvest safety hazards in incoming fish and shellfish and Table 5.3 summarises possible safety hazards introduced in the post harvest and further processing of fish and shellfish.

It is important to identify potential hazards and defects in the operation from the point of view of plant construction, equipment used in the plant and hygienic practices, including those which may be associated with the use of ice and water. This is covered by the pre-requisite programme and is used to denote hazards that are common to almost any point in the process.

Table 5.2 Examples of Pre-harvest and Harvest Hazards in Incoming Fish & Shellfish

	Biological		Chemical		Physical
Parasites	Parasites of public health significance: Trematodes Nematodes, Cestodes	Chemicals	Pesticides, herbicides, algicides, fungicides, anti-oxidants (added in feeds)	Foreign Matter	fish hooks
Pathogenic bacteria	Salmonella, Shigella, E. coli, Vibrio cholerae, Vibrio parahaemolyticus, Vibrio vulnificus	Veterinary drug residues	Antibiotics, growth promoters (hormones), other veterinary drugs and feed additives		
Enteric Viruses	Norwalk virus	Heavy metals	Metals leached from marine sediments and soil, from industrial wastes, from sewage or animal manures		
Biotoxins	Biotoxins, Scombrotoxin				
		Miscellaneous	Petroleum		

Table 5.3 Examples of Hazards Introduced in the Post Harvest and Further Processing of Fish & Shellfish*

	Biological		Chemical		Physical
Pathogenic bacteria	Listeria monocytogenes, Clostridium botulinum Staphylococcus aureus	Chemicals:	Disinfectants, Sanitizers or Lubricants (Misapplication)	Foreign Matter	Metal fragments; hard or sharp objects
Enteric Viruses	Hepatitis A, Rotovirus		Disinfectants, Sanitizers or Lubricants (nonapproved)		
Biotoxins	Scombrotoxin, Staph. Enterotoxin, botulinum toxin				
		Ingredients and Additives	Misapplication and non-approved		

Note: For biological hazards, environmental factors (for example: temperature, oxygen availability, pH and AW) play a major role in their activity and growth, therefore the type of processing the fish or shellfish will undergo, and its subsequent storage, will determine their risk to human health and inclusion in a food safety management plan. In addition, some hazards may show a certain degree of overlap between the two levels of operation through their existence and manifestation into the water supply.

* For hazards relating to specific products see the relevant processing section.

For the example on canned tuna developed in this section, the following essential potential hazards can be identified:

Table 5.4 An example of potential hazards for canned tuna

	In raw materials (frozen tuna)	During processing or storage or transportation
Biological	Presence of *Cl. botulinum*, Presence of scombrotoxin	Contamination by *Cl. Botulinum*, Growth of *Cl. Botulinum*, Survival of spores of *Cl. Botulinum*, Contamination and growth of *Staphylococcus aureus* Microbial recontamination after heat processing Production of scombrotoxin during processing, Production of staphylotoxin
Chemical	Presence of heavy metals	Recontamination by metals coming from the cans Recontamination by cleaning agents, by the brine, by mechanical grease, …
Physical	Presence of foreign material	Recontamination during processing (pieces of knives, by the cans, …)

For the example on canned tuna developed in this section, the following potential defects can be identified:

Table 5.5 An example of potential defects of canned tuna

	In raw materials (frozen tuna)	During processing or storage or transportation
Biological	Decomposition	Decomposition, survival of micro-organisms responsible of decomposition, …
Chemical		oxidation during storage, …
Physical		Objectionable matters (viscera, scales, skin, …), formation of struvite crystals, container defects (panelled container, …)
Others	species substitution	abnormal flavours, incorrect weight, incorrect coding, incorrect labelling

5.3.3.1.1 Hazards

It is equally important to consider, naturally occurring food safety hazards in the environment from which fish or shellfish are harvested. In general, risks to consumer health from seafood captured in unpolluted marine environments are low, provided these products are handled in line with principles of Good Manufacturing Practice. However, as with all foods, there are some health risks associated with the consumption of certain products, which may be increased when the catch is mishandled after harvest. Fish from some marine environments, such as tropical reef fish, can pose a consumer risk from natural marine toxins, such as ciguatera. The risk of adverse health effects from certain hazards might be increased under certain circumstances in products from aquaculture when compared with fish and crustacean from the marine environment. The risks of foodborne disease associated with products from aquaculture are related to inland and coastal ecosystems, where the potential of environmental contamination is greater when compared to capture fisheries. In some parts of the world, where fish or shellfish are consumed either raw or partially cooked, there is an increased risk of foodborne parasitic or bacterial disease. In order to perform a hazard analysis as part of the process of developing a HACCP plan, processors must have scientific information on potential hazards associated with raw material and products for further processing.

5.3.3.1.2 Defects

Potential defects are outlined in the essential quality, labelling and composition requirements described in the Codex Standards listed in Appendix XII*. Where no Codex Standard exists regard should be made to national regulations and/or commercial specifications.

End product specifications outlined in Appendices II - XI*, describe optional requirements which are intended to assist buyers and sellers in describing those provisions which are often used in commercial transactions or in designing specifications for final products. These requirements are intended for voluntary application by commercial partners and not necessarily for application by governments.

5.3.3.2 Significance of Hazards and Defects

One of the most important activities, which must be performed in a processing facility as part of the food safety management system is to determine if an identified hazard or defect is significant. The two primary factors that determine whether a hazard or defect is significant for HACCP purposes are probability of occurrence of an adverse health effect and the severity of the effect. A hazard that has a high severity of effect, such as death from *Clostridium botulinum* toxin, may impose a socially unacceptable risk at very low probability of occurrence, and thus warrant the application of HACCP controls (i.e., be a significant hazard for purposes of HACCP). Thus, in the processed canned tuna, *Clostridium botulinum* should be considered a significant hazard to be controlled through the application of a validated thermal process schedule. On the other hand, a hazard with a relatively low severity, such as mild gastroenteritis, might not warrant the HACCP controls at the same very low probability of occurrence, and thus not be significant for purposes of HACCP.

Information gathered during the product description exercise (refer to Section 5.3.1 - Describe Product) could also help facilitate the determination of significance since the likelihood of occurrence of hazard or defect can be affected by factors such as how the consumer will likely use the product (e.g., to consumed or cooked raw); the types of consumers who will likely consume it (e.g., immuno-compromised, elderly, children, etc.) and the method of storage and distribution (e.g., refrigerated or frozen).

Once significant hazard and defects have been identified, consideration needs to be given to assess their potential to be introduced or controlled at each step of the process. The use of a flow diagram (refer to Section 5.3.2 - Flow Diagram) is beneficial for this purpose. Control measures must be considered for significant hazard (s) or defect (s) associated with each step with the aim of eliminating its possible occurrence or to reduce it to an acceptable level. A hazard or defect may be controlled by more that one control measure. For illustrative purposes, tables 5.6 and 5.7 demonstrate an approach to listing significant hazards and defects and the related control measures for the processing step, " Heat Processing".

Table 5.6 An example of the significant hazard survival of *Cl. Botulinum* at the step of heat processingfor canned tuna

Processing step	Potential hazard	Is the potential hazard significant?	Justification	Control measures
12. Heat processing	*Cl. botulinum* viable spores	Yes	An insufficient heat processing may result in survival of *C. botulinum* spores and therefore, possibility of toxin production. A product must be commercially sterile	Ensure adequate heat applied for proper time at retort

Table 5.7 An example of the significant defect rancidity during the storage of frozen tuna for canned tuna

Processing step	Potential defect	Is the potential defect significant?	Justification	Control measures
2. Storage of frozen tuna	Persistent and distinct objectionable odours or flavours indicative of rancidity	Yes	Product does not meet quality or customer requirements	Controlled temperature in the storage premises Stock management procedure Maintenance procedure of the refrigeration system Personnel training and qualification

Table 5.8 A schematic example of a hazard analysis with corresponding control measures and the application of the Codex decision tree for the determination of a critical control point at processing step 12 of the example process as set out in Figure 5.2

Processing Step N° 12 Heat processing		Application of Codex Decision Tree			
Potential Hazards	Control Measures				
Cl. botulinum viable spores	Ensure adequate heat applied for proper time at retort	Q1: Do control measures exist If yes - go to Q2. If no - consider whether control measures are available or necessary within the process. Proceed to next identified hazard	Q2: Is the step specifically designed to eliminate or reduce the likely occurrence of *Cl. botulinum* to an acceptable level If yes - this step is a CCP. If no - go to Q3	Q3: Could contamination occur in excess of acceptable levels or could these increase to unacceptable levels If yes - go to Q4. If no - not a CCP	Q4: Will a subsequent step eliminate or reduce the hazard to an acceptable level If yes - not a CCP. If no - CCP. *What about consideration of a previous step*
		A: Yes: a heat processing procedure (schedule, method) is clearly defined	A: Yes, this step was specifically designed to eliminate spores		
		Decision: Processing step N°12 《Heat processing》is a Critical Control Point			

5.3.4 Determine Critical Control Points and Defect Action Points

A thorough and concise determination of Critical Control Points and Defect Action Points in a process is important in ensuring food safety and compliance with elements related to essential quality, composition and labelling provisions of the appropriate Codex standard. The Codex decision tree (Figure 5.1, step 7) is a tool, which can be applied, to the determination of CCPs and a similar approach may be used for DAPs. Using this decision tree, a significant hazard or defect at a step can be assessed through a logical sequence of questions. Where CCPs and DAPs have been identified at a step, that point in the process must be controlled to prevent, reduce or eliminate the likely occurrence of the hazard or defect to an acceptable level. For illustrative purposes, an example of the application of the Codex decision tree to a hazard and defect using the canned tuna fish processing line, are shown in Tables 5.8 & 5.9, respectively.

Table 5.9 A schematic example of a defect analysis with corresponding control measures and the application of the Codex decision tree for the determination of a defect action point at processing step 2 of the example process as set out in Figure 5.2

Processing Step N° 12 Heat processing		Application of Codex Decision Tree			
Potential Defects	Control Measures				
Persistent and distinct objectionable odours or flavours indicative of rancidity	Controlled temperature in storage premises. Stock management procedure	Q1: Do control measures exist? If yes - go to Q2. If no - consider whether control measures are available or necessary within the process. Proceed to next identified hazard	Q2: Is the step specifically designed to eliminate or reduce the likely occurrence of rancidity to an acceptable level If yes - this step is a DAP. If no - go to Q3	Q3: Could rancidity occur in excess of acceptable levels or could it increase to unacceptable levels If yes - go to Q4. If no - not a DAP	Q4: Will a subsequent step eliminate rancidity or reduce its likely occurrence to acceptable level If yes - not a DAP. If no - DAP. *What about consideration of a previous step?*
		A: Yes, the storage temperature is controlled, procedures exist	A: No	A : Yes, if the storage time is too long and/or the storage temperature is too high	A : No
		Decision: Processing Step N°2 ? Storage of frozen tuna is a Defect Action Point			

5.3.5 Establish Critical Limits

For each CCP and DAP, critical limits for the control of the hazard or defect must be specified. For any given hazard or defect, it may be necessary to have more than one critical limit designated for each control measure. The establishment of critical limits should be based on scientific evidence and validated by appropriate technical experts to ensure its effectiveness in controlling the hazard or defect to the determined level. Table 5.10 illustrates critical limits for a CCP and a DAP using a canned tuna fish processing line as an example.

5.3.6 Establish Monitoring Procedures

Any monitoring system developed by the multi-disciplinary team should be designed to detect loss of control at a CCP or DAP relative to its critical limit. The monitoring activity of a CCP or DAP should be documented in a concise fashion providing details regarding the individual responsible for the observation or measurement, the methodology used, the parameter (s) being monitored and the frequency of the inspections. The complexity of the monitoring procedure should also be carefully considered. Considerations include optimising the number of individuals performing the measurement and selection of appropriate methods, which will produce rapid results (for example: time, temperature, pH) . For CCPs, records of monitoring should be acknowledged and dated by a responsible person for verification.

Because each process is unique for each product, it is possible only to present, for illustrative purposes, an example of a monitoring approach for a CCP and DAP using the canned tuna fish processing line.

This example is shown in Table 5. 10.

5. 3. 7 Establish Corrective Action

An effective HACCP or DAP plan is anticipatory by nature and it is recognised that corrective action may be necessary from time to time. A documented corrective action programme should be established to deal with instances where the critical limit has been exceeded and loss of control has occurred at a CCP or DAP. The goal of this plan is to ensure that comprehensive and specific controls are in place and can be implemented to prevent the affected lot (s) from reaching the consumer. For example, fish and shellfish should be held and rejected if they are known to contain harmful substances and/or defects which would not be eliminated or reduced to an acceptable level by normal procedures of sorting or preparation. Of equal importance, is an assessment by plant management and other appropriate personnel to determine the underlying reason (s) why control was lost. For the latter, a modification to HACCP and DAP plans may be necessary. A record of investigation results and actions taken should be documented by a responsible person for each instance where loss of control occurred at a CCP or DAP. The record should demonstrate that control of the process has been re-established, that appropriate product disposition has occurred and that preventative action has been initiated. An example of a corrective action approach for a CCP and DAP using a canned tuna fish processing line is illustrated in Table 5. 10.

5. 3. 8 Establish Verification Procedures

A processing facility should establish a verification procedure carried out by qualified individuals, to periodically assess if the HACCP and DAP plans are adequate, implemented and working properly. This step will help determine if CCPs and DAPs are under control. Examples of verification activities include: validation of all components of the HACCP plan including: a paper review of HACCP system, its procedures and records; review of corrective actions and product disposition actions when critical limits are not met and validation of established critical limits. The latter is particularly important when an unexplained system failure has occurred, when a significant change to the process, product or packaging is planned or when new hazards or defects have been identified. Observation, measurement and inspection activities within the processing facility should also be incorporated as a part of the verification procedure, where applicable. Verification activities should be carried out by qualified competent individuals. The verification frequency of the HACCP and DAP plans should be sufficient to provide assurance that their design and implementation will prevent food safety problems as well as issues associated with essential quality, composition and labelling provisions of the appropriate Codex standard to enable problems to be detected and dealt with in a timely manner. For illustration purposes, an example of a verification procedure approach for a CCP and DAP using the canned tuna fish processing line is shown in Table 5. 10.

5. 3. 9 Establish Documentation and Record Keeping Procedures

Documentation may include Hazard Analysis, CCP determination, critical limit determination, and procedures for monitoring, corrective action and verification.

A current, accurate and concise record keeping system will greatly enhance the effectiveness of a HACCP programme and facilitate in the verification process. Examples of the elements of a HACCP plan that should be documented have been provided in this section for illustrative purposes. Inspection and corrective action records should be practical and collect all the appropriate data necessary to demonstrate "real-time" control or deviation control of a CCP. Records are recommended but not required for a DAP except where a loss of control occurred. For illustration purposes, an example of a record keeping approach for a CCP and DAP using the canned tuna fish processing line is shown in Table 5. 10.

5.3.10 Review of HACCP and DAP Plans

Upon completion of all the steps for the development of HACCP and DAP plans as outlined in Figure 1 a full review of all components should be conducted. The purpose of these reviews is to verify that the plans are capable of meeting their objectives.

Table 5.10 An example of the results of the application of HACCP principles to the two specific steps in the canned tuna process (Tables 5.8 & 5.9), for a CCP & a DAP, respectively

CCP

Processing Step No. 12 : Heat Processing
Hazard: *Clostridium botulinum* viable spores

Critical Limit	Monitoring Procedure	Corrective Action	Records	Verification
Those specific parameters associated with heat processing	Who: Qualified person assigned to heat processing What: All parameters Frequency: every batch How: Checks of sterilisation schedule and other factors	Who: qualified personnel What: Personnel retraining New heat processing or batch destruction Corrective maintenance of equipment Hold product until safety can be evaluated. Who: Appropriate trained personnel	Monitoring records, corrective action records, product evaluation records, calibration records, validation records, audit records, HACCP plan review record	Validation, finished product evaluation, internal audit, review of records, calibration of machinery (may be a prerequisite), review of HACCP plan, external audit

DAP

Processing Step No. 2 : Storage of frozen tuna
Defect: Persistent and distinct objectionable odours or flavours indicative of rancidity

Critical Limit:	Monitoring Procedure	Corrective Action	Records	Verification
Number of rancid sample units cannot exceed acceptance number of established sampling plan. Storage temperature and time	Who: Appropriate trained personnel How: Organoleptic examination Chemical tests Checking of the storage premise temperature Checking of stock forms What: fish quality and acceptability based on product Codex standard. Frequency: as required	What: Application of an intensified monitoring According to the results of this intensified inspection, immediate processing, sorting or reject of frozen tuna exceeding the critical limits Adjust storage temperature Personnel retraining Who: Appropriate trained personnel	Analysis results Stock forms Temperature records	On-site audit Review of monitoring and corrective action reports

5.4 Conclusion

Section 5 has demonstrated the principles of HACCP and how they should be applied to a process to ensure safe product. The same principles can be used to determine the points in a process where it is necessary to control defects. Since every facility and each processing line is different it is possible within this Code only to demonstrate the types of potential hazards and defects that must be considered. Furthermore, because of the nature of the significance of hazards and defects it is not possible to categorically determine which steps in a process will be CCPs and/or DAPs without actually assessing the process, the objectives of the process, its environment and expected outcomes. The example of the canned tuna processing line is intended to illustrate how to apply the principles, given the outcome of a commercially sterile product, and why a HACCP and DAP plan will be unique to each operation.

The remaining Sections in the Code concentrate on aquaculture and molluscan shellfish production and to the handling and processing of fish, shellfish and their products and attempt to illustrate the potential hazards and defects at the various stages in a wide range of processes. In developing a HACCP or DAP plan it will be necessary to consult Sections 3 & 5 before turning to the appropriate processing section for specific advice. It should also be noted that Section 8 refers to processing of fresh, frozen and minced fish and will provide useful guidance for most of the other processing operations.

SECTION 6 AQUACULTURE PRODUCTION

Preamble

Aquaculture establishments should operate in a responsible way such that they comply with the recommendations of the Code of Conduct for Responsible Fisheries (FAO, Rome, 1995) in order to minimize any adverse impact on human health and environment including any potential ecological changes. Fish farms should operate effective fish health and welfare management. Fry and fingerlings should be disease free and should comply with the OIE Codes of Practice (International Aquatic Animal Health Code, 6th Edition, 2003). Growing fish should be monitored for disease. When using chemicals at fish farms, special care should be exercised so that these substances are not released into the surrounding environment.

Whilst the fish health, environment, and ecological aspects are important considerations in aquaculture activities, this section focuses on food safety and quality aspects.

This Section of the Code applies to industrialised and commercial aquaculture production, producing all aquatic animals, except mammalian species, aquatic reptiles and amphibians for direct human consumption, but excluding bivalve molluscs covered in Section 7 of the code, hereafter referred to as "fish that are intended for direct human consumption. Such intensive or semi-intensive aquaculture systems use higher stocking densities, stock from hatcheries, use mainly formulated feeds and may utilise medication and vaccines. This Code is not intended to cover extensive fish farming systems that prevail in many developing countries or integrated livestock and fish culture systems. This section of the code covers the feeding, growing, harvesting and transport stages of aquaculture production. Further handling and processing of fish are covered elsewhere in the code.

In the context of recognising controls at individual processing steps, this section provides examples of potential hazards and defects and describes technological guidelines, which can be used to develop control measures and corrective action. At a particular step only the hazards and defects, which are likely to be introduced or controlled at that step, are listed. It should be recognised that in preparing a HACCP and/or DAP plan it is essential to consult Section 5 which provides guidance for the application of the principles of HACCP and DAP analysis. However, within the scope of this Code of Practice it is not possible to give details of critical limits, monitoring, record keeping and verification for each of the steps since these are specific to particular hazards and defects.

The Example flow diagram will provide guidance to some of the common steps in aquaculture production.

This flow chart is for illustrative purpose only. For implementation of HACCP principles, a complete and

This flow chart is for illustrative purpose only. For implementation of HACCP principles, a complete and comprehensive flow chart has to be drawn up for each product. References correspond to relevant Sections of the Code.

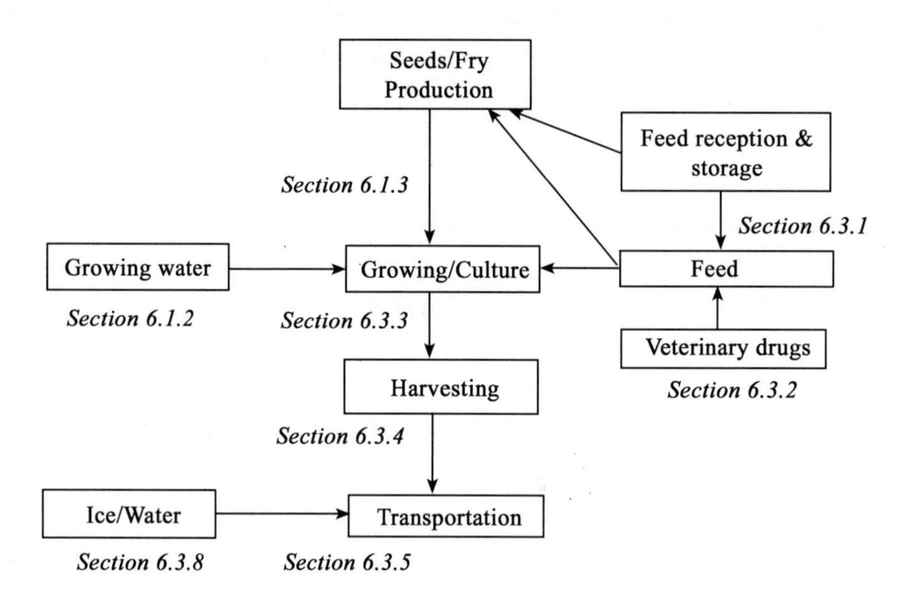

Figure 6.1 Example of a flow chart for aquaculture production

comprehensive flow chart has to be drawn up for each product. References correspond to relevant Sections

6.1 GENERAL

The general principles in Section 3 apply to aquaculture production, in addition to the following:

6.1.1 Site selection

- The siting, design and construction of fish farms should follow principles of good aquaculture practice, appropriate to species;
- The physical environment with regard to temperature, current, salinity and depth should also be considered since different species have different environmental requirements. Closed recirculation systems should be able to adapt the physical environment to the environment requirements of the farmed fish species;
- Fish farms should be located in areas where the risk of contamination by chemical, physical or microbiological hazards is minimal and where sources of pollution can be controlled;
- Soil for the construction of earthen ponds should not contain such concentrations of chemicals and other substances, which may lead to the presence of unacceptable levels of contamination in fish;
- Ponds should have separated inlets and discharge canals, so that water supplies and effluent are not

mixed;

· Adequate facility for treatment of effluent should be provided to allow sufficient time for sediments and organic load settlement before used water is discharged into the public water body;

· Water inlets and outlets to ponds should be screened to prevent the entrance of unwanted species;

· Fertilizers, liming materials or other chemicals and biological materials, should be used in accordance with good aquaculture practice;

· All sites should be operated so as to not adversely impact human health from the consumption of the fish in farm.

6.1.2 Growing Water Quality

· The water in which fish are raised should be suitable for the production of products which are safe for human consumption;

· The water quality should be monitored regularly such that the health and sanitation of the fish is continuously maintained to ensure aquaculture products are safe for human consumption;

· Fish farms should not be sited where there is a risk of contamination of the water in which fish are reared;

· Appropriate design and construction of fish farms should be adopted to ensure control of hazards and prevention of water contamination.

6.1.3 Source of Fry and Fingerlings

· The source of postlarvae, fries and fingerlings should be such to avoid the carryover of potential hazards into the growing stocks.

6.2 IDENTIFICATION OF HAZARDS AND DEFECTS

Consumption of fish and fishery products can be associated with a variety of human health hazards. Broadly the same hazards are present in aquaculture products as in corresponding varieties caught in the wild (Section 4.1). The risk of harm from a particular hazard might be increased, under some circumstances, in aquaculture products compared with fish caught in the wild - for instance if the withdrawal time for residues of veterinary drugs has not been observed. High stocking densities, compared with the natural situation, might increase the risk of cross-infection of pathogens within a population of fish and might lead to deterioration of water quality. On the other hand, farmed fish can also present a lower risk of harm. In systems where the fish receive formulated feeds, the risks associated with transmission of hazards through the food consumed by the fish could be reduced. For example, infection with nematode parasites is absent from, or very much reduced in, farmed salmon compared with salmon caught in the wild. Raising fish in cages in the marine environment poses few hazards and low risks. In closed recirculation systems hazards are even further reduced. In those systems, the water is constantly refreshed and reused and water quality is controlled within safe measures.

6.2.1 Hazards

Aquaculture products possess broadly the same hazards that are present in corresponding varieties caught in the wild (Section 5.3.3.1). Potential hazards that are specific to aquaculture products include: residues of veterinary drugs in excess of recommended guidelines and other chemicals used in aquaculture production, contamination of faecal origin where the facilities are close to human habitation or animal husbandry.

6.2.2 Defects

The same defects are present in aquaculture products as in corresponding varieties caught in the wild (Section 5.3.3.1). A defect which may occur is objectionable odours/flavours. During transport of live fish, it is important to reduce stress, as stressing fish can lead to deterioration in quality. Also, care should be taken to minimise physical damage to fish as this can lead to bruising.

6.3 PRODUCTION OPERATIONS

6.3.1 Feed Supply

Feeds used in aquaculture production should comply with the Codex Recommended Code of Practice on Good Animal Feeding (CAC/RCP-54 (2004)).

Potential Hazards: Chemical contamination, mycotoxins and microbiological contamination.

Potential Defects: Decomposed feeds, fungal spoilage.

Technical Guidance:

· Feed and fresh stocks should be purchased and rotated and used prior to the expiry of their shelf life.

· Dry fish feeds should be stored in cool and dry areas to prevent spoilage, mould growth and contamination. Moist feed should be properly refrigerated according to manufacturers instructions.

· Feed ingredients should not contain unsafe levels of pesticides, chemical contaminants, microbial toxins, or other adulterating substances.

· Industrially produced complete feeds and industrially produced feed ingredients should be properly labelled. Their composition must fit the declaration on the label and they should be hygienically acceptable.

· Ingredients should meet acceptable, and if applicable, statutory standards for levels of pathogens, mycotoxins, herbicides, pesticides and other contaminants which may give rise to human health hazards.

· Only approved colours of the correct concentration should be included in the feed.

· Moist feed or feed ingredients should be fresh and of adequate chemical and microbiological quality.

· Fresh or frozen fish should reach the fish farm in an adequate state of freshness.

· Fish silage and offal from fish, if used, should be properly cooked or treated to eliminate potential hazards to human health.

· Feed which is compounded industrially or at the fish farm, should contain only such additives, growth promoting substances, fish flesh colouring agents; anti-oxidising agents, caking agents or veterinary drugs which are permitted for fish by the official agency having jurisdiction.

· Products should be registered with the relevant national authority as appropriate.

· Storage and transport conditions should conform to the specifications on the label.

· Veterinary drug and other chemical treatments should be administered in accordance with recommended practices and comply with national regulations.

· Medicated feeds should be clearly identified in the package and stored separately, in order to avoid errors.

· Farmers should follow manufacturers' instructions on the use of medicated feeds.

· Product tracing of all feed ingredients should be assured by proper record keeping.

6.3.2 Veterinary Drugs

Potential Hazards: Residues of veterinary drugs.

Potential Defects: Unlikely.

Technical Guidance:

· All veterinary drugs for use in fish farming should comply with national regulations and international guidelines (in accordance with the Recommended International Code of Practice for Control of the Use of Veterinary Drugs (CAC/RCP 38-1993) and the Codex Guidelines for the Establishment of a regulatory programme for control of veterinary drugs residues in foods (CAC/GL 16-1993)).

· Prior to administering veterinary drugs, a system should be in place to monitor the application of the drug to ensure that the withdrawal time for the batch of treated fish can be verified.

· Veterinary drugs or medicated feeds should be used according to manufacturers' instructions, with particular attention to withdrawal periods.

· Products should be registered with the appropriate national authority.

· Products should only be prescribed or distributed by personnel authorised under national regulations.

· Storage and transport conditions should conform to the specifications on the label.

· Control of diseases with drugs should be carried out only on the basis of an accurate diagnosis.

· Records should be maintained for the use of veterinary drugs in aquaculture production.

· For those fish which tested with drug residue concentrations above the MRL (or in some countries, by an industry imposed lower level), harvest of the batch should be postponed until the batch complies with the MRL. After an assessment of the Good Aquaculture Practices regarding pre-harvest measures, appropriate steps should be taken to modify the drug residue control system.

· A post harvest control should reject all fish that do not comply with the requirements set for veterinary drug residues by the relevant national authority.

6.3.3 Growing

Potential Hazards: Microbiological and chemical contamination.

Potential Defects: Abnormal colour, muddy flavour, physical damage.

Technical Guidance:

· Source of postlarvae, fries and fingerlings should be controlled to assure healthy stock.

· Stocking densities should be based on culture techniques, fish species, size and age, carrying capacity of the fish farm, anticipated survival and desired size at harvesting.

· Diseased fish should be quarantined when necessary and appropriate and dead fish should be disposed immediately in a sanitary manner that will discourage the spread of disease and the cause of death should be investigated.

· Good water quality should be maintained by using stocking and feeding rates that do not exceed the carrying capacity of the culture system.

· Growing water quality should be monitored regularly, so as to identify potential hazards and defects.

· The fish farm should have a management plan that includes a sanitation programme, monitoring and corrective actions, defined fallowing periods, appropriate use of agrochemicals, verification procedures for fish farming operations and systematic records.

· Equipment such as cages and nets should be designed and constructed to ensure minimum physical damage of the fish during the growing stage.

· All equipment and holding facilities should be easy to clean and to disinfect and should be cleaned and disinfected regularly and as appropriate.

6.3.4 Harvesting

Potential Hazards: Unlikely.

Potential Defects: Physical damage, physical/biochemical change due to stress of live fish.

Technical Guidance:

· Appropriate harvesting techniques should be applied to minimise physical damage and stress.

· Live fish should not be subjected to extremes of heat or cold or sudden variations in temperature and salinity.

· Fish should be free from excessive mud and weed soon after being harvested by washing it with clean seawater or fresh water under suitable pressure.

· Fish should be purged, where necessary, to reduce gut contents and pollution of fish during further processing.

· Fish should be handled in a sanitary manner according to the guidelines in Section 4 of the Code.

· Harvesting should be rapid so that fish are not exposed unduly to high temperatures.

· All equipment and holding facilities should be easy to clean and to disinfect and should be cleaned and disinfected regularly and as appropriate.

6.3.5 Holding and Transportation

Potential Hazards: microbiological and chemical contamination.

Potential Defects: physical damage, physical/biochemical change due to stress of live fish.

Technical Guidance:

· Fish should be handled in such a way as to avoid unnecessary stress.

· Fish should be transported without undue delay.

· Equipment for the transport of live fish should be designed for rapid and efficient handling without causing physical damage or stress.

· All equipment and holding facilities should be easy to clean and to disinfect and should be cleaned and disinfected regularly and as appropriate.

· Records for transport of fish should be maintained to ensure full product tracing.

· Fish should not be transported with other products which might contaminate them.

6.3.6 Storage and transport of live fish

This section is designed for the storage and transportation of live fish originating from aquaculture or capture.

Potential Hazards: microbiological contamination, biotoxins, chemical contamination (e.g. oil, cleaning and disinfecting agents).

Potential Defects: Dead fish, physical damage, off flavours, physical/biochemical change due to stress of live fish.

Technical Guidance:

· Only healthy and undamaged fish should be chosen for live storage and transport. Damaged, sick and dead fish should be removed before introduction to the holding or conditioning tanks.

· Holding tanks should be checked regularly during storage and transportation. Damaged, sick and dead fish should be removed immediately when found.

· Clean water utilised to fill holding tanks, or to pump fish between holding tanks, or for conditioning fish, should be similar in properties and composition to the water from where the fish was originally taken to reduce fish stress.

· Water should not be contaminated with either human sewage or industrial pollution. Holding tanks and transportation systems should be designed and operated in a hygienic way to prevent contamination of water and equipment.

· Water in holding and conditioning tanks should be well aerated before fish is transferred into them.

· Where seawater is used in holding or conditioning tanks, for species prone to toxic algae contamination, seawater containing high level of cell concentrations should be avoided or filtered properly.

· No fish feeding should occur during storage and transport of live fish. Feeding will pollute water of holding tanks very quickly and, in general, fish should not be fed 24 hours before transporting.

· Material of holding and conditioning tanks, pumps, filters, piping, temperature control system, intermediate and final packaging or containers should not be harmful to fish or present hazards to humans.

· All equipment and facilities should be cleaned and disinfected regularly and as needed.

6.3.6.1 Live fish stored and transported at ambient temperature

Potential Hazards: *microbiological contamination, biotoxins, chemical contamination (e.g. oil, cleaning and disinfecting agents)*

Potential Defects: *Dead fish, physical damage, off flavours, physical/biochemical change due to stress of live fish*

Technical Guidance:

· Depending on the source of water, requirements of the species and time of storage and/or transport, it could be necessary to re-circulate the water and filter it through mechanical and/or biofilters.

· Water intake of holding tanks on board of vessels should be located so as to avoid contamination from vessel's sewage, waste and engine cooling discharge. Pumping of water should be avoided when the vessel comes into harbour or sailing through waters near sewage or industrial discharges. Equivalent precautions should be adopted for water intake on land.

· Facilities for storing and transportation (holding tanks) of live fish should be capable to:

— maintain the oxygenation of water in the holding tanks through either, continuous water flow, direct oxygenation (with oxygen or air bubbling), or regularly and as needed changing of the water of the holding tank;

— maintain the temperature of storage and transport, for species sensitive to temperature fluctuations. It may be necessary to insulate the holding tanks and install a temperature control system;

— keep water in reserve which might be needed in case the holding tank should drain. The volume in fixed facilities (storage) should be at least of the same volume of the total holding tanks in operation. The volume in land transport facilities should be at least capable to compensate water for evaporation, leakage, purges, filter cleaning and eventual mixing of water for control purposes;

· For species known to exhibit strong territoriality or cannibalism or hyperactivity when under stress, these fish should be separated in individual tanks or appropriately secured/banned to prevent damage (an alternative method is reduction of temperature).

6.3.6.2 Live fish stored and transported at low temperatures

Potential Hazards: *microbiological contamination, biotoxins, chemical contamination (e.g. oil, cleaning and disinfecting agents)*

Potential Defects: *Dead fish, physical damage, off flavours, physical/biochemical change due to stress of live fish*

Technical Guidance:

· Conditioning should aim at reducing the metabolic rate of fish in order to minimize the stress to them. Conditioning of the fish at low temperatures should be done according to the characteristics of the species (minimum temperature, cooling rate, water/humidity requirements, packaging conditions). Conditioning is a biological operation to reduce the metabolic rate of the fish minimising the stress to them.

· The level of temperature to be reached should be in accordance with the species, transport and packaging conditions. There is a range of temperature in which fish do not exhibit or have reduced physical activity. The limit is attained at the temperature at which the metabolic rate of the fish is minimised without causing adverse effects to them (basal metabolic rate).

· When performing conditioning, only approved anaesthetics and procedures accepted by the regulations should be used.

· Conditioned fish should be packed without delay in proper insulated containers.

· Remaining water or water for use with packaging material for conditioned fish should be clean, of similar composition and pH to the water the fish was taken from, but to the temperature of storage.

· Water absorbent pads, shredded wood, wood shavings or sawdust and tying material that may be utilised for packaging conditioned fish should be clean, first use, free of possible hazards and be wet right at the time of packaging.

· Conditioned and packed fish should be stored or transported under conditions that assure proper temperature control.

SECTION 8 PROCESSING OF FRESH, FROZEN AND MINCED FISH

In the context of recognising controls at individual processing steps, this section provides examples of potential hazards and defects and describes technological guidelines, which can be used to develop control measures and corrective action. At a particular step only the hazards and defects, which are likely to be introduced or controlled at that step, are listed. It should be recognised that in preparing a HACCP and/or DAP plan it is essential to consult Section 5 which provides guidance for the application of the principles of HACCP and DAP analysis. However, within the scope of this Code of Practice it is not possible to give details of critical limits, monitoring, record keeping and verification for each of the steps since these are specific to particular hazards and defects.

In general, the processing of fresh, frozen fish and minced fish, will range in sophistication. In its simplest form, the processing of fresh and frozen fish may be presented in a raw state such as dressed, fillets, and minced to be distributed in markets and institutions or used in processing facilities. For the latter, the processing of fresh, frozen and minced fish is often an intermediate step to the production of value added products (for example, smoked fish as described in section 12, canned fish as described in section 16, frozen breaded or battered fish as described in section 15). Traditional methods often prevail in the design of a process. However, modern scientific food technology is having an increasingly important role in enhancing the preservation and shelf-stability of a product. Regardless of the complexity of a particular process, the fabrication of the desired product relies on the consecutive execution of individual steps. As stressed by this Code, the application of appropriate elements of the pre-requisite programme (Section 3) and HACCP principles (Section 5) at these steps will provide the processor with reasonable assurance that the essential quality, composition and labelling provisions of the appropriate Codex standard will be maintained and food safety issues controlled.

The example of the flow diagram (Figure 8.1) will provide guidance to some of the common steps involved in a fish fillet preparation line, and three examples of final product types: modified atmosphere packaging (MAP), minced and frozen fish. As in the further processing of fresh fish in a MAP product, or minced or frozen fish, the section labelled "Fish Preparation" is used as the basis for all the other fish processing operations (Sections 9-16)[1], where appropriate.

8.1 FINFISH PREPARATION

The hygienic conditions and technical manner in which fish are prepared is similar and is not influenced greatly by its intended purpose (for direct distribution or for further processing). However, variations will exist in the form in which the fresh fish flesh is to be utilised. The forms may include, but not limited to, dressed, fillets or steaks.

8.1.1 Raw, Fresh or Frozen Fish Reception (Processing Steps 1)

Potential Hazards: Microbiological pathogens, viable parasites, biotoxins, scombrotoxin, chemicals (including veterinary drug residues) and physical contamination.

Potential Defects: Decomposition, parasites, physical contamination.

Technical Guidance:

• for raw fish material, product specifications could include the following characteristics:

— organoleptic characteristics such as appearance, odour, texture, etc;

— chemical indicators of decomposition and/or contamination, for example, TVBN, histamine, heavy metals, pesticide residues, nitrates etc;

— microbiological criteria, in particular for intermediate raw materials, to prevent the processing of raw material containing microbial toxins;

— foreign matter;

— physical characteristics such as size of fish;

— species homogeneity.

• training in species identification and communication in product specification should be provided to fish handlers and appropriate personnel to ensure a safe source of incoming fish where written protocols exist. Of special consideration, are the reception and sorting of fish species that poses a risk of biotoxins such as ciguatoxin in large carnivorous tropical and sub-tropical reef fish or scombrotoxin in scombroid species or parasites;

• skills should be acquired by fish handlers and appropriate personnel in sensory evaluation techniques to ensure raw fish meet essential quality provisions of the appropriate Codex standard;

• fish requiring gutting on arrival at the processing facility should be gutted efficiently, without undue delay and with care to avoid contamination (see Section 8.1.5 - Washing & Gutting);

• fish should be rejected if it is known to contain harmful, decomposed or extraneous substances, which will not be reduced or eliminated to an acceptable level by normal procedures of sorting or preparation;

• information about the harvesting area.

8.1.1.1 Sensory Evaluation of Fish

The best method of assessing the freshness or spoilage of fish is by sensory evaluation techniques[2]. It

---

[1] Sections 10 – 15 under elaboration.

[2] *Guidelines for Sensory Evaluation of Fish and Shellfish in Laboratories* (CAC/GL 31-1999).

This flow chart is for illustrative purposes only For in factory HACCP implementation a complete and comprehensive flow chart has to be drawn up for each process. References correspond to relevant Sections of the Code.

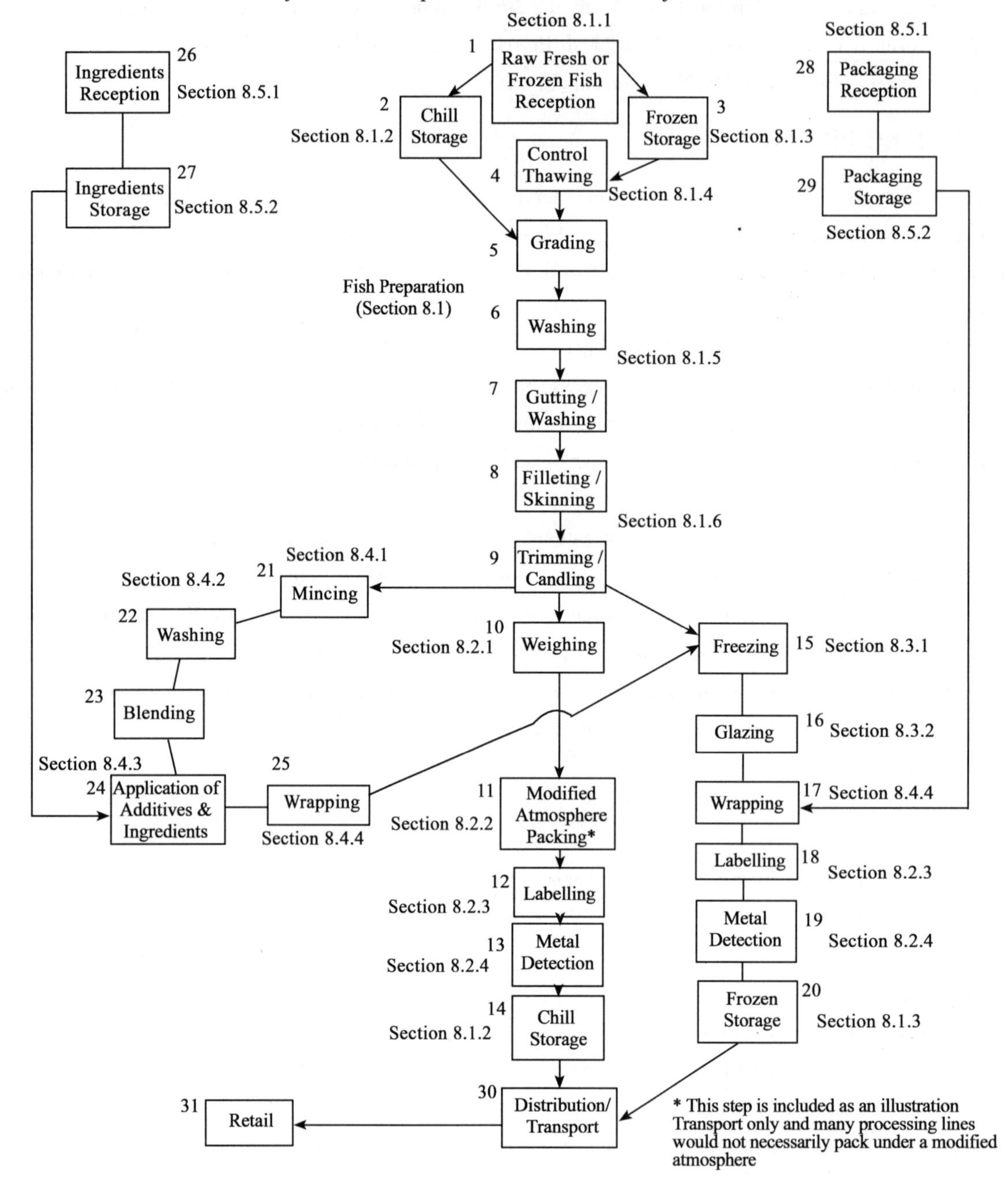

Figure 8.1 Example of a flow chart of a fish fillet preparation line, including MAP, mincing and freezing operations

is recommended that appropriate sensory evaluation criteria be used to evaluate the acceptability of fish and to eliminate fish showing loss of essential quality provisions of the appropriate Codex standards. As an example, fresh white fish species are considered unacceptable when showing the following characteristics:

Skin/Slime	dull, gritty colours with yellow brown dotting slime
Eyes	Concave, opaque, sunken discoloured
Gills	grey - brown or bleached, slime opaque yellow, thick or clotting
Odour	flesh odour amines, ammonia, milky lactic, sulphide, faecal, putrid, rancid

8.1.2 Chilled Storage (Processing Steps 2 & 14)

Potential Hazards: Microbiological pathogens, biotoxin, and scombrotoxin.

Potential Defects: Decomposition, physical damage.

Technical Guidance:

- fish should be moved to the chill storage facility without undue delay;
- the facility should be capable of maintaining the temperature of the fish between 0 ~ +4℃;
- the chill room should be equipped with a calibrated indicating thermometer. Fitting of a recording thermometer is strongly recommended;
- stock rotation plans should ensure proper utilisation of the fish;
- the fish should be stored in shallow layers and surrounded by sufficient finely divided ice or with a mixture of ice and of water before processing;
- fish should be stored such that damage will be prevented from over-stacking or over-filling of boxes;
- where appropriate replenish ice supply on the fish or alter temperature of the room.

8.1.3 Frozen Storage (Processing Steps 3 & 20)

Potential Hazards: Microbiological pathogens, toxins, viable parasites

Potential Defects: Dehydration, rancidity, loss of nutritional quality

Technical Guidance:

- the facility should be capable of maintaining the temperature of the fish at or colder than -18℃, and with minimal temperature fluctuations;
- the store should be equipped with a calibrated indicating thermometer. Fitting of a recording thermometer is strongly recommended;
- a systematic stock rotation plan should be developed and maintained;
- product should be glazed and/or wrapped to protect it from dehydration;
- fish should be rejected if known to contain defects, which subsequently cannot be reduced or eliminated to an acceptable level by re-working. An appropriate assessment should be carried out to determine the reason (s) for loss of control and the DAP plan modified where necessary;
- for killing of parasites harmful to human health, the freezing temperature and monitoring of duration of freezing should be combined with good inventory control to ensure sufficient cold treatment.

8.1.4 Control Thawing (Processing Step 4)

Potential Hazards: Microbiological pathogens, biotoxins and scombrotoxin.

Potential Defects: Decomposition.

Technical Guidance:

- the thawing method should be clearly defined and should address the time and temperature of thawing, temperature measuring instrument used and placement of device for measurement. The thawing schedule (time and temperature parameters) should be carefully monitored. Selection of the thawing method should take into account in particular the thickness and uniformity of size of the products to be thawed;
- thawing time and temperature and fish temperature critical limits should be selected so as to control

the development of micro-organisms, histamine, where high risk species are concerned or persistent and distinctive objectionable odours or flavours indicative of decomposition or rancidity;

· where water is used as the thawing medium, it should be of potable quality;

· where recycling of water is used, care should be taken to avoid the build up of microorganisms;

· where water is used, circulation should be sufficient to produce even thawing;

· during thawing, according to the method used, products should not be exposed to excessively high temperatures;

· particular attention should be paid to controlling condensation and drip from the fish. An effective drainage should be made;

· after thawing, fish should be immediately processed or refrigerated and kept at the adequate temperature (temperature of melting ice);

· the thawing schedule should be reviewed as appropriate and amended where necessary.

8.1.5 Washing and Gutting (Processing Steps 6 & 7)

Potential Hazards: Microbiological pathogens, biotoxins and scombrotoxin.

Potential Defects: Presence of viscera, bruising, off-flavours, cutting faults.

Technical Guidance:

· gutting is considered complete when the intestinal tract and internal organs have been removed;

· an adequate supply of clean sea water or potable water should be available for washing of:

— whole fish to remove foreign debris and reduce bacterial load prior to gutting;

— gutted fish to remove blood and viscera from the belly cavity;

— surface of fish to remove any loose scales;

— gutting equipment and utensils to minimise build-up of slime and blood and offal;

· depending on the vessel or processing facility product flow pattern and where a prescribed critical limit for staging time and temperature regime has been established for the control of histamine or a defect, the gutted fish should be drained and well iced or appropriately chilled in clean containers and stored in specially designated and appropriate areas within the processing facility;

· separate and adequate storage facilities should be provided for the fish roe, milt and livers, if these are saved for later utilisation.

8.1.6 Filleting, Skinning, Trimming and Candling (Processing Steps 8 & 9)

Potential Hazards: Viable parasites, microbiological pathogens, biotoxins and scombrotoxin, presence of bones.

Potential Defects: Parasites, presence of bones, objectionable matter (e.g. skin, scales, etc.), decomposition.

Technical Guidance:

· to minimise time delays, the design of the filleting line and candling line, where applicable, should be continuous and sequential to permit the uniform flow without stoppages or slowdowns and removal of waste;

· an adequate supply of clean sea water or potable water should be available for washing of:

— fish prior to filleting or cutting especially fish that have been scaled;

— fillets after filleting or skinning or trimming to remove any signs of blood, scales or viscera;

— filleting equipment and utensils to minimise build-up of slime and blood and offal;

— for fillets to be marketed and designated as boneless, fish handlers should employ appropriate inspection techniques and use the necessary tools to remove bones not meeting Codex standards[1][2] or commercial specifications.

· The candling of skinless fillets by skilled personnel, in a suitable location which optimises the illuminating effect, is an effective technique in controlling parasites (in fresh fish) and should be employed when implicated fish species are being used;

· the candling table should be frequently cleaned during operation in order to minimise the microbial activity of contact surfaces and the drying of fish residue due to heat generated from the lamp;

· where a prescribed critical limit for staging time and temperature regime has been established for the control of histamine or a defect, the fish fillets should be well iced or appropriately chilled in clean containers, protected from dehydration and stored in appropriate areas within the processing facility.

8.2 PROCESSING OF VACUUM OR MODIFIED ATMOSPHERE PACKED FISH

This section is designed to augment the processing of fresh fish section with additional operation steps pertaining specifically to the modified atmosphere packing of fish (see also Appendix I).

8.2.1 Weighing (Processing Step 10)

Potential Hazards: *Unlikely.*

Potential Defects: *Incorrect net weight.*

Technical Guidance:

· weigh scales should be periodically calibrated with a standardised mass to ensure accuracy.

8.2.2 Vacuum or Modified Atmosphere Packaging (Processing Step 11)

Potential Hazards: *Subsequent microbiological pathogens and biotoxins, physical contamination (metal).*

Potential Defects: *Subsequent decomposition.*

Technical Guidance:

The extent to which the shelf-life of the product can be extended by vacuum or MAP will depend on the species, fat content, initial bacterial load, gas mixture, type of packaging material and, especially important, the temperature of storage. Refer to Appendix I for process control issues in modified atmosphere packaging.

· modified atmosphere packaging should be strictly controlled by:

— monitoring the gas to product ratio;

— types and ratio of gas mixtures used;

— type of film used;

— type and integrity of the seal;

— temperature control of product during storage.

· occurrence of adequate vacuum and package;

· fish flesh should be clear of the seam area;

· packaging material should be inspected prior to use to ensure that it is not damaged or contaminated;

---

[1] Codex Standard for Quick Frozen Blocks of Fish Fillet, Minced Fish Flesh and Mixtures of Fillets and Minced Fish Flesh (Codex Stan. 165-1989, Rev. 1-1995).

[2] Codex Standard for Quick Frozen Fish Fillets (Codex Stan. 190-1995).

· packaging integrity of the finished product should be inspected at regular intervals by an appropriately trained personnel to verify the effectiveness of the seal and the proper operation of the packaging machine;

· following sealing, MAP or vacuumed products should be transferred carefully and without undue delay to chilled storage;

· Ensure that adequate vacuum is attained, and the package seals are intact.

8.2.3 Labelling (Processing Steps 12 & 18)

Potential Hazards: Unlikely.

Potential Defects: Incorrect labelling.

Technical Guidance:

· prior to their application, labels should be verified to ensure that all information declared meet, where applicable, the Codex General Standard for the Labelling of Pre-packaged Foods[1], labelling provisions of the appropriate Codex Standard for products and/or other relevant national legislative requirements;

· in many cases it will be possible to re-label incorrectly labelled products. An appropriate assessment should be carried out to determine the reason (s) for incorrect labelling and the DAP plan should be modified where necessary;

8.2.4 Metal Detection (Processing Steps 13 & 19)

Potential Hazards: Metal contamination.

Potential Defects: Unlikely.

Technical Guidance:

· it is important that line speeds are adjusted to allow for the proper functioning of a metal detector;

· routine procedures should be initiated to ensure product rejected by the detector is investigated as to the cause of the rejection;

· metal detectors, if used, should be periodically calibrated with a known standard to ensure proper operation;

8.3 PROCESSING OF FROZEN FISH

This section is designed to augment the processing of fresh fish section with additional operation steps pertaining specifically to the processing of frozen fish.

8.3.1 Freezing Process (Processing Step 15)

Potential Hazards: Viable parasites.

Potential Defects: Texture deterioration, development of rancid odours, freezer burn.

Technical Guidance:

The fish product should be subjected to a freezing process as quickly as possible since unnecessary delays before freezing will cause temperature of the fish products to rise, increasing the rate of quality deterioration and reducing shelf-life due to the action of micro-organisms and undesirable chemical reactions.

· a time and temperature regime for freezing should be established and should take into consideration the freezing equipment and capacity; the nature of the fish product including thermal conductivity, thickness, shape and temperature and the volume of production, to ensure that the range of temperature of maximum crystallisation is passed through as quickly as possible;

---

[1] Codex General Standard for the Labelling of Pre-packaged Foods (CODEX STAN 1-1985, Rev. 2-1991).

· the thickness, shape and temperature of fish product entering the freezing process should be as uniform as possible;

· processing facility production should be geared to the capacity of freezers;

· frozen product should be moved to the cold storage facility as quickly as possible;

· the core temperature of the frozen fish should be monitored regularly for completeness of the freezing process;

· frequent checks should be made to ensure correct operation of freezing;

· accurate records of all freezing operations should be kept;

· for killing of parasites harmful to human health, the freezing temperature and monitoring of duration of freezing should be combined with good inventory control to ensure sufficient cold treatment.

8.3.2 Glazing (Processing Step 16)

Potential Hazards: Microbiological pathogens.

Potential Defects: Subsequent dehydration, incorrect net weight.

Technical Guidance:

· glazing is considered complete when the entire surface of the frozen fish product is covered with a suitable protective coating of ice and should be free of exposed areas where dehydration (freezer-burn) can occur;

· if additives are used in the water for glazing, care should be taken to ensure its proper proportion and application with product specifications;

· where the labelling of a product is concerned, information on the amount or proportion of glaze applied to a product or a production run should be kept and used in the determination of the net weight which is exclusive of the glaze;

· where appropriate monitoring should ensure that spray nozzles do not become blocked;

· where dips are used for glazing it is important to replace the glazing solution periodically to minimise the bacterial load and build-up of fish protein, which can hamper freezing performance.

8.4 PROCESSING OF MINCED FISH

This section is designed to augment the processing of fresh fish section (prior to mincing) and processing of frozen fish section (after mincing) with additional operation steps pertaining specifically to the processing of minced fish.

8.4.1 Mincing Fish Using Mechanical Separation Process (Processing Step 21)

Potential Hazards: Microbiological pathogens, biotoxins and scombrotoxin, physical contamination (metal, bones, rubber from separator belt, etc.).

Potential Defects: Incorrect separation (i.e. objectionable matter), decomposition, presence of defect bones, parasites.

Technical Guidance:

· the separator should be fed continuously but not excessively;

· candling is recommended for fish suspected of high infestation with parasites;

· split fish or fillets should be fed to the separator so that the cut surface contacts the perforated surface;

· fish should be fed to the separator in a size that it is able to handle;

· in order to avoid time-consuming adjustments of the machinery and variations in quality of the finished product, raw materials of different species and types should be segregated and processing of separate batches should be carefully planned;

· the perforation sizes of the separator surface as well as the pressure on the raw material should be adjusted to the characteristics desired in the final product;

· the separated residual material should be carefully removed on a continuous or near-continuous basis to the next processing stage;

· temperature monitoring should ensure undue temperature rises of the product are avoided.

8.4.2 Washing of Minced Fish (Processing Step 22)

Potential Hazards: Microbiological pathogens and scombrotoxin.

Potential Defects: Poor colour, poor texture, excess of water.

Technical Guidance:

· if necessary the mince should be washed and should be adequate for the type of product desired;

· stirring during washing should be carried out with care, but it should be kept as gentle as possible in order to avoid excessive disintegration of the minced flesh which will reduce the yield due to the formation of fines;

· the washed minced fish flesh may be partially de-watered by rotary sieves or centrifugal equipment and the process completed by pressing to appropriate moisture content if necessary, and depending on eventual end-use, the de-watered mince should be either strained or emulsified;

· special attention should be taken to ensure mince being strained is kept cool;

· the resulting waste water should be disposed of in a suitable manner.

8.4.3 Blending and Application of Additives and Ingredients to Minced Fish (Processing Steps 23 & 24)

Potential Hazards: Physical contamination, non-approved additives and/or ingredients.

Potential Defects: Physical contamination, incorrect addition of additives.

Technical Guidance:

· if fish, ingredients and /or additives are to be added, they should be blended in the proper proportions to achieve the desired sensory quality;

· additives should comply with the requirements of the Codex General Standard for Food Additives;

· the minced fish product should be packaged and frozen immediately after preparation; if it is not frozen or used immediately after preparation it should be chilled.

8.4.4 Wrapping and Packing (Processing Steps 17 & 25)

Potential Hazards: Microbiological pathogens.

Potential Defects: Subsequent dehydration, decomposition.

Technical Guidance:

· packaging material should be clean, sound, durable, sufficient for its intended use and of food grade material;

· the packaging operation should be conducted to minimise the risk of contamination and decomposition;

· products should meet appropriate standards for labelling and weights.

8.5 PACKAGING, LABELS & INGREDIENTS

8.5.1 Reception – Packaging, Labels & Ingredients (Processing Steps 26 & 28)

Potential Hazards: Microbiological pathogens, chemical and physical contamination.

Potential Defects: *Misdescription.*

Technical Guidance:

• only ingredients, packaging material and labels complying with the processors' specification should be accepted into the processing facility;

• labels which are to be used in direct contact with the fish should be fabricated of a nonabsorbent material and the ink or dye used on that label should be approved by the official agency having jurisdiction;

• ingredients and packaging material not approved by the official agency having jurisdiction should be investigated and refused at reception.

8.5.2 Storage - Packaging, Labels & Ingredients (Processing Steps 27 & 29)

Potential Hazards: *Microbiological pathogens, chemical and physical contamination.*

Potential Defects: *Loss of quality characteristics of packaging materials or ingredients.*

Technical Guidance:

• ingredients and packaging should be stored appropriately in terms of temperature and humidity;

• a systematic stock rotation plan should be developed and maintained to avoid out of date materials;

• ingredients and packaging should be properly protected and segregated to prevent crosscontamination;

• defective ingredients and packaging should not be used.

SECTION 9 PROCESSING OF FROZEN SURIMI

In the context of recognising controls at individual processing steps, this section provides examples of potential hazards and defects and describes technological guidelines, which can be used to develop control measures and corrective action. At a particular step only the hazards and defects, which are likely to be introduced or controlled at that step, are listed. It should be recognised that in preparing a HACCP and/or DAP plan it is essential to consult Section 5 which provides guidance for the application of the principles of HACCP and DAP analysis. However, within the scope of this Code of Practice it is not possible to give details of critical limits, monitoring, record keeping and verification for each of the steps since these are specific to particular hazards and defects.

Frozen surimi is an intermediate food ingredient made from myofibrillar fish protein isolated from other constituent fish protein by repeated washing and de-watering of minced fish. Cryoprotectants are added so that the mince can be frozen and will retain the capacity to form gel when heat-treated after thawing. Frozen surimi is usually blended with other components and further processed into surimi-based products such as kamaboko or crab analogs (imitation crab) that utilise its gel forming ability.

The main emphasis of this section of the code is to give guidance to the manufacture of frozen surimi processed from marine groundfish such as Alaska Pollock and Pacific Whiting by mechanised operations that are common in Japan, the United States and some other country in which there are processors under mechanised operation.

The vast majority of frozen surimi is processed from marine groundfish such as Alaska Pollock and Pacific Whiting. However, technological advances and the change of main raw fish species for frozen surimi production will necessitate periodic revision of this section of the Code of Practice.

Frozen surimi is manufactured using various methods, but this flow chart shows the most typical procedure. This flow chart is for illustrative purpose only For in-factory HACCP implementation a complete and comprehensive flow chart has to be draw up for each process.
References correspond to relevant Sections of the Code.

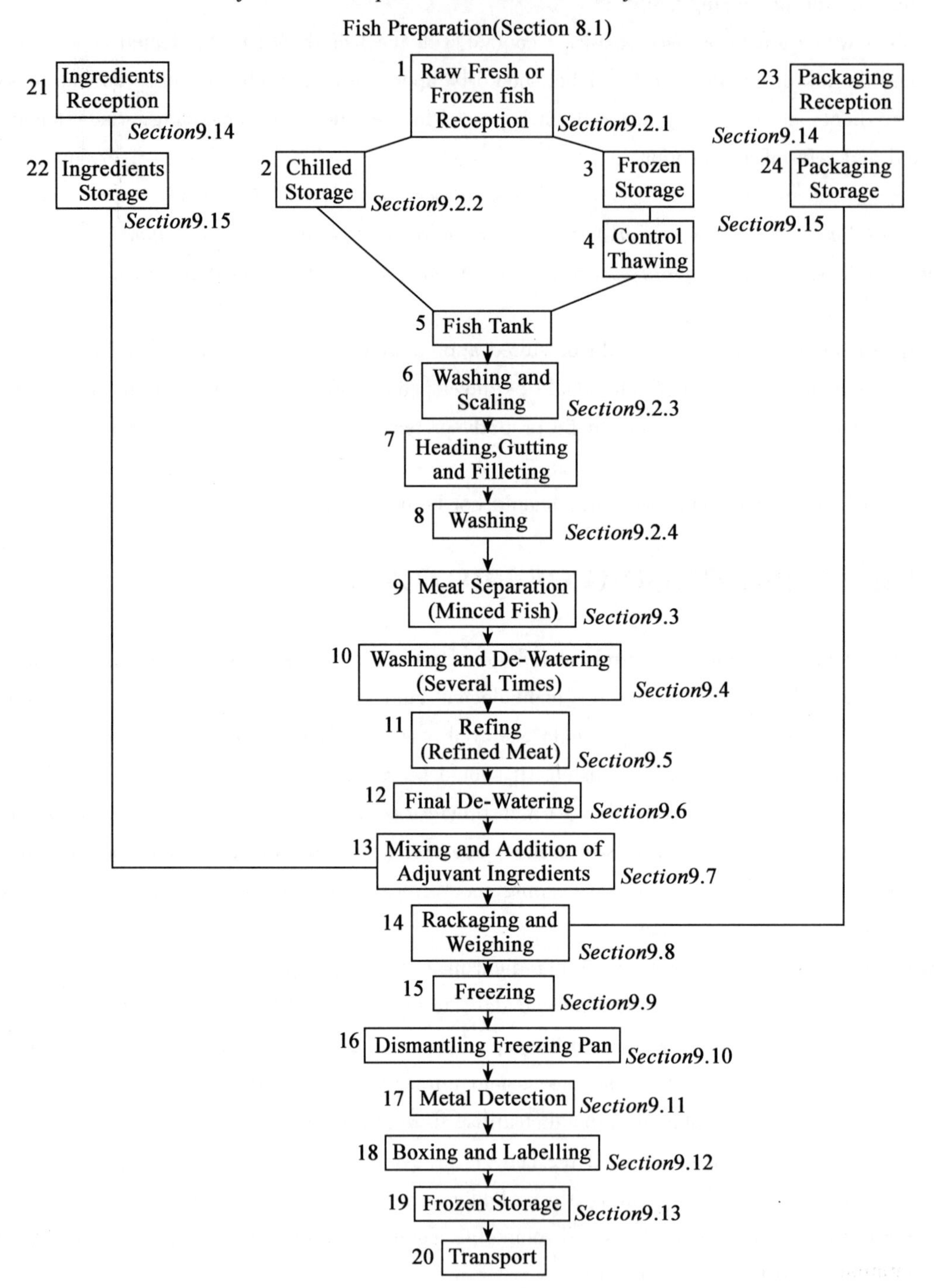

Figure 9. 1 Example of a flow chart of a frozen surimi production process

9.1 GENERAL CONSIDERATIONS OF HAZARDS AND DEFECTS FOR FROZEN SURIMI PRODUCTION

9.1.1 Hazards

Frozen surimi is an intermediate ingredient that will be further processed into surimi-based products such as kamaboko and crab analogs. Many of the potential food safety hazards will be controlled during subsequent processing. For example, pathogenic bacteria such as Listeria monocytogenes and toxin formers such as *Clostridium botulinum* (that becomes a hazard due to modified atmosphere packaging of the end product) should be controlled during the cooking or pasteurising steps of final processing. Possible *Staphylococcus aureus* contamination that produces heat-stable enterotoxins should be adequately controlled by the prerequisite programme. Parasites will not be a hazard since the final product will be cooked or pasteurised.

If scombrotoxin-forming fish such as tuna or mackerel or tropical reef fish that may accumulate ciguatera toxin are utilised for surimi, appropriate controls for these hazards should be developed. Likewise, due to the highly mechanised nature of surimi processing, appropriate controls should be instituted to assure that metal fragments (e.g., bearings, bolts, washers, and nuts) are excluded or eliminated in the end product.

In countries that produce frozen surimi by traditional non-mechanised methods from locally available fish species for local consumption, extensive consideration should be given to pre-requisite programmes described in section 3.

9.1.2 Defects

Certain quality attributes of frozen surimi is important for the successful manufacture of surimi-based products such as kamaboko and crab analogs that meet consumer expectations of quality. Some of these important factors are colour, moisture content, pH or gel strength. These and others are described in more detail in Appendix X of the code entitled Optional Final Product Requirements for Frozen Surimi[1].

Myxosporidia is a parasite that is common in marine groundfish such as Pacific Whiting. This organism contains protease enzymes that chemically separates proteins that can ultimately affect the gel strength of surimi even at very low incidence. If species are used that are known to contain this parasite, a protease inhibitors such as beef plasma protein or egg whites may be needed as additives to attain the necessary gel strength capabilities for kamaboko or crab analogs production.

Decomposed fish should not be used as raw material for frozen surimi production. The sensory qualities will not be sufficient to produce acceptable kamaboko or crab analog end products. It also necessary to note that decomposed fish should not be used as raw material for production of frozen surimi, because proliferation of spoilage bacteria that cause decomposition of the end product will cause negative effect on the gel forming ability of frozen surimi by denaturing salt soluble protein.

The washing and de-watering cycle should be sufficient to achieve separation of the water-soluble protein from the myofibrillar proteins. If water-soluble proteins remain in the product it will negatively affect the gel forming ability and the long term frozen storage shelf life.

Objectionable matter such as small bones, scales and black belly lining should be minimised as it negatively affects the usability of frozen surimi for processing into end products.

Due to the comminuted nature of raw surimi, the use of food additive may be necessary to achieve the

[1] Under elaboration.

level of quality that is desired. These additives should be introduced to surimi in accordance to appropriate regulations and manufacturer's recommendation in order to avoid quality problems and regulatory actions.

Consideration should be given to the thermal stability of fish proteins. At normal room temperatures most fish proteins will undergo denaturing that will inhibit the gel forming ability of the product. Alaska Pollock and other cold water marine fish should not be subjected to temperatures above 10℃ during processing. Warm water fishes may denature at a slower rate and may not be as temperature sensitive.

In countries that produce frozen surimi by traditional non-mechanised methods from locally available fish species for local consumption, special consideration should be given to several defects. Since the growth of spoilage bacteria that cause decomposition and protein denaturation increases with temperature, the conditions that the raw and processed product is subjected to should be carefully monitored.

9.2 FISH PREPARATION (Processing Steps 1 to 8)

Refer to Section 8.1 steps 1 through 8 for information regarding preparation of fish for processing. For frozen surimi processing, consideration should be given to the following for each step:

9.2.1 Raw Fresh and Frozen Fish Reception (Processing Step 1)

Potential Hazards: unlikely when using marine ground fish as the raw material

Potential Defects: decomposition, protein denaturation

Technical Guidance:

· harvested fish intended for frozen surimi processing should preferably be kept at 4℃ or below;

· consideration should be given to the age and condition of fish used for surimi processing as the factors will affect the final gel strength capability. Especially, care should be taken to raw fish received many hours after harvest. For example acceptable period after harvest should be as follows, but processing as fast as possible after harvest will better retain adequate quality of frozen surimi:

— round; within 14 days of harvest, when stored at 4℃ or below;

— dressed; within 24 hours after dressing, when stored at 4℃ or below.

· date, time of harvesting, origin and harvester or vendor of products received should be properly recorded and identified;

· presence of decomposition in raw product should not be allowed, as it will negatively affect the gel strength capability of the end product. Harvested fish in poor condition may not result in specified colour characteristics;

· Fish that is used for frozen surimi processing should have a flesh for adequate gel strength capability. For example an aggregate flesh for Alaska Pollock (*Theragra chalcogramma*) should have pH of 7.0 ± 0.5

· fish that is crushed and suffocated due to abnormally big tow size and duration during harvesting should be deleted from the line in order to avoid negative effect to gel forming ability.

9.2.2 Chilled Storage (Processing Step 2)

Potential Hazards: unlikely

Potential Defects: protein denaturation

Technical Guidance:

· chilled storage at the processing facility should be minimised with prompt processing in order to minimise protein denaturation and loss of gel strength capability;

· raw fish should be preferably stored at 4℃ or below and the dates of harvesting and the time of receipt of the fish should identify the lot of fish used for processing.

9.2.3 Washing and Scaling (Processing Step 6)

Potential Hazards: unlikely

Potential Defects: protein denaturation, colour, objectionable matter

Technical Guidance:

· the epidermis (slime layer), scales and loose pigment should be removed before heading and gutting. This will lessen the level of impurities and extraneous material that can negatively affect the gel strength capability and colour of the end product.

9.2.4 Washing (Processing Step 8)

Potential Hazards: unlikely

Potential Defects: impurities, extraneous materials

Technical Guidance:

· headed and gutted fish should be re-washed. This will lessen the level of impurities and extraneous material that can negatively affect the gel strength capability and colour of the end product.

9.3 MEAT SEPARATION PROCESS (Processing Step 9)

Potential Hazards: metal fragments

Potential Defects: impurities

Technical Guidance:

· fish flesh is minced using mechanical separation process, therefore metal detection equipment that is capable of sensing product that has become contaminated with metal fragments of the size likely to cause human injury should be installed at the most appropriate place in the process to eliminate the hazard;

· procedures should be established to assure that chemical contamination of the product is not likely;

· separated minced meat should be immediately spread into water and transferred to the washing and de-watering step to prevent blood from congealing and causing loss of gel strength capability.

9.4 WASHING AND DE-WATERING PROCESS (Processing Step 10)

Potential Hazards: pathogenic microbial growth

Potential Defects: decomposition, protein denaturation, residual water-soluble protein

Technical Guidance:

· temperature of the water and minced fish flesh in the rotating sieve or wash water should be adequately controlled to prevent the growth of pathogenic microbes;

· wash water should be 10℃ or below for adequate separation of water-soluble proteins. Wash water for Pacific Whiting should be lower than 5℃ since this species will usually have a high protease activity. Some warm water species may be processed at temperatures up to 15℃;

· product should be processed promptly to minimise possible pathogenic microbial growth;

· minced fish should be spread uniformly in the water to assure dilution of the water- soluble components and effect proper separation from the myofibrillar protein;

· consideration should be given to the specific design of the washing and de-watering step in regards to the desired yield, quality and fish species;

· a sufficient amount of potable water should be available for washing;

· the pH of wash water should be near 7.0. Wash water should preferably have a total hardness of

100ppm or below in terms of converted $CaCO_3$;

· salt or other de-watering aids can be added (less than 0.3% salt) in the final stage of washing to enhance dehydration efficiency;

· food additives should be added in accordance with national regulations and manufacturer's instructions, if use in this process;

· wastewater should be disposed of in a suitable manner;

· wash water should not be recycled unless there are appropriate controls on its microbial quality.

9.5 REFINING PROCESS (Processing Step 11)

Potential Hazards: *pathogenic microbial growth*, *metal fragments*

Potential Defects: *objectionable matter*, *protein denaturation*

Technical Guidance:

· temperature of the minced fish flesh in the refining process should be adequately controlled to prevent the growth of pathogenic bacteria;

· for preventing protein denaturation, temperature of minced fish flesh should not exceed 10℃ in the refining process;

· product should be processed promptly to minimise possible pathogenic microbial growth;

· metal detection equipment that is capable of sensing product that has become contaminated with metal fragments of the size likely to cause human injury should be installed at the most appropriate place in the process to eliminate the hazard;

· objectionable matter such as small bones, black membranes, scales, bloody flesh and connective tissue should be removed from washed flesh with appropriate refining equipment before final de-watering;

· equipment should be properly adjusted to effect efficient product throughput;

· refined product should not be allowed to accumulate on sieve screens for long periods of time.

9.6 FINAL DE-WATERING PROCESS (Processing Step 12)

Potential Hazards: *pathogenic microbial growth*

Potential Defects: *decomposition*, *protein denaturation*

Technical Guidance:

· temperature of the refined fish flesh in the final de-watering process should be adequately controlled to prevent the growth of pathogenic bacteria;

· temperature of refined fish flesh should not exceed 10℃ for cold water fish species, such as Alaska Pollock. For Pacific Whiting the temperature should not be exceed 5℃, since this species usually will have a high protease activity. Some warm water species may be processed at temperatures up to 15℃;

· product should be processed promptly to minimise possible pathogenic microbial growth;

· the moisture level of refined product should be controlled to specified levels with appropriate de-watering equipment (e.g., centrifuge, hydraulic press, screw press);

· consideration should be given to variations in moisture levels due to the age, condition or mode of capture of the raw fish. In some cases dehydration should be performed before refining.

9.7 MIXING AND ADDITION OF ADJUVANT INGREDIENTS PROCESS (Processing Step 13)

Potential Hazards: *pathogenic microbial growth, metal fragments*

Potential Defects: *improper use of food additives, protein denaturation*

Technical Guidance:

· temperature of the product in the mixing process should be adequately controlled to avoid the growth of pathogenic bacteria;

· temperature of dehydrated fish flesh during mixing should not exceed 10℃ for cold water fish species such as Alaska Pollock. For Pacific Whiting the temperature should not exceed 5℃ since this species usually will have a high protease activity. Some warm water species may be processed at temperatures up to 15℃;

· product should be processed promptly to minimise possible pathogenic microbial growth;

· metal detection equipment that is capable of sensing product that has become contaminated with metal fragments of the size likely to cause human injury should be installed at the most appropriate place in the process to eliminate the hazard;

· food additives should be the same and comply with Codex General Standard for Food additives;

· food additives should be mixed homogeneously;

· Cryoprotectants should be used in frozen surimi. Sugars and/or polyhydric alcohols are commonly used to prevent protein denaturation in the frozen state;

· food grade enzyme inhibitors (e. g. egg white, beef protein plasma) should be used for species that exhibit high levels of proteolytic enzyme activity such as Pacific Whiting that reduce the gel forming ability of surimi during kamaboko or crab analogs processing. The use of protein plasma should be appropriately labelled.

9.8 PACKAGING AND WEIGHING (Processing Step 14)

Potential hazards: *pathogenic microbial growth*

Potential defects: *foreign matter (packaging), incorrect net weight, incomplete packaging, denaturation of protein*

Technical Guidance:

· temperature of the product should be adequately controlled during packaging to avoid the growth of pathogenic bacteria;

· product should be packaged promptly to minimise possible pathogenic microbial growth;

· the packaging operation should have procedures established that make possible cross contamination unlikely;

· product should be stuffed into clean plastic bags or packaged into clean containers that have been properly stored;

· product should be appropriately shaped;

· packaging should be conducted rapidly to minimise the risk of contamination or decomposition;

· packaged products should not contain voids;

· the product should meet appropriate standards for net weight.

See also Section 8.2.1 "Weighing" and Section 8.4.4 "Wrapping and Packing".

9.9 FREEZING OPERATION (Processing Step 15)

Refer to Section 8.3.1 for general considerations for freezing fish and fishery products.

Potential Hazards: unlikely

Potential Defects: protein denaturation, decomposition

Technical Guidance:

- after packaging and weighing the product should be promptly frozen to maintain the quality of the product;
- procedures should be established that specifies maximum time limits from packaging to freezing.

9.10 DISMANTLING FREEZING PAN (Processing Step 16)

Potential Hazards: unlikely

Potential Defects: damage to plastic bag and product

Technical Guidance:

- care should be taken to avoid breakage of plastic bag and the product itself in order to refrain from deep dehydration during long-term cold storage.

9.11 METAL DETECTION (Processing Step 17)

Refer to Section 8.2.4 "Metal Detection" for general information.

Potential Hazards: metal fragments

Potential Defects: unlikely

Technical Guidance:

- Metal detection equipment that is capable of sensing product that has become contaminated with metal fragments of the size likely to cause human injury should be installed at the most appropriate place in the process to eliminate the hazard.

9.12 BOXING AND LABELLING (Processing Step 18)

Refer to Section 8.2.3 "Labelling" and Section 8.4.4 "Wrapping and Packing".

Potential Hazards: unlikely

Potential Defects: incorrect label, damage to packaging

Technical Guidance:

- boxing should be clean, durable and suitable for the intended use;
- the boxing operation should be conducted to avoid the damage of packaging materials;
- product in damaged boxing should be re-boxed so that it is properly protected.

9.13 FROZEN STORAGE (Processing Step 19)

Refer to Section 8.1.3 "Frozen Storage" for general information concerning fish and fishery products.

Potential Hazards: unlikely

Potential Defects: decomposition, protein denaturation

Technical Guidance:

- frozen surimi should be stored at -20℃ or colder to prevent protein denaturation from taking place. Quality and shelf life will be maintained more adequately if the product is stored at -25℃ or colder;

· stored frozen product should have adequate air circulation to assure that it remains properly frozen. This includes preventing product from being stored directly on the floor of the freezer.

9.14 RAW MATERIAL RECEPTION - PACKAGING AND INGREDIENTS (Processing Steps 21 and 22)

Refer to Section 8.5.1 "Raw Material Reception - Packaging, Labels and Ingredients".

9.15 RAW MATERIAL STORAGE - PACKAGING AND INGREDIENTS (Processing Steps 23 and 24)

Refer to Section 8.5.2 "Raw Material Storage - Packaging, Labels and Ingredients".

SECTION 10 PROCESSING OF QUICK-FROZEN COATED FISH PRODUCTS

In the context of recognising controls at individual processing steps, this section provides examples of potential hazards and defects and describes technological guidelines, which can be used to develop control measures and corrective action. At a particular step only the hazards and defects, which are likely to be introduced or controlled at that step, are listed. It should be recognised that in preparing a HACCP and/or DAP plan it is essential to consult Section 5 which provides guidance for the application of the principles of HACCP and DAP analysis. However, within the scope of this Code of Practice it is not possible to give details of critical limits, monitoring, record keeping and verification for each of the steps since these are specific to particular hazards and defects.

10.1 GENERAL ADDITION TO PRE-REQUISITE PROGRAMME

· conveyor systems used to transport uncoated and coated fish should be designed and constructed to prevent damaging and contamination of the products;

· shims sawn for formed fish production and held for tempering should be kept at temperatures that will prevent deterioration of the essential quality of the product;

· if the whole process is run continuously an adequate number of processing lines should be available to avoid interruptions and batch-wise processing. If the process has to be interrupted, intermediate products have to be stored under deep-frozen conditions until being further processed;

· pre-frying baths, freezing cabinets used for re-freezing should be equipped with permanent temperature and belt speed control device;

· the proportion of sawdust should be minimised by using appropriate sawing equipment;

· sawdust should be kept well separated from fish cores used for coated products, should be temperature controlled, not stay too long at ambient temperature and should be stored preferably in frozen state prior to further processing into suitable products.

10.2 IDENTIFICATION OF HAZARDS AND DEFECTs

Refer also to Section 5.3.3 and Appendix XI.

This Section describes the main hazards and defects specific to QF coated fish and shellfish.

This flow chart is for illustrative purposes only For in-factory HACCP implementation a complete and comprehensive flow chart has to be drawn up for each process.
References correspond to relevant Sections of the Code.

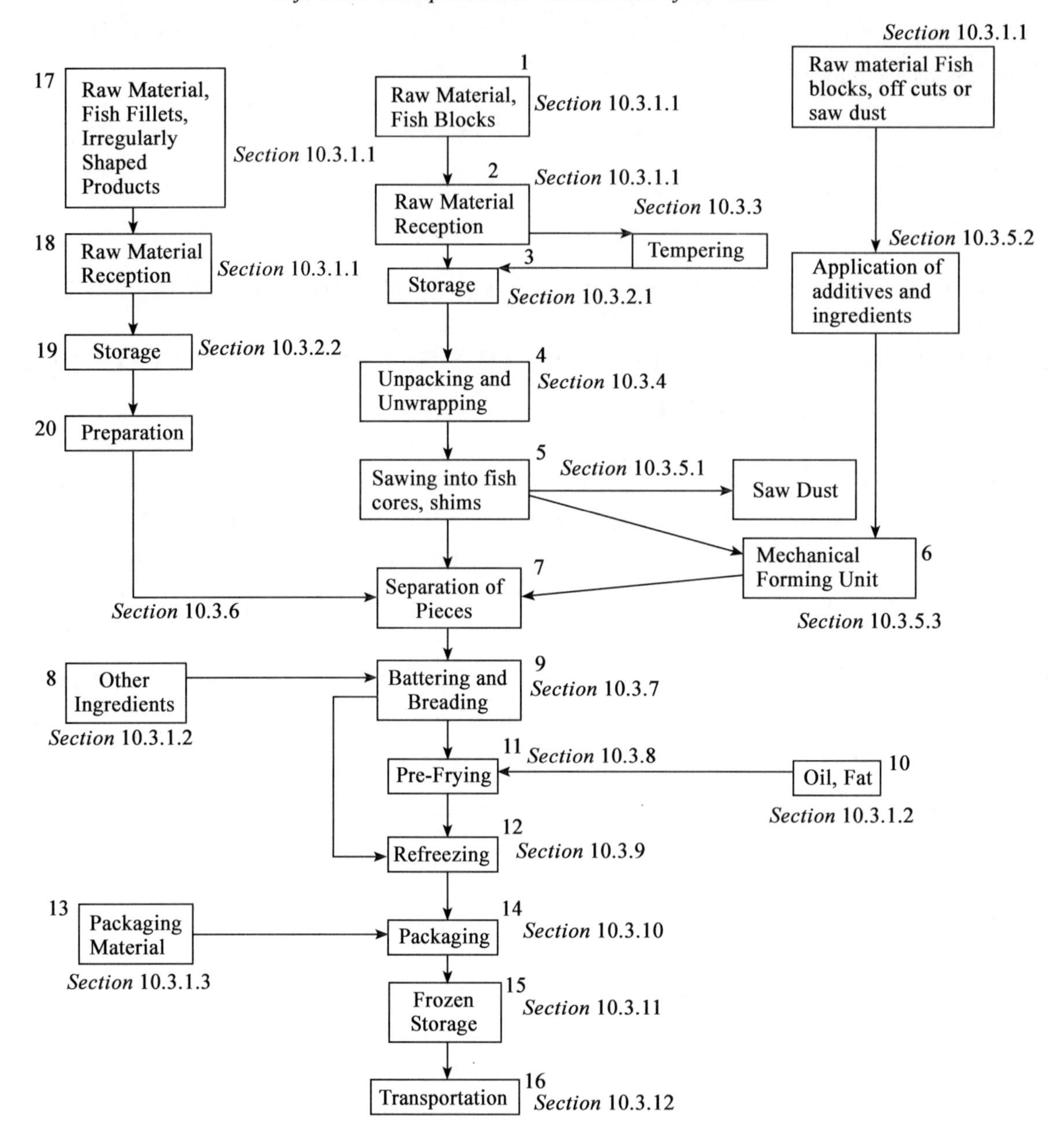

Figure 10. 1　Example of a flow chart for the processing of coated fish products

10. 2. 1　Hazards

Refer also to Section 5. 3. 3. 1.

The production and storage of batter for application to fish portions, fillets, etc. , may involve either rehydratation of a commercial batter mix or preparation from raw ingredients. During the preparation of this batter and its use, the potential hazard for the possible growth and toxins production of *Staphylococcus aureus* and *Bacillus cereus* must be controlled.

10. 2. 2　Defects

Potential defects are outlined in the essential quality, labelling and composition requirements described in the relevant Codex Standard for Quick Frozen Fish Sticks (Fish Fingers), Fish Portions and Fish Fillets - Breaded or in Batter (CODEX STAN. 166-1989).

End product specifications outlined in Appendix XI describe optional requirements specific to QF coated fishery products.

10.3 PROCESSING OPERATIONS

Refer to figure 10.1 for an example of a flow chart for coated fish product processing.

10.3.1 Reception

10.3.1.1 Fish

Potential Hazards: *chemical and biochemical contamination, histamine.*

Potential Defects: *tainting, block irregularities, water and air pockets, packaging material, foreign matter, parasites, dehydration, decomposition.*

Technical Guidance:

- Temperatures of all incoming lots should be recorded;
- Packaging material of frozen products should be examined for dirt, tearing and evidence of thawing;
- Cleanliness and suitability of the transport vehicle to carry frozen fish products should be examined;
- Use of temperature recording devices with the shipment is recommended;
- Representative samples should be taken for further examination for possible hazards and defects.

10.3.1.2 Other Ingredients

Potential Hazards: *chemical, biochemical and microbiological contamination.*

Potential Defects: *mould, colour deviations, filth, sand.*

Technical Guidance:

- breading and batter should be inspected for broken packaging material, signs of rodent and insect infestations and other damage such as dirt on packaging materials and wetness;
- cleanliness and suitability of the transport vehicle to carry food products should be examined;
- representative samples of the ingredients should be taken and examined to ensure that the product is not contaminated and meets specifications for use in the end product;
- ingredients should be shipped on transportation vehicles that are suitable for handling food products and ingredients. Vehicles that have previously hauled potentially unsafe or hazardous material should not be used for hauling food products or ingredients.

10.3.1.3 Packaging Materials

Potential Hazards: *foreign matter.*

Potential Defects: *tainting of products.*

Technical Guidance:

- packaging material used should be clean, sound, durable, sufficient for its intended use and of food grade material;
- for pre-fried products it should be impermeable for fat and oil;
- cleanliness and suitability of the transport vehicle to carry food packaging material should be examined;
- pre-printed labelling and packaging material should be examined for accuracy.

10.3.2 Storage of Raw Material, Other Ingredients and Packaging Materials

10.3.2.1 Fish (Frozen Storage)

Refer to Section 8.1.3

10. 3. 2. 2 Fish (chilled storage)

For storage of nonfrozen fish, refer to section 8. 1. 2.

10. 3. 2. 3 Other Ingredients and Packaging Materials

Potential Hazards: biological, physical and chemical contamination.

Potential Defects: loss of quality and characteristics of ingredients, rancidity.

Technical Guidance:

· all other ingredients and packaging material should be stored in a dry and clean place under hygienic conditions;

· all other ingredients and packaging material should be stored appropriately in terms of temperature and humidity;

· a systematic stock rotation plan should be developed and maintained to avoid out of date materials;

· ingredients should be protected from insects, rodents and other pests;

· defective ingredients and packaging material should not be used.

10. 3. 3 Frozen Fish Block/Fillet tempering

Potential Hazards: Unlikely.

Potential Defects: Incorrect dimension due to sawing of over softened fish flesh (applies to fish sticks).

Technical Guidance:

· Depending on the use of the fish, the tempering of frozen fish blocks/fillets should be carried out in a manner which will allow the temperature of the fish to rise without thawing.

· Tempering block/fillets of frozen fish in chilled storage is a slow process that usually requires at least 12 hours or more.

· Over softening of the outer layers is undesirable (poor performance during sawing) and should be avoided. It could be avoided if facilities used for tempering are maintained at a temperature of 0 ~ 4℃ and if fish blocks/fillets are stacked in layers.

· microwave tempering is an alternate method but should also be controlled to prevent softening of outer layers.

10. 3. 4 Unwrapping, Unpacking

Potential Hazards: Microbiological contamination.

Potential Defects: remaining undetected packaging material, contamination by filth.

Technical Guidance:

· during unwrapping and unpacking of fish blocks care should be given not to contaminate the fish;

· special attention has to be given to cardboard and/or plastic material partly or fully embedded in the blocks;

· all packaging material should be disposed of properly and promptly;

· Protect wrapped, unwrapped and unpacked fish blocks when cleaning and sanitizing processing lines during breaks and between shifts if the production process is interrupted.

10. 3. 5 Production of Fish Core

10. 3. 5. 1 Sawing

Potential Hazards: foreign material (metal or plastic parts of saws).

Potential Defects: irregularly shaped pieces or portions.

Technical Guidance:

· sawing instruments should be kept in clean and hygienic conditions;

· saw-blades must be inspected regularly, to avoid tearing of the product and breakage;

· saw dust must not collect on the saw-table and must be collected in special containers if used for further processing;

· sawn shims used to form irregularly shaped fish cores by mechanical pressure should be kept in clean, hygienic conditions until further manufacturing.

10.3.5.2 Application of additives and Ingredients

Also refer to Section 8.4.3

Potential Hazards: foreign material, microbiological contamination.

Potential Defects: Incorrect addition of additives.

Technical Guidance:

· The temperature of the product in the mixing process should be adequately controlled to avoid the growth of pathogenic bacteria.

10.3.5.3 Forming

Potential Hazards: foreign material (metal or plastic from machine) and/or microbiological contamination (fish mixture only).

Potential Defects: poorly formed fish cores, cores subject to too much pressure (mushy, rancid).

Technical Guidance:

· Forming of fish cores is a highly mechanised method of producing fish cores for battering and breading. It utilises either hydraulic pressure to force shims (sawn portions of fish blocks) into moulds that are ejected onto the conveyor belt or mechanical forming of fish mixtures;

· forming machines should be kept in hygienic conditions;

· formed fish cores should be examined closely for proper shape, weight and texture.

10.3.6 Separation of Pieces

Potential Hazards: Unlikely.

Potential Defects: adhering pieces or portions.

Technical Guidance:

· the fish flesh cores cut from the blocks or fish fillets or other irregular shaped QF fish material must be well separated from each other and should not adhere to each other;

· fish cores that are touching each other going through the wet coating step should be removed and placed back on the conveyor in order to get a uniform batter coat and a uniform breading pick-up;

· cored fish should be monitored for foreign material and other hazards and defects before coating;

· Remove from production any broke, misshape or out of specification peaces.

10.3.7 Coating

In industrial practice the order and the number of coating steps may differ and may therefore deviate considerably from this scheme.

10.3.7.1 Wet Coating

Potential Hazards: Microbiological contamination.

Potential Defects: Insufficient cover or excessive cover of coating.

Technical Guidance:

· fish pieces must be well coated from all sides;

· surplus liquid, which should be reused, must be re-transported under clean and hygienic conditions;

· surplus liquid on fish pieces should be removed by clean air;

· viscosity and temperature of hydrated batter mixes should be monitored and controlled within certain parameters to effect the proper amount of breading pick-up;

· to avoid microbiological contamination of the hydrated batter, appropriate means should be adopted to ensure that significant growth does not take place, such as temperature control, dumping liquid contents and regular or scheduled clean-ups and/or sanitation during the manufacturing shift.

10. 3. 7. 2 Dry Coating

Potential Hazards: *microbiological contamination.*

Potential Defects: *insufficient coating or excessive coating.*

Technical Guidance:

· dry coating must cover the whole products and should stick well on the wet coating;

· surplus coating is removed by blowing away with clean air and/or by vibration of conveyors and must be removed in a clean and hygienic way if further use is intended;

· flow of breading from the application hopper should be free, even and continuous;

· coating defects should be monitored and be in accordance to Codex Standard for Frozen Fish Fingers, Fish Portions and Fish Fillets - Breaded or in Batter (Codex Standard 166-1989);

· the proportion of breading and fish core should be in accordance to Codex Standard for Frozen Fish Fingers, Fish Portions and Fish Fillets - Breaded or in Batter (Codex Standard 166-1989).

10. 3. 8 Pre-Frying

There are some variations in industrial production for the frying process in so far that QF coated products are completely fried including fish core and re-frozen later. For this case alternative hazards and defects have to be described and not all statements in this section apply. In some regions it is common practice to manufacture raw (not pre-fried) coated fish products.

Potential Hazards: *Unlikely.*

Potential Defects: *over-oxidised oil, insufficient frying, loosely adhering coating, burnt pieces and portions.*

Technical Guidance:

· frying oil should have a temperature between approx. 160℃ and 195℃;

· coated fish pieces should remain in frying oil for sufficient time depending on the frying temperature to get a satisfying colour, flavour, and structure to adhere firmly to the fish core, but core should be kept frozen throughout the whole time;

· frying oil has to be exchanged when colour becomes too dark or when concentration of fat degradation products exceeds certain limits;

· remains from coating which concentrate at the bottom of the frying bath have to be removed regularly to avoid partial dark coloration on coated products caused by upwelling of oil;

· excessive oil should be removed from coated products after pre-frying by a suitable device.

10. 3. 9 Re-freezing- Final Freezing

Potential Hazards: *foreign material.*

Potential Defects: *Insufficient freezing leads to sticking of units together or to walls of freezing equipment and facilitates mechanical removal of breading/batter.*

Technical Guidance:

· re-freezing to -18℃ or lower of the whole product should take place immediately after prefrying;

· products should be allowed to stay sufficient time in freezer cabinet to assure core temperature of products of -18℃ or lower;

· cryogenic freezers should have sufficient compressed gas flow to effect proper freezing of the product;

· processors that utilise blast freezers may package the product in the consumer containers before freezing.

10. 3. 10 Packing and Labelling

Refer to Section 8. 2. 3 "Labelling", Section 8. 4. 4 "Wrapping and Packing" and Section 8. 2. 1. "Weighing" .

Potential Hazards: Microbiological contamination.

Potential Defects: Under- or over-packing, improper sealed containers, wrong or misleading labelling.

Technical Guidance:

· packaging should be made without delay after refreezing under clean and hygienic conditions. If packaging is made later (e. g. batch processing) re-frozen products should be kept under deep frozen conditions until being packed;

· packages should be checked regularly by weight control, end products should be checked by a metal detector and/or other detection methods if applicable;

· packaging of cartons or plastic bags to master shipping containers should be done without delay and under hygienic conditions;

· both consumer packages and shipping containers should be appropriately lot coded for product tracing in the event of a product recall.

10. 3. 11 Storage of End Products

Also refer to Section 8. 1. 3.

Potential Hazards: Unlikely.

Potential Defects: texture and Flavour deviations due to fluctuations in temperature, deep freezer burn, cold store flavour, cardboard flavour.

Technical Guidance:

· all end products should be stored at frozen temperature in a clean, sound and hygienic environment;

· severe fluctuations of storage temperature (greater than 3℃) has to be avoided;

· too long storage time (depending on fat content of species used and type of coating) should be avoided;

· products should be properly protected from dehydration, dirt and other forms of contamination;

· all end products should be stored in the freezer to allow proper air circulation.

10. 3. 12 Transport of End Product

Also refer to Section 3. 6. " Transportation" and Section 17 " Transport" under elaboration

Potential Hazards: Unlikely.

Potential Defects: thawing of frozen product.

Technical Guidance:

· during all transportation steps deep-frozen conditions should be maintained −18℃ (maximum fluctuation ± 3℃) until final destination of product is reached;

· cleanliness and suitability of the transport vehicle to carry frozen food products should be examined;

· use of temperature recording devices with the shipment is recommended.

SECTION 14 PROCESSING OF SHRIMPS AND PRAWNS

Scope: Shrimp frozen for further processing may be whole, head-off or deheaded or raw headless, peeled, peeled and de-veined or cooked on board harvest or processing vessels or at on shore processing plants.

In the context of recognising controls at individual processing steps, this section provides examples of potential hazards and defects and describes technological guidelines, which can be used to develop control measures and corrective action. At a particular step only the hazards and defects, which are likely to be introduced or controlled at that step, are listed. It should be recognised that in preparing a HACCP and/or DAP plan it is essential to consult Section 5 which provides guidance for the application of the principles of HACCP and DAP analysis. However, within the scope of this Code of Practice it is not possible to give details of critical limits, monitoring, record keeping and verification for each of the steps since these are specific to particular hazards and defects.

14.1 FROZEN SHRIMPS AND PRAWNS – GENERAL

· shrimps for frozen product originate from a wide variety of sources as varied as deep cold seas to shallow tropical inshore waters and rivers through to aquaculture in tropical and semi tropical regions.

· the methods of catching, or harvesting and processing are as equally varied. Species in northern regions may be caught by freezer vessels, cooked, individually quick frozen and packed on board in their final marketing form. More often however, they will be raw IQF on board for further processing at on-shore plants, or even landed chilled on ice. Shrimps of these species are invariably pre-cooked at onshore plants through in-line integrated process lines, followed by mechanical peeling, cooking, freezing, glazing and packing. A much larger product line is produced in tropical and sub-tropical countries from wild caught and cultivated *Penaeus* species: whole, headless (head off), peeled, peeled and deveined raw and/or cooked products presented in different marketing forms (easy-peel, tail-on, tail-off, butterfly, stretched, sushi shrimp). This wide range of products is prepared in shrimp processing plants that may be small and use manual techniques or large dimensions fully mechanised equipments. Cooked shrimp products are generally peeled after cooking.

· warm water shrimps may also be subject to further added value processes such as marinading and batter and crumb coatings.

· since some raw shrimp products, as well as cooked ones, may be consumed without further processing safety considerations are paramount.

· the processes described above are captured on the flow chart, but it must be appreciated that because of the diverse nature of production methods individual HACCP/DAP plans must be devised for each product.

· Other than the previous description of on-board cooking, there is no reference to processing of shrimps at sea or in farms. It is assumed that product will be correctly handled and processed in line with the relevant sections in the code of practice and that where appropriate some element of pre-preparation, such as de-heading, will have taken place prior to receipt at processing plants.

Figure 14.2 *This flow chart is for illustrative purposes only. For in-factory HACCP implementation a complete and comprehensive flow chart has to be drawn up for each process.*

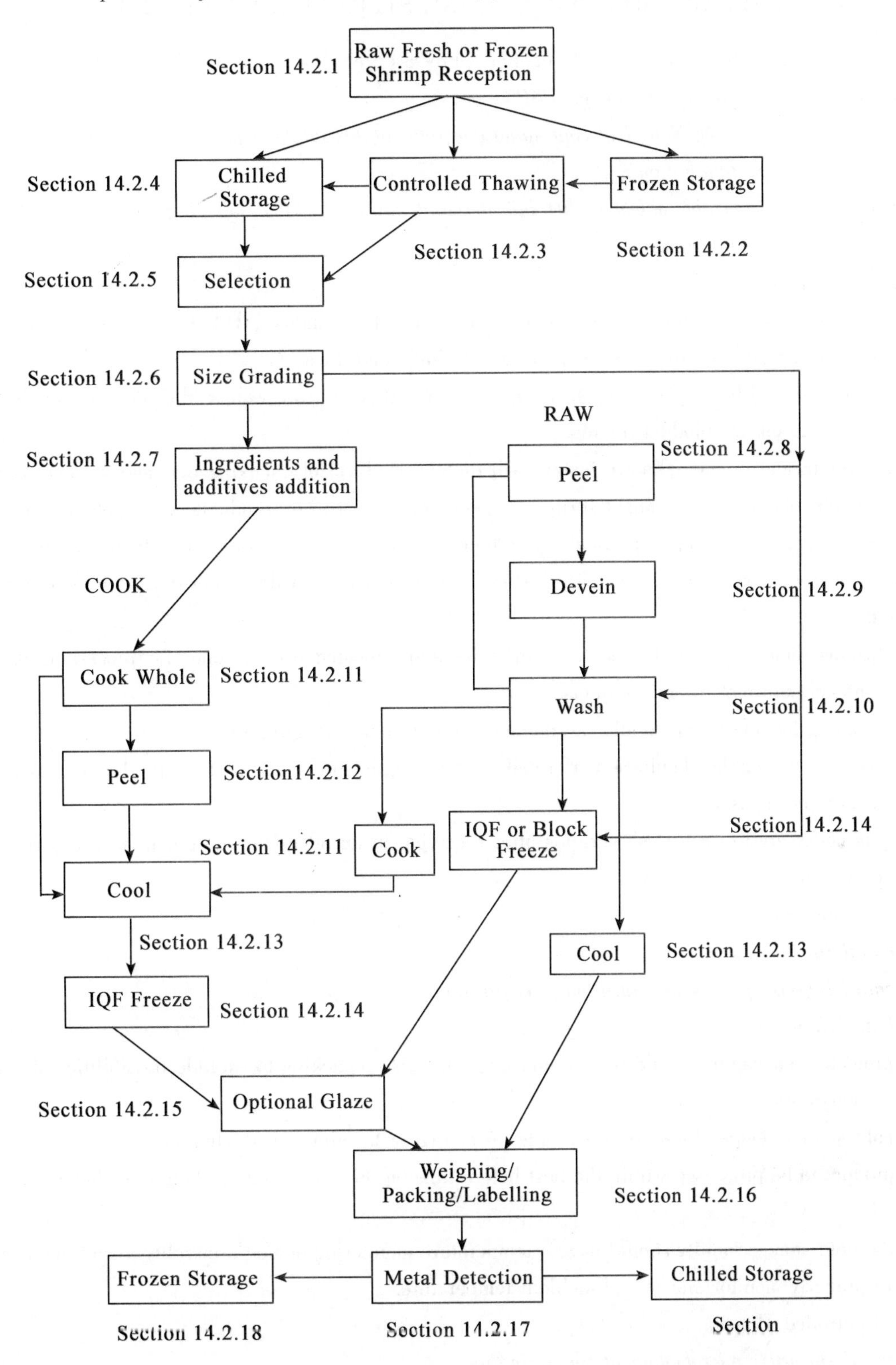

Figure 14. 1 Example of a flow chart of a shrimp and prawn processing line

14.2 SHRIMP PREPARATION (PROCESSING STEPS 14.2.1 TO 14.2.18)

14.2.1 Raw Fresh and Frozen Shrimp Reception (Process Steps)

Potential Hazards: *phytotoxins (e.g. PSP)*

microbiological contamination antioxidants sulphites pesticides fuel oil (chemical contamination)

Potential Defects: *variable batch quality mixed species taints blackspot softening from head enzymes decomposition*

Technical Guidance:

· inspection protocols should be devised to cover identified quality , HACCP and DAP plan parameters together with appropriate training for inspectors to undertake these tasks.

· shrimps should be inspected upon receipt to ensure that they are well iced or deep frozen and properly documented to ensure product tracing.

· the origin and previous known history will dictate the level of checking that may be necessary for, for example, phytotoxins in sea caught shrimps (specifically for head on products), for potential antibiotics presence in aquaculture shrimps, particularly if there is no supplier assurance certification. In addition, other chemical indicators for heavy metals, pesticides and indicators of decomposition such as TVBN's may be applied.

· shrimps should be stored in suitable facilities and allocated use-by times for processing to ensure quality parameters are met in end products.

· incoming lots of shrimp should be monitored for sulphites at harvesting.

· a sensory evaluation should be performed on incoming lots to ensure that the product is of acceptable quality and not decomposed.

· it is necessary to wash fresh shrimps after receiving in an adequate equipment with a series of low velocity sprays with chilled clean water.

14.2.2 Frozen Storage

Potential Hazards: *unlikely.*

Potential Defects: *protein denaturation*, *dehydration.*

Technical Guidance:

· protective packaging should be undamaged, otherwise repacking to exclude possibilities of contamination and dehydration.

· cold storage temperatures to be suitable for storage with minimum fluctuation.

· product to be processed within the best before time on the packaging, or before as dictated at reception.

· the cold storage facility should have a temperature monitoring device preferably a continuous recording unit to properly monitor and record ambient temperature.

14.2.3 Controlled Thawing

Potential Hazards: *microbiological contamination*

contamination from wrapping

Potential Defects: *decomposition*

Technical Guidance:

· thawing processes may be undertaken from block frozen or IQF shrimps depending on the raw material source. The outer and inner packaging should be removed prior to defrosting to prevent contamination and extra care should be taken on block frozen prawns where inner wax or polyethylene packaging may be entrapped with blocks.

· thawing tanks should be purpose designed and allow for ‘counter current’ water defrosting where necessary to maintain lowest temperatures possible. However water re-use is discouraged.

· Clean sea water or water and ice of potable quality should be used for thawing with a water temperature no higher than 20℃ (68oF) by use of additional ice to achieve a defrosted product at a temperature cooler than 4℃.

· thawing should be achieved as quickly as possible to maintain quality.

· it is desirable for the exit conveyor, leading from the defrost tanks, to be equipped with a series of low velocity sprays to wash the shrimps with chilled clean water.

· immediately after thawing, the shrimps should be re-iced or held in chill to avoid temperature abuse before further processing.

14.2.4 Chilled Storage

Refer to Section 8.1.2 “Chilled Storage” for general information concerning fish and fishery products.

Potential Hazards: *microbiological contamination*

Potential Defects: *decomposition*

Technical Guidance:

· chilled storage, preferably under ice in chill rooms at less than 4℃ after reception.

· the chilled storage facility should have a temperature monitoring device (preferably a continuous recording unit) to properly monitor and record ambient temperatures.

· Unnecessary delays should be avoided during chilled storage in order to prevent quality deterioration.

14.2.5 Selection

Potential Hazards: *unlikely*

Potential Defects: *decomposition*

Technical Guidance:

· shrimps may be selected for different quality grades according to specification requirements. This should be undertaken with minimum of delay followed by re-icing of the shrimps

14.2.6 Size Grading

Potential Hazards: *microbiological contamination*

Potential Defects: *decomposition*

Technical Guidance:

· Size grading of shrimps is undertaken through mechanical graders of various degrees of sophistication and manually. There is a possibility of shrimps becoming trapped in the bars of the graders so that regular inspection is required to prevent ‘carry over’ of old prawns and bacteriological contamination.

· Shrimp should be re-iced and stored in chill prior to further processing.

· The grading process should be carried out promptly to prevent unnecessary microbiological growth and product decomposition.

14.2.7 Addition of Ingredients and Use of Additives

Potential Hazards: *chemical and microbiological contamination*

sulphites

Potential Defects: *decomposition*

improper use of additives

Technical Guidance:

· according to specification and legislation, certain treatments may be applied to shrimps to improve organoleptic quality, preserve yield or preserve them for further processing.

· examples would including sodium metabisulphite to reduce shell blackening, sodium benzoate to extend shelf-life between processes and sodium polyphosphates to maintain succulence through processing and prevent black spot after peeling, whilst common salt would be added as brine for flavour.

· these ingredients and additives can be added at various stages, for instance common salt and sodium polyphosphates at defrost stages or chilled brine as a flume conveyor between cooking and freezing, or as glaze.

· at whatever stage ingredients and additives are added, it is essential to monitor the process and product to ensure that any legislative standards are not exceeded, quality parameters are met and that where dip baths are used, the contents are changed on a regular basis according to drawn up plans.

· chill conditions to be maintained throughout.

· sulphites used to prevent blackspot formation autolysis should be used in accordance with manufacturer's instructions and Good manufacturing Practice

14.2.8 Full and Partial Peeling

Potential Hazards: *microbiological cross contamination*

Potential Defects: *decomposition*

shell fragments

foreign matter

Technical Guidance:

· this process applies mainly to warm water prawns and could be as simple as inspecting and preparing whole large prawns for freezing and down-grading blemished prawns for full peeling.

· other peeling stages could including full peeling or partial peeling leaving tail swimmers intact.

· whatever the process, it is necessary to ensure that the peeling tables are kept clear of contaminated shrimps and shell fragments with water jets and the shrimps are rinsed to ensure no carry over of shell fragments.

14.2.9 Deveining

Potential Hazards: *microbiological cross contamination*

metal contamination

Potential Defects: *objectionable matter*

decomposition

foreign matter

Technical Guidance:

· the vein is the gut which may appear as a dark line in the upper dorsal region of prawn flesh. In large warm water prawns, this may be unsightly, gritty and a source of bacterial contamination.

· removal of the vein is by razor longitudinally cutting along the dorsal region of the shrimp with a razor slide and removal of the vein by pulling. This may be partially achieved with head-off shell-on shrimps as well.

· this operation is considered to be a mechanical though labour intensive process so that:

· cleaning and maintenance schedules should be in place and cover the need for cleaning before, after and during processing by trained operatives.

· further, it is essential to ensure that damaged and contaminated shrimps are removed from the line and that no debris build up is allowed.

14. 2. 10 Washing

Potential Hazards: *microbiological contamination*

Potential Defects: *decomposition*

foreign matter

Technical Guidance:

· washing of peeled and deveined shrimps is essential to ensure that shell and vein fragments are removed.

· shrimps should be drained and chilled without delay prior to further processing.

14. 2. 11 Cooking Processes

Potential Hazards: *survival of pathogenic micro-organisms due to insufficient cooking*

microbiological cross contamination

Potential Defects: *over cooking*

Technical Guidance:

· the cooking procedure, in particular time and temperature, should be fully defined according to the specification requirements of the final product, for example whether it is to be consumed without further processing and the nature and origin of the raw shrimp and uniformity of size grading.

· the cooking schedule should be reviewed before each batch and where continuous cookers are in use, constant logging of process parameters should be available.

· only potable water should be used for cooking, whether in water or via steam injection.

· the monitoring methods and frequency should be appropriate for the critical limits identified in the scheduled process.

· maintenance and cleaning schedules should be available for cookers and all operations should only be undertaken by fully trained staff.

· adequate separation of cooked shrimps exiting the cooking cycle utilising different equipment is essential to ensure no cross contamination.

14. 2. 12 Peeling of Cooked Shrimps

Potential Hazards: *microbiological cross contamination*

Potential Defects: *presence of shell*

Technical Guidance:

· cooked shrimps have to be properly peeled through mechanical or manual peeling in line with cooling and freezing processes.

· cleaning and maintenance schedules should be available, implemented by fully trained staff to ensure efficient and safe processing are essential.

14. 2. 13 Cooling

Potential Hazards: *microbiological cross contamination and toxin formation*

Potential Defects: *unlikely*

Technical Guidance:

· cooked shrimps, should be cooled as quickly as possible to bring the temperature of the product to a temperature range limiting bacteria proliferation or toxin production

· cooling schedules should enable the time-temperature requirements to be met and maintenance and cleaning schedules should be in place and complied with by fully trained operatives.

· only cold/iced potable water or clean water should be used for cooling and should not be used for further batches, although for continuous operations a top-up procedure and maximum run-length will be defined.

· raw/cooked separation is essential.

· after cooling and draining, the shrimps should be frozen as soon as possible, avoiding any environmental contamination.

14.2.14 Freezing Processes

Potential Hazards: *microbiological contamination*

Potential Defects: *slow freezing - textural quality and clumping of shrimps*

Technical Guidance:

· the freezing operation will vary tremendously according to the type of product. At its simplest, raw whole or head-off shrimps may be block or plate frozen in purpose-designed cartons into which potable water is poured to form a solid block with protective ice.

· cooked and peeled *Pandalus* cold water prawns, at the other extreme, tend to be frozen through fluidised bed systems, whilst many of the warm water shrimp products are IQF frozen either on trays in blast freezers or in continuous belt freezers.

· whichever the freezing process, it is necessary to ensure that the freezing conditions specified are met and that for IQF products, there is no clumping, i.e. pieces frozen together. Putting product into a blast freezer before it is at operating temperature may result in glazed, slow frozen product and contamination.

· freezers are complex machines requiring cleaning and maintenance schedules operated by fully trained staff.

14.2.15 Glazing

Potential Hazards: *microbiological cross-contamination*

Potential Defects: *inadequate glaze*, *too much glaze*, *spot welding*, *incorrect labelling.*

Technical Guidance:

· glazing is applied to frozen shrimps to protect against dehydration and maintain quality during storage and distribution.

· ice block frozen shrimps is the simplest form of glazing, followed by dipping and draining frozen shrimps in chilled potable water. A more sophisticated process is to pass frozen size graded shrimps under cold-water sprays on vibratory belts such that the shrimps pass at a steady rate to receive an even and calculable glaze cover.

· ideally, glazed shrimps should receive a secondary re-freezing prior to packing, but if not, they should be packaged as quickly as possible and moved to cold storage. If this is not achieved, the shrimps may freeze together and 'spot weld' or clump as the glaze hardens.

· there are Codex methods for the determination of glaze.

14.2.16 Weighing, Packing and Labelling of All Products

Refer to Section 8.4.4 "Wrapping and Packing" and Section 8.5. "Packaging, Labels & Ingredients".

Potential Hazards: *sulphites*

Potential Defects: *incorrect labelling*, *decomposition*

Technical Guidance:

· all wrappings for products and packaging including glues and inks should have been specified to be food grade, odourless with no risk of substances likely to be harmful to health being transferred to the packed food.

· all food products should be weighed in packaging with scales appropriately tared and calibrated to ensure correct weight.

· where products are glazed, checks should be carried out to ensure the correct compositional standards to comply with legislation and packaging declarations.

· ingredients lists on packaging and labelling should declare presence of ingredients in the food product in descending order by weight, including any additives used and still present in the food.

· all wrapping and packaging should be carried out in a manner to ensure that the frozen products remain frozen and that temperature rises are minimal before transfer back to frozen storage.

· sulphites should be used in accordance with manufacturer's instructions and Good manufacturing Practice.

· where sulphites were used in the process, care should be taken that they are properly labelled.

14. 2. 17 Metal Detection

Potential Hazard: *presence of metal*

Potential Defect: *unlikely*

Technical Guidance:

· products should be metal detected in final pack through machines set to the highest sensitivity possible.

· larger packs will be detected at a lower sensitivity than smaller packs so that consideration should be given to testing product prior to packing. However, unless potential re-contamination prior to packing can be eliminated, it is probably still better to check in-pack.

14. 2. 18 Frozen Storage of End Product

Refer to Section 8. 1. 3. "Frozen storage" for general information concerning fish and fishery products.

Potential Hazard: *unlikely*

Potential Defects: *texture and flavour deviations due to fluctuations in temperature*, *deep freezer burn*, *cold store flavour*, *cardboard flavour*

Technical Guidance:

· frozen products should be stored at frozen temperature in a clean, sound and hygienic environment.

· the facility should be capable of maintaining the temperature of the shrimp at or below minus 18℃ with minimal temperature fluctuations (+ or −3℃) .

· the storage area should be equipped with a calibrated indicating thermometer. Fitting of a recording thermometer is strongly recommended.

· a systematic stock rotation plan should be developed and maintained.

· products should be properly protected from dehydration, dirt and other forms of contamination.

· all end products should be stored in the freezer to allow proper air circulation.

SECTION 15 PROCESSING OF CEPHALOPODS

In the context of recognising controls at individual processing steps, this section provides examples of potential hazards and defects and describes technological guidelines, which can be used to develop control measures and corrective action. At a particular step only the hazards and defects, which are likely to be introduced or controlled at that step, are listed. It should be recognised that in preparing a HACCP and/or DAP plan it is essential to consult Section 5 which provides guidance for the application of the principles of HACCP and DAP analysis. However, within the scope of this Code of Practice it is not possible to give details of critical limits, monitoring, record keeping and verification for each of the steps since these are specific to particular hazards and defects.

This section applies to fresh and processed cephalopods including cuttlefish (Sepia and Sepiella), squid (*Alloteuthis*, *Berryteuthis*, *Dosidicus*, *Ilex*, *Lolliguncula*, *Loligo*, *Loliolus*, *Nototodarus*, *Ommastrephes*, *Onychoteuthis*, *Rossia*, *Sepiola*, *Sepioteuthis*, *Symplectoteuthis* and *Todarodes*) and octopuses (*Octopus* and *Eledone*) intended for human consumption.

Fresh Cephalopods are extremely perishable and should be handled at all times with great care and in such a way as to prevent contamination and inhibit the growth of micro-organisms. Cephalopods should not be exposed to direct sunlight or to the drying effects of winds, or any other harmful effects of the elements, but should be carefully cleaned and cooled down to the temperature of melting ice, 0℃ (32℉), as quickly as possible.

This section shows an example of a cephalopod process. Figure 15. 1 lists the steps associated with receiving and processing fresh squid. It should be noted that there are a variety of processing operations for cephalopods and this process is being used for illustrative purposes only.

15. 1 RECEPTION OF CEPHALOPODS (Processing Step 1)

Potential Hazards: Microbiological contamination, chemical contamination, parasites

Potential Defects: Damaged products, foreign matter

Technical Guidance:

· The processing facility should have in place a programme for inspecting cephalopods on catching or arrival at the factory. Only sound product should be accepted for processing.

· Product specifications could include:

— organoleptic characteristics such as appearance , odour, texture etc. which can also used as indicators of fitness of consumption;

— chemical indicators of decomposition and / or contamination e. g. TVBN, heavy metals (cadmium);

— microbiological criteria;

— parasites e. g. *Anisakis* foreign matter;

— the presence of lacerations, breakages and discolouration of the skin, or a yellowish tinge spreading from the liver and digestive organs inside the mantle, which are indicative of product deterioration.

· Personnel inspecting product should be trained and experienced with the relevant species in order to recognise any defects and potential hazards.

Further information can be found on Section 8 "Processing of Fresh, Frozen and Minced Fish" and Co-

This flow chart is for illustrative purposes only.For in-factory HACCP implementation a complete and comprehensive flow chart has to be drawn up for each process.

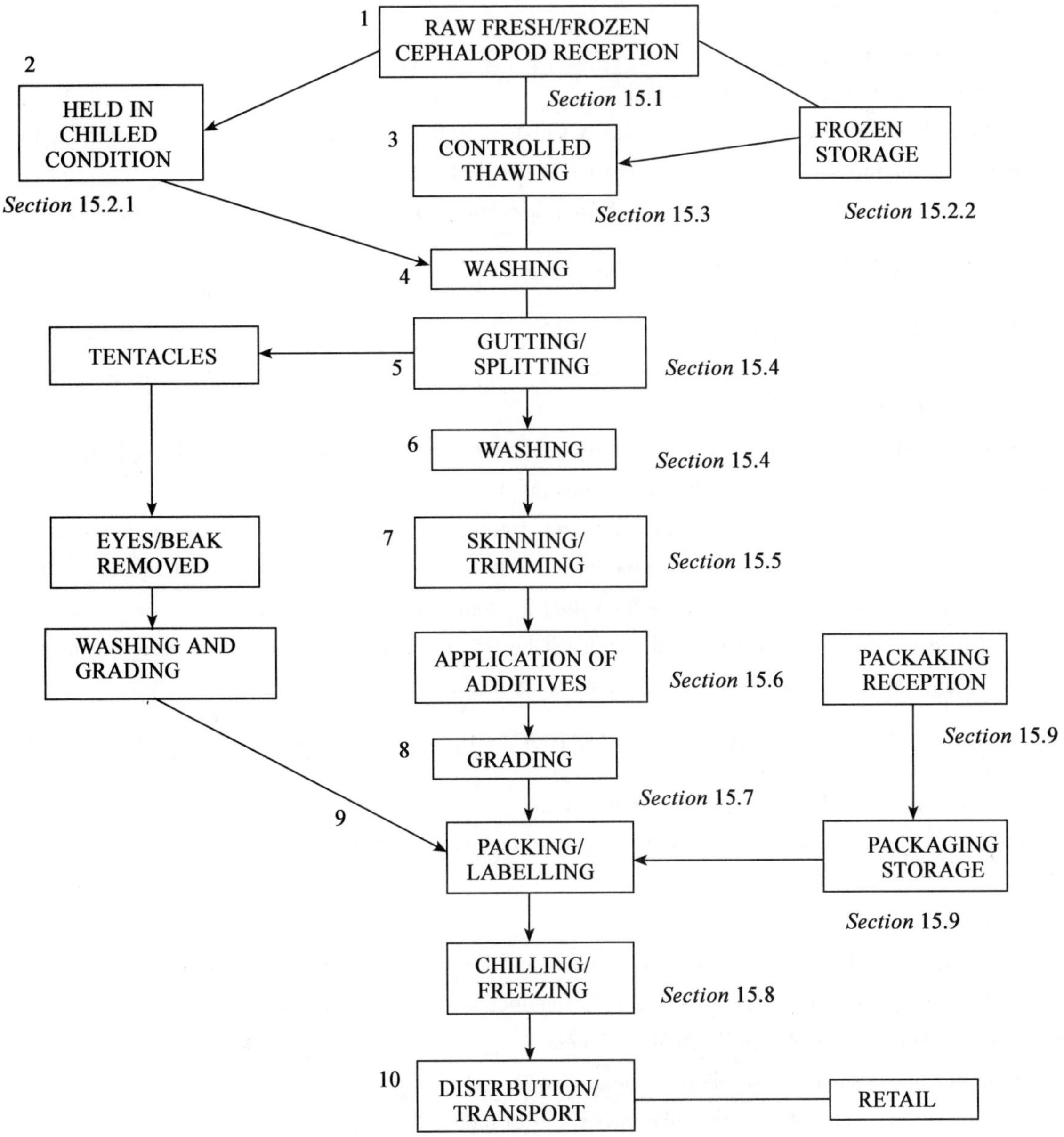

1 RAW FRESH/FROZEN CEPHALOPOD RECEPTION

Section 15.1

2 HELD IN CHILLED CONDITION

Section 15.2.1

FROZEN STORAGE

Section 15.2.2

3 CONTROLLED THAWING

Section 15.3

4 WASHING

5 GUTTING/ SPLITTING

Section 15.4

TENTACLES

6 WASHING

Section 15.4

EYES/BEAK REMOVED

7 SKINNING/ TRIMMING

Section 15.5

WASHING AND GRADING

APPLICATION OF ADDITIVES

Section 15.6

PACKAKING RECEPTION

Section 15.9

8 GRADING

Section 15.7

9 PACKING/ LABELLING

PACKAGING STORAGE

Section 15.9

CHILLING/ FREEZING

Section 15.8

10 DISTRBUTION/ TRANSPORT

RETAIL

Figure 15. 1 Example of a possible squid processing line

dex Guidelines for Sensory Evaluation of Fish and Shellfish in Laboratories.

15. 2 STORAGE OF CEPHALOPODS

15. 2. 1 Chilled storage (Processing steps 2 and 10)

Potential Hazards: *Microbiological contamination*

Potential Defects: *Decomposition*, *physical damage*

Technical Guidance:

Refer to Section 8. 1. 2 "Chilled Storage"

15.2.2 Frozen Storage (Processing steps 2 & 10)

Potential Hazards: Heavy metals e.g. cadmium migration from the gut.

Potential Defects: Freezer-burn

Technical Guidance:

Refer to Section 8.1.3 "Frozen Storage".

· Consideration needs to be given to the fact that when there are high cadmium levels in the gut contents there may be migration of this heavy metal into the flesh.

· Products should be properly protected from dehydration by sufficient packaging or glaze.

15.3 CONTROLLED THAWING (Processing step 3)

Potential Hazards: Microbiological contamination

Potential Defects: Decomposition, discoloration

Technical Guidance:

· The thawing parameters should be clearly defined and include time and temperature. This is important to prevent the development of pale pink discoloration.

· Critical limits for the thawing time and temperature of the product should be developed. Particular attention should be paid to the volume of product being thawed in order to control discoloration.

· If water is used as the thawing medium then it should be of potable quality

· If re-circulated water is used then care must be taken to avoid the build up of micro organisms

For further guidance refer to Section 8.1.4 "Control Thawing".

15.4 SPLITTING, GUTTING AND WASHING (Processing Steps 4, 5, 6, 11, 12 &13)

Potential Hazards: Microbiological contamination

Potential Defects: Presence of gut contents, parasites, shells, ink discolouration, beaks, decomposition.

Technical Guidance:

· Gutting should remove all intestinal material and the cephalopod shell and beaks if present.

· Any by-product of this process which is intended for human consumption e.g. tentacles, mantle should be handled in a timely and hygienic manner.

· Cephalopods should be washed in clean seawater or potable water immediately after gutting to remove any remaining material from the tube cavity and to reduce the level of micro-organisms present on the product.

· An adequate supply of clean seawater or potable water should be available for the washing of whole cephalopods and cephalopod products

15.5 SKINNING, TRIMMING (Processing Step 7)

Potential Hazards: Microbiological contamination

Potential Defects: presence of objectionable matter, bite damage, skin damage, decomposition

Technical Guidance:

· The method of skinning should not contaminate the product nor should it allow the growth of micro-organisms e.g. enzymatic skinning or hot water techniques should have defined time/temperature parameters to prevent the growth of micro-organisms.

· Care should be taken to prevent waste material from cross contaminating the product.

· An adequate supply of clean seawater or potable water should be available for the washing or product during and after skinning.

15. 6 APPLICATION OF ADDITIVES

Potential Hazards: *Physical contamination, non approved additives, non fish allergens*

Potential Defects: *Physical contamination, additives exceeding their regulatory limits*

Technical Guidance:

· Mixing and application of appropriate additives should be carried out by trained operators

· It is essential to monitor the process and product to ensure that regulatory standards are not exceeded and quality parameters are met

· Additives should comply with requirements of the Codex general Standard for Food Additives.

15. 7 Grading/Packing/Labelling (Processing Steps 8 & 9)

Refer to Section 8. 2. 3 "Labelling" .

Potential Hazards: *chemical or physical contamination from packaging*

Potential Defects: *incorrect labelling, incorrect weight, dehydration*

Technical Guidance:

· Packaging material should be clean, be suitable for it's intended purpose and manufactured from food grade materials;

· Grading and packing operations should be carried out with minimal delay to prevent deterioration of the cephalopod;

· Where sulphites were used in the process, care should be taken that they are properly labelled.

15. 8 FREEZING (Processing Step 10)

Potential Hazards: *parasites*

Potential Defects: *freezer burn, decomposition, loss of quality due to slow freezing.*

Technical Guidance:

Cephalopods should be frozen as rapidly as possible to prevent deterioration of the product and a resulting reduction in shelf life due to microbial growth and chemical reactions.

· The time/temperature parameters developed should ensure rapid freezing of product and should take into consideration the type of freezing equipment, capacity, the size and shape of the product, and production volume. Production should be geared to the freezing capacity of the processing facility;

· If freezing is used as a control point for parasites, then the time/temperature parameters need to ensure that the parasites are no longer viable need to be established;

· The product temperature should be monitored regularly to ensure the completeness of the freezing operation as it relates to the core temperature;

· Adequate records should be kept for all freezing and frozen storage operations.

For further guidance refer to Section 8. 3. 1 "Freezing Process" and to Annex 1 on Parasites.

15. 9 PACKAGING, LABELS AND INGREDIENTS – RECEPTION AND STORAGE

Consideration should be given to the potential hazards and defects associated with packaging, labelling and ingredients. It is recommended that users of this code consult Section 8. 5 "Packaging, Labels and Ingredients" .

SECTION 16 PROCESSING OF CANNED FISH, SHELLFISH AND OTHER AQUATIC INVERTEBRATES

This section applies to fish, shellfish, cephalopods and other aquatic invertebrates.

In the context of recognising controls at individual processing steps, this section provides examples of potential hazards and defects and describes technological guidelines, which can be used to develop control measures and corrective action. At a particular step only the hazards and defects, which are likely to be introduced or controlled at that step, are listed. It should be recognised that in preparing a HACCP and/or DAP plan it is essential to consult Section 5 (Hazard Analysis Critical Control Point (HACCP) and Defect Action Point (DAP) Analysis) which provides guidance for the application of the principles of HACCP and DAP analysis. However, within the scope of this Code of Practice it is not possible to give details of critical limits, monitoring, record keeping and verification for each of the steps since these are specific to particular hazards and defects.

This section concerns the processing of heat processed sterilised canned fish and shellfish products which have been packed in hermetically sealed containers[1] and intended for human consumption.

As stressed by this Code, the application of appropriate elements of the pre-requisite programme (Section 3) and HACCP principles (Section 5) at these steps will provide the processor with reasonable assurance that the essential quality, composition and labelling provisions of the appropriate Codex standard will be maintained and food safety issues controlled. The example of the flow diagram (Figure 16. 1) will provide guidance to some of the common steps involved in a canned fish or shellfish preparation line.

16. 1 GENERAL - ADDITION TO PRE-REQUISITE PROGRAMME

Section 3 (Pre-requisite programme) gives the minimum requirements for good hygienic practices for a processing facility prior to the application of hazard and defect analyses.

For fish and shellfish canneries, additional requirements to the guidelines described in Section 3 are necessary due to the specific technology involved. Some of them are listed below, but reference should also be made to the Recommended International Code of Hygienic Practice for Low-Acid and Acidified Low-Acid Canned Food (CAC/PRC 23-1979, Rev. 2 (1993)) for further information

· design, working and maintenance of baskets and handling and loading devices aimed at retorting should be appropriate for the kind of containers and materials used. These devices should prevent any excessive abuse to the containers.

· an adequate number of efficient sealing machines should be available to avoid undue delay in processing;

· retorts should have a suitable supply of energy, vapour, water and/or air so as to maintain in it sufficient pressure during the heat treatment of sterilisation ; their dimensions should be adapted to the production to avoid undue delays;

· every retort should be equipped with an indicating thermometer, a pressure gauge and a time and

---

[1] Aseptic filling is not covered by this Code. Reference of the relevant code is made in Appendix XII.

This flow chart is for illustrative purpose only For in-factory implementation of HACCP principles, a complete and comprehensive flow chart has to be drawn up for each product. References correspond to relevant Sections of the Code.

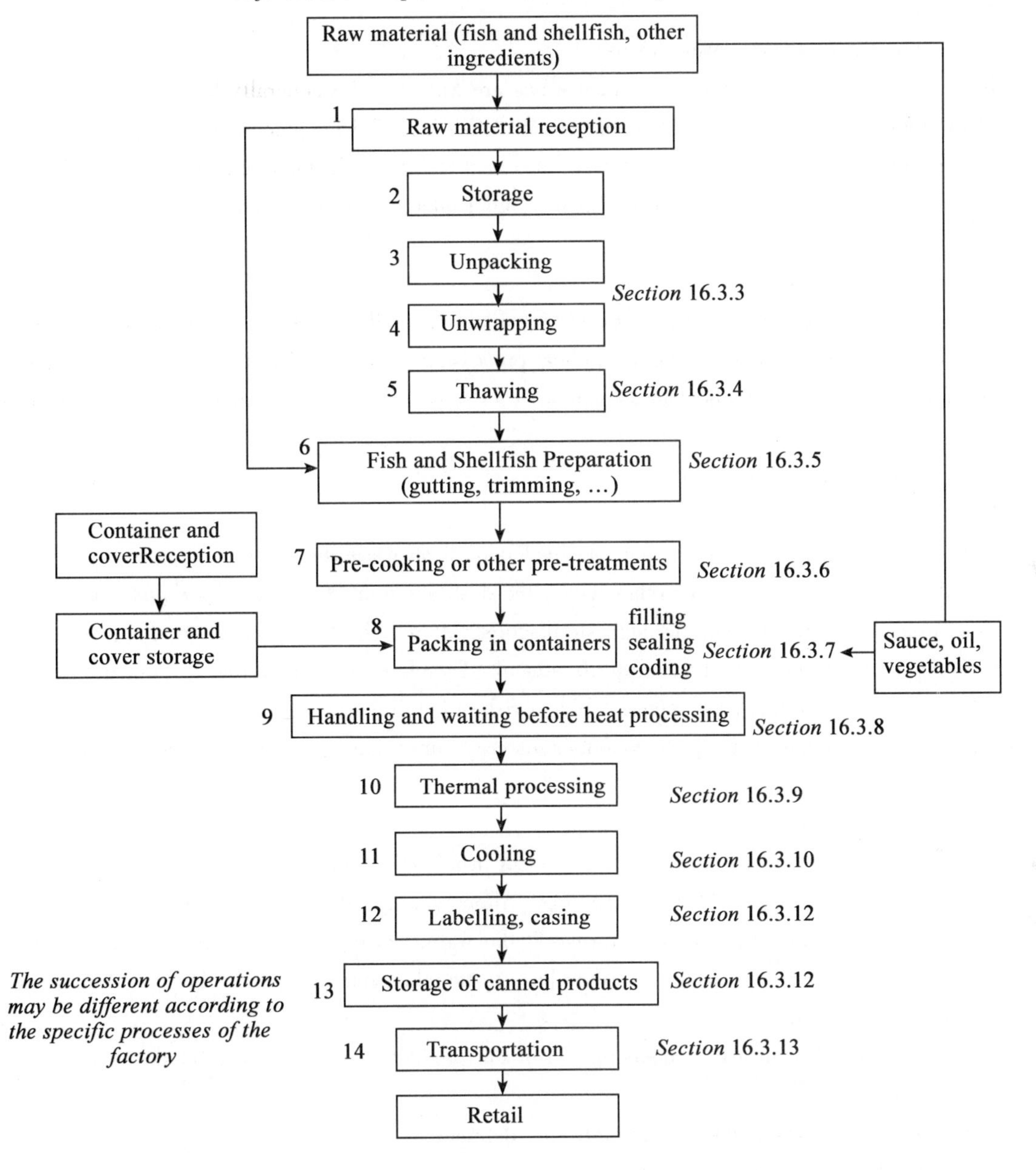

Figure 16. 1 Example of a flow chart for the processing of canned fish and shellfish

temperature recorder, an accurate clearly visible clock should be installed in the retorting room;

- canneries using steam retorts should consider installing automatic steam controller valves;
- Instruments used to control and to monitor in particular the thermal process should be kept in good condition and should be regularly verified or calibrated. Calibration of instruments used to measure temperature should be made in comparison with a reference thermometer. This thermometer should be regularly calibrated. Records concerning the calibration of instruments should be established and kept.

16. 2 IDENTIFICATION OF HAZARDS AND DEFECTS

Refer also to Section 4. 1 (Potential Hazards Associated with Fresh Fish and Shellfish)

This section describes the main potential hazards and defects specific to canned fish and shellfish.

16.2.1 Hazards

A Biological Hazards

A1 Naturally occurring marine toxins

Biotoxins such as tetrodotoxines or ciguatoxines are known to be generally heat-stable, so the knowledge of the identity of the species and/or the origin of fish intended for processing is important.

Phycotoxins such as DSP, PSP or ASP are also heat stable, so it important to know the origin and the status of the area of origin of molluscan shellfish or other affected species intended for processing.

A2 Scombrotoxins

Histamine

Histamine is heat-stable, and so its toxicity remains practically intact in containers. Good practices for the conservation and handling from capture to heat processing are essential to prevent the histamine production. The Codex Commission adopted in its standards for some fish species maximum levels tolerated for histamine.

A3 Microbiological toxins

Clostridium botulinum

The botulism risk usually appears after an inadequate heat processing and inadequate container integrity. The toxin is heat-sensitive, on the other hand, the destruction of *Clostridium botulinum* spores, in particular from proteolytic strains, requires high sterilisation values. The heat processing effectiveness depends on the contamination level at the time of the treatment. Therefore, it is advisable to limit proliferation and the contamination risks during processing. A higher risk of botulinum could result from any of the following: inadequate heat processing, inadequate container integrity, unsanitary post process cooling water and unsanitary wet conveying equipment.

Staphylococcus aureus

Toxins from *Staphylococcus aureus* can be present in a highly contaminated raw material or can be produced by bacterial proliferation during processing. After canning, there is also the potential risk of post process contamination with *Staphylococcus aureus* if the warm wet containers are handled in an unsanitary manner. These toxins are heat-resistant, so they have to be taken into account in the hazard analysis.

B Chemical Hazards

Care should be taken to avoid contamination of the product from components of the containers (e.g. lead)

and chemical products (lubricants, sanitizers, detergents).

C Physical Hazards

Containers prior to filling may contain materials such as metal or glass fragments.

16.2.2 Defects

Potential defects are outlined in the essential quality, labelling and composition requirements described in the relevant Codex Standards listed in Appendix XII. Where no Codex Standard exists regard should be made to national regulations and/or commercial specifications.

End product specifications outlined in Appendix IX describe optional requirements specific to canned products.

16.3 PROCESSING OPERATIONS

Processors can also refer to the Recommended International Code of Hygienic Practice for Low-Acid and Acidified Low-Acid Canned Foods (CAC/RCP 23-1979, Rev. 2 (1993)) in order to obtain detailed advice on canning operations.

16. 3. 1 Reception of raw material, containers, covers and packaging material and other ingredients

16. 3. 1. 1 Fish and shellfish (Processing step 1)

Potential Hazards: Chemical and biochemical contamination (DSP, PSP, scombrotoxine, heavy metals...)

Potential Defects: Species substitution, decomposition, parasites

Technical Guidance:

Refer to section 8. 1. 1 (Raw Fresh or Frozen Fish Reception) and to other relevant sections; and also:

· When live shellfish (crustaceans) are received for canning processing, inspection should be carried out in order to discard dead or badly damaged animals.

16. 3. 1. 2 Container, cover and packaging materials (Processing step 1)

Potential Hazards: Subsequent microbiological contamination

Potential Defects: Tainting of the product

Technical Guidance:

Refer to section 8. 5. 1 (Raw Material Reception - Packaging, Labels & Ingredients); and also:

· Containers, cover and packaging materials should be suitable for the type of product, the conditions provided for storage, the filling, sealing and packaging equipment and the transportation conditions;

· the containers in which fish and shellfish products are canned should be made from suitable material and constructed so that they can be easily closed and sealed to prevent the entry of any contaminating substance;

· containers and cover for canned fish and shellfish should meet the following requirements:

— they should protect the contents from contamination by micro-organisms or any other substance;

— their inner surfaces should not react with the contents in any way that would adversely affect the product or the containers ;

— their outer surfaces should be resistant to corrosion under any likely conditions of storage;

— they should be sufficiently durable to withstand the mechanical and thermal stresses encountered during the canning process and to resist physical damage during distribution.

16. 3. 1. 3 Other ingredients (Processing step 1)

Refer to section 8. 5. 1 (Raw Material Reception - Packaging, Labels & Ingredients) .

16. 3. 2 Storage of raw material, containers, covers and packaging materials

16. 3. 2. 1 Fish and shellfish (Processing step 2)

Refer to sections 8. 1. 2 (Chilled storage), 8. 1. 3 (Frozen storage and 7. 6. 2 Conditioning and storage of molluscan shellfish in sea water tanks, basins, etc.)

16. 3. 2. 2 Containers and packaging (Processing step 2)

Potential Hazards: Unlikely

Potential Defects: Foreign matters

Technical Guidance:

Refer to section 8. 5. 2 (Raw Material Storage - Packaging, Labels & Ingredients); and also:

· all materials for containers or packages should be stored in satisfactory clean and hygienic conditions;

· during storage, empty containers and covers should be protected from dirt, moisture and temperature fluctuations, in order to avoid condensations on containers and in the case of tin cans, the development of corrosion;

· during loading, stowing, transportation and unloading of empty containers, any shock should be avoided. Containers shouldn't be stepped on. These precautions become more imperative when containers are put in bags or on pallets. Shocks can deform the containers (can body or flange), that can compromise tightness (shocks on the seam, deformed flange) or be prejudicial to appearance.

16.3.2.3 Other ingredients (Processing step 2)

Refer to section 8.5.2 (Raw Material Storage - Packaging, Labels & Ingredients).

16.3.3 Unwrapping, unpacking (Processing steps 3 and 4)

Potential Hazards: Unlikely

Potential Defects: Foreign matter

Technical Guidance:

· During unwrapping and unpacking operations, precautions should be taken in order to limit product contamination and foreign matters introduction into the product. To avoid microbial proliferation, waiting periods before further processing should be minimised.

16.3.4 Thawing (Processing step 5)

Refer to section 8.1.4 (Control Thawing)

16.3.5 Fish and shellfish preparatory processes (Processing step 6)

16.3.5.1 Fish preparation (gutting, trimming...)

Potential Hazards: Microbiological contamination biochemical development (histamine)

Potential Defects: Objectionable matters (viscera, skin, scales, ... in certain products), off flavours, presence of bones, parasites...

Technical Guidance:

Refer to sections 8.1.5 (Washing and Gutting) and 8.1.6 (Filleting, Skinning, Trimming and Candling); and also:

· when skinning of fish is operated by soaking in soda solution, a particular care should be taken to carry out an appropriate neutralisation.

16.3.5.2 Preparation of molluscs and crustaceans

Potential Hazards: Microbiological contamination, hard shell fragments

Potential Defects: Objectionable matters

Technical Guidance:

Refer to sections 7.7 (Heat Treatment/Heat Shocking of Molluscan Shellfish in Establishment; and also:

· when live shellfish are used, inspection should be carried out in order to discard dead or badly damaged animals;

· particular care should be taken to ensure that shell fragments are removed from shellfish meat.

16.4 PRE-COOKING AND OTHER TREATMENTS

16.4.1 Pre-Cooking

Potential hazards : chemical contamination (polar components of oxidised oils), microbiological or biochemical (scombrotoxin) growth.

Potential defects : water release in the final product (for products canned in oil), abnormal flavours.

Technical guidance:

16.4.1.1 General Considerations

· methods used to pre-cook fish or shellfish for canning should be designed to bring about the desired effect with a minimum delay and a minimum amount of handling; the choice of method is usually strongly influenced by the nature of the treated material. For products canned in oil such as sardines or tunas, pre-cooking should be sufficient in order to avoid excessive release of

water during heat processing;

· means should be found to reduce the amount of handling subsequent to pre-cooking, wherever practical;

· if eviscerated fish is used, then the fish should be arranged in the belly down position for precooking to allow for the drainage of fish oils and juices which may accumulate and affect product quality during the heating process;

· where appropriate, molluscan shellfish, lobsters and crabs, shrimps and prawns and cephalopods should be pre- cooked according to technical guidance laid down in sections 7 (Processing of Molluscan Shellfish), 13 (Processing of Lobsters and Crabs), 14 (Processing of Shrimps and Prawns) and 15 (Processing of Cephalopods);

· care should be taken to prevent temperature abuse of scombrotoxic species before pre-cooking.

16.4.1.1.1 Pre-cooking Schedule

· the pre-cooking method, in particular, in terms of time and temperature, should be clearly defined. The pre-cooking schedule should be checked;

· fish pre-cooked together in batches should be very similar in size. It also follows that they should all be at the same temperature when they enter the cooker.

16.4.1.1.2 Control of Quality of Pre-cooking Oils and Other Fluids

· only good quality vegetable oils should be used in pre-cooking fish or shellfish for canning (see Codex Standard for Named Vegetable Oils (CODEX STAN 210-1999), Codex Standard for Olive Oils and Olive Pomace Oils (CODEX STAN 33-1981, Rev. 2-2003) and Codex Standard for Fats and Oils not Covered by Individual Standards CODEX STAN 19-1981, Rev. 1 1999);

· cooking oils should be changed frequently in order to avoid the formation of polar compounds. Water used for pre-cooking should also be changed frequently in order to avoid contaminants;

· care must be taken that the oil or the other fluids used such as vapour or water do not impart an undesirable flavour to the product.

16.4.1.1.3 Cooling

· except for products, which are packed when still hot, cooling of pre-cooked fish or shellfish should be done as quickly as possible to bring the product temperatures in a range limiting proliferation or toxin production, and under conditions where contamination of the product can be avoided;

· where water is used to cool crustacea for immediate shucking, it should be potable water or clean seawater. The same water should not be used for cooling more than one batch.

16.4.1.2 Smoking

· refer to section 12 (Processing of smoked fish)

16.4.1.3 Use of Brine and Other Dips

Potential hazards: *microbiological and chemical contamination by the dip solution.*

Potential defects: *adulteration (additives), abnormal flavours.*

Technical guidance:

· Where fish or shellfish are dipped or soaked in brine or in solutions of other conditioning or flavouring agents or additives in preparation for canning, solution strength and time of immersion should both be carefully controlled to bring about the optimum effect;

· dip solutions should be replaced and dip tanks and other dipping apparatus should be thoroughly cleaned at frequent intervals;

· care should be taken to ascertain whether or not the ingredients or additives used in dips would be permitted in canned fish and shellfish by the related Codex Standards and in the countries where the product will be marketed.

16.4.2 Packing in Containers (Filling, Sealing and Coding) (Processing Step 8)

16.4.2.1 Filling

Potential hazards: *microbiological growth (waiting period), microbiological survival growth and recontamination after heat processing due to incorrect filling or faulty containers, foreign material.*

Potential defects: *incorrect weight, foreign matter.*

Technical guidance:

· a representative number of containers and covers should be inspected immediately before delivery to the filling machines or packing tables to ensure that they are clean, undamaged and without visible flaws;

· if necessary, empty containers should be cleaned. It is also a wise precaution to have all containers turned upside down to make certain that they do not contain any foreign material before they are used;

· care should also be taken to remove faulty containers, because they can jam a filling or sealing machine, or cause trouble during heat processing (bad sterilisation, leaks);

· empty containers should not be left on the packing tables or in conveyor systems during clean up of premises to avoid contamination or splashes;

· where appropriate, to prevent microbial proliferation, containers should be filled with hot fish and shellfish (> 63℃, for example for fish soups) or should be filled quickly (the shortest possible waiting period) after the end of the pre-treatments;

· if the fish and shellfish must be held for a long time before packing into containers, they should be chilled;

· containers of canned fish and shellfish should be filled as directed in the scheduled process;

· mechanical or manual filling of containers should be checked in order to comply with the filling rate and the headspace specified in the adopted sterilisation schedule. A regular filling is important not only for economical reasons, but also because the heat penetration and the container integrity can be affected by excessive filling changes;

· the necessary amount of headspace will depend partly on the nature of the contents. The filling should also take into account the heat processing method. Headspace should be allowed as specified by the container manufacturer;

· furthermore, containers should be filled such as the end product meets the regulatory provisions or the accepted standards concerning weight of contents;

· where canned fish and shellfish is packed by hand, there should be a steady supply of fish, shellfish and eventually other ingredients. Build-up of fish and shellfish, as well as filled containers at the packing table should be avoided;

· the operation, maintenance, regular inspection, calibration and adjustment of filling machines should received particular care. The machine manufacturers' instructions should be carefully followed;

· the quality and the amount of other ingredients such as oil, sauce, vinegar... should be carefully controlled to bring about the optimum desired effect;

· if fish has been brine-frozen or stored in refrigerated brine, the amount of salt absorbed should be taken into consideration when salt is added to the product for flavouring;

· filled containers should be inspected:

— to ensure that they have been properly filled and will meet accepted standards for weight of contents

— and to verify product quality and workmanship just before they are closed;

— manual filled products such as small pelagic fish should be carefully checked by the operators to verify that container flanges or closure surface have not any product residues, which could impede the formation of a hermetic seal. For automatic filled products, a sampling plan should be implemented.

16.4.2.2 Sealing

Sealing the container and covers are one of the most essential processes in canning.

Potential hazards: subsequent contamination due to a bad seam.

Potential defects: unlikely.

Technical guidance:

· the operation, maintenance, regular inspection and adjustment of sealing machines should received particular care. The sealing machines should be adapted and adjusted for each type of container and each closing method which are used. Whatever the type of sealing equipment, the manufacturers or equipment supplier's instructions should be followed meticulously;

· seams and other closures should be well formed with dimensions within the accepted tolerances for the particular container;

· qualified personnel should conduct this operation;

· if vacuum is used during packing, it should be sufficient to prevent the containers from bulging under any condition (high temperature or low atmospheric pressure) likely to be encountered during the distribution of the product. This is useful for deep containers or glass containers. It is difficult and hardly necessary to create a vacuum in shallow containers that have relatively large flexible covers;

· excessive vacuum may cause the container to panel, particularly if the headspace is large, and may also cause contaminants to be sucked into the container if there is a slight imperfection in the seam;

· to find the best methods to create vacuum, competent technologists should be consulted;

· regular inspections should be made during production to detect potential external defects on containers. At intervals sufficiently close to each other in order to guarantee a closure in accordance with specifications, the operator, the supervisor of the closure or any other competent person should examine the seams or the closure system for the other types of containers, which are used. Inspections should consider for example vacuum measurements and seam teardown. A sampling plan should be used for the checks;

· in particular, at each start of the production line and at each change in container dimensions, after a jamming, a new adjustment or a restarting after a prolonged stop of the sealing machine, a check should be carried out;

· all appropriate observations should be recorded.

16.4.2.3 Coding

Potential hazards: subsequent contamination due to damaged containers.

Potential defects: loss of traceability due to an incorrect coding.

Technical guidance.

· each container of canned fish and shellfish should bear indelible code markings from which allimportant details concerning its manufacture (type of product, cannery where the canned fish or shellfish was produced, production date, etc.) can be determined coding equipment must be carefully adjusted so that the containers are not damaged and the code remains legible;

· coding may sometimes be carried out after the cooling step.

16.4.3 Handling of Containers After Closure - Staging Before Heat Processing (Processing Step 9)

Potential hazards: microbiological growth (waiting period), subsequent contamination due to damaged containers.

Potential defects: Unlikely.

Technical guidance.

· containers after closure should always be handled carefully in such a way as to prevent every damage capable to cause defects and microbiological recontamination;

· if necessary, filled and sealed metal containers should be thoroughly washed before heat processing to remove grease, dirt and fish or shellfish stains on their outside walls;

· to avoid microbial proliferation, the waiting period should be as short as possible;

· if the filled and sealed containers must be held for a long time before heat processing, the product should be held at temperature conditions which minimise microbial growth;

· every cannery should develop a system, which will prevent non heat-processed canned fish and shellfish from being accidentally taken past the retorts into the storage area.

16.4.4 Thermal Processing (Processing Step 10)

Heat processing is one of the most essential operations in canning.

Canners can refer to the Recommended International Code of Hygienic Practice for Low-Acid and Acidified Low-Acid Canned Foods (CAC/RCP 23-1979, rev. 2 in 1993) in order to obtain detailed advice on heat processing. In this Section, only some essential elements are pointed out.

Potential hazards: survival of spores of Clostridium botulinum.

Potential defects: survival of micro-organisms responsible of decomposition.

Technical guidance.

16.4.4.1 Sterilisation Schedule

· to determine the sterilisation schedule, at first, the heat process required to obtain the commercial sterility should be established taking into account some factors (microbial flora, dimensions and nature of the container, product formulation, etc.). A sterilisation schedule is established for a certain product in a container of a given size;

· Proper heat generation and temperature distribution should be carried out. Standard heat processing procedures and experimentally established sterilisation schedules should be checked and validated by an expert to confirm that the values are appropriate for each product and retort;

· before any changes in operations (initial temperature of filling, product composition, size of containers, fullness of the retort, etc.) are made, competent technologists should be consulted as to the need for re-evaluation of the process.

16. 4. 4. 2 Heat Processing Operation

• only qualified and properly trained personnel should operate retorts. Therefore it is necessary that retort operators control the processing operations and ensure the sterilisation schedule is closely followed, including meticulous care in timing, monitoring temperatures and pressures, and in maintaining records;

• it is essential to comply with the initial temperature described in the schedule process to avoid under-processing. If the filled containers were held at refrigerated temperatures because of a too long waiting period before heat processing, the sterilisation schedule should take into account these temperatures;

• in order that the heat processing is effective and process temperature is controlled, air must be evacuated from the retort through a venting procedure that is deemed efficient by a competent technologist. Container size and type, retort installation and loading equipment and procedures should be considered;

• the timing of the heat processing should not commence until the specified heat processing temperature has been reached, and the conditions to maintain uniform temperature throughout the retort achieved, in particular, until the minimum safe venting time has elapsed;

• for other types of retorts (water, steam/air, flame, etc.) refer to the Recommended International Code of Hygienic Practice for Low-Acid and Acidified Low-Acid Canned Foods (CAC/RCP 23-1979, rev. 2 in 1993);

• if canned fish and shellfish in different size containers are processed together in the same retort load care must be taken to ensure the process schedule used is sufficient to provide commercial sterility for all container sizes processed;

• when processing fish and shellfish in glass containers, care must be taken to ensure that the initial temperature of the water in the retort is slightly lower than that of the product being loaded. The air pressure should be applied before the water temperature is raised.

16. 4. 4. 3 Monitoring of Heat Processing Operation

• during the application of heat processing, it is important to ensure that the sterilisation process and factors such as container filling, minimal internal depression at closing, retort loading, initial product temperature, etc. are in accordance with the sterilisation schedule;

• retort temperatures should always be determined from the indicating thermometer, never from the temperature recorder;

• permanent records of the time, temperature and other pertinent details should be kept concerning each retort load;

• the thermometers should be tested regularly to ensure that they are accurate. Calibration records should be maintained; the recording thermometer readings should never exceed the indicating thermometer reading;

• inspections should be made periodically to ensure that retorts are equipped and operated in a manner that will provide thorough and efficient heat processing, that each retort is properly equipped, filled and used, so that the whole load is brought up to processing temperature quickly and can be maintained at that temperature throughout the whole of the processing period;

• the inspections should be made under the guidance of a competent technologist.

16. 4. 5 Cooling (Processing Step 11)

Potential hazards: recontamination due to a bad seam and contaminated water.

Potential defects: formation of struvite crystals, buckled containers, scorch.

Technical guidance:

· after heat processing, canned fish and shellfish should, wherever practical, be water cooled under pressure to prevent deformations, which could result in a loss of tightness. In case of recycling, potable water should always be chlorinated (or other appropriate treatments used) for this purpose. The residual chlorine level in cooling water and the contact time during cooling should be checked in order to minimise the risk of post-processing contamination. The efficiency of the treatment other than chlorination should be monitored and verified;

· in order to avoid organoleptic defects of the canned fish and shellfish, such as scorch or overcooking, the internal temperature of containers should be lowered as quickly as possible;

· for glass containers, the temperature of the coolant in the retort should be, at the beginning, lowered slowly in order to reduce the risks of breaking due to thermal shock;

· where canned fish and shellfish products are not cooled in water after heat processing, they should be stacked in such a way that they will cool rapidly in air.

· heat processed canned fish and shellfish should not be touched by hand or articles of clothing unnecessarily before they are cooled and thoroughly dry. They should never be handled roughly or in such a way that their surfaces, and in particular their seams, are exposed to contamination;

· rapid cooling of canned fish and shellfish avoids the formation of struvite crystals ;

· every cannery should develop a system to prevent unprocessed containers being mixed with processed containers.

16.4.5.1 Monitoring After Heat Processing and Cooling

· canned fish and shellfish should be inspected for faults and for quality assessment soon after it is produced and before labelling;

· representative samples from each code lot should be examined to ensure that the containers do not exhibit external defects and the product meets the standards for weight of contents, vacuum, workmanship and wholesomeness. Texture, colour, odour, flavour and condition of the packing medium should be assessed;

· if desired, stability tests could be made in order to verify in particular the heat processing;

· this examination should be made as soon as practical after the canned fish and shellfish have been produced, so that if there are any faults due to failings on the part of cannery workers or canning equipment, these failings can be corrected without delay. Segregating and properly disposing of all defective units or lots that are unfit for human consumption should be ensured.

16.4.6 Labelling, Casing and Storage of Finished Products (Processing steps 12 and 13)

Refer to Section 8.2.3 "Labelling"

Potential hazards : *subsequent recontamination due to the damage of containers or to an exposition to extreme conditions*

Potential defects : *incorrect labelling*

Technical guidance

· the materials used for labelling and casing canned fish and shellfish should not be conducive to corrosion of the container. Cases should have an adequate size in order that the containers fit them and are not damaged by any move inside. Cases and boxes should be the correct size and strong enough to protect the canned fish and shellfish during distribution;

· code marks appearing on containers of canned fish and shellfish should also be shown on the cases in which they are packed;

- storage of canned fish and shellfish should be made in order not to damage the containers. In particular, pallets of finished products should not be stacked excessively high and the forklift trucks used for the storage should be used in a proper manner;
- canned fish and shellfish should be so stored that they will be kept dry and not exposed to extremes of temperature.

16.4.7 Transportation of Finished Products (Processing step 14)

Potential hazards: *subsequent recontamination due to the damage of containers or to an exposition to extreme conditions*

Potential defects: *Unlikely*

Technical guidance

Refer to section 17 (Transportation); and also:

- transportation of canned fish and shellfish should be made in order not to damage the containers. In particular, the forklift trucks used during the loading and unloading should be used in a proper manner;
- cases and boxes should be completely dry. In fact, moisture has effects on the mechanical characteristics of boxes and the protection of containers against damages during transportation couldn't be sufficient;
- metal containers should be kept dry during transportation in order to avoid corroding and/or rust.

SECTION 17 TRANSPORT

Refer to the Recommended International Code of Practice-General Principles of Food Hygiene, Section Ⅷ - Transportation, CAC/RCP 1969, Rev. 4 (2003) and the Code of Hygienic Practice for the Transport of Food in Bulk and Semi-Packaged Food (CAC/RCP 47-2001).

Transportation applies to all sections and is a step of the flow diagram which needs specific skills. It should be considered with the same care as the other processing steps. This section provides examples of potential hazards and defects and describes technological guidelines, which can be used to develop control measures and corrective action. At a particular step only the hazards and defects, which are likely to be introduced or controlled at that step, are listed. It should be recognised that in preparing a HACCP and/or DAP plan it is essential to consult Section 5 which provides guidance for the application of the principles of HACCP and DAP analysis. However, within the scope of this Code of Practice it is not possible to give details of critical limits, monitoring, record keeping and verification for each of the steps since these are specific to particular hazards and defects.

It is particularly important throughout the transportation of fresh, frozen or refrigerated fish, shellfish and their products that care is taken to minimise any rise in temperature of the product and that the chill or frozen temperature, as appropriate, is maintained under controlled conditions. Moreover, appropriate measures should be applied to minimize damage to products and also their packaging.

17.1 FOR FRESH, REFRIGERATED AND FROZEN PRODUCTS

Refer to 3.6 Transportation.

Potential Hazards: *Biochemical development (histamine). Microbial growth and contamination*

Potential Defects: *Decomposition, physical damage. Chemical contamination (fuel).*

Technical Guidance:

· check temperature of product before loading;

· avoid unnecessary exposure to elevated temperatures during loading and unloading of fish, shellfish and their products;

· load in order to ensure a good air flow between product and wall, floor and roof panels ; load stabilizer devices are recommended;

· monitor air temperatures inside the cargo hold during transportation; the use of a recording thermometer is recommended;

· during transportation:

— frozen products should be maintained at -18℃ or below (maximum fluctuation +3℃)

— fresh fish, shellfish and their products should be kept at a temperature as close as possible to 0℃. Fresh whole fish should be kept in shallow layers and surrounded by finely divided melting ice; adequate drainage should be provided in order to ensure that water from melted ice does not stay in contact with the products or melted water from one container does not cross contaminate products in other containers.

— transportation of fresh fish in containers with dry freezer bags and not ice should be considered where appropriate.

— transportation of fish in an ice slurry, chilled sea water or refrigerated sea water (e. g. pelagic fish) should be considered where appropriate. Chilled sea water or refrigerated sea water should be used under approved conditions.

— refrigerated processed products should be maintained at the temperature specified by the processor but generally should not exceed 4℃.

— provide fish, shellfish and their products with adequate protection against contamination from dust, exposure to higher temperatures and the drying effects of the sun or wind.

17. 2 FOR LIVE FISH AND SHELLFISH

refer to the specific provisions laid down in the relevant sections of the code.

17. 3 FOR CANNED FISH AND SHELLFISH

refer to the specific provisions laid down in section 16.

17. 4 FOR ALL PRODUCTS

· before loading, the cleanliness, suitability and sanitation of the cargo hold of the vehicles should be verified;

· loading and transportation should be made in order to avoid damage and contamination of the products and to ensure the packaging integrity;

· after unloading, the accumulation of waste should be avoided and should be disposed of in a proper manner.

SECTION 18 RETAIL

In the context of recognising controls at individual processing steps, this section provides examples of potential hazards and defects and describes technological guidelines, which can be used to develop control measures and corrective action. At a particular step only the hazards and defects, which are likely to be introduced or controlled at that step, are listed. It should be recognised that in preparing a HACCP and/or DAP plan it is essential to consult Section 5 which provides guidance for the application of the principles of HACCP and DAP analysis. However, within the scope of this Code of Practice it is not possible to give details of critical limits, monitoring, record keeping and verification for each of the steps since these are specific to particular hazards and defects.

Fish, shellfish and their products at retail should be received, handled, stored and displayed to consumers in a manner that minimizes potential food safety hazards and defects and maintains essential quality. Consistent with the HACCP and DAP approaches to food safety and quality, products should be purchased from known or approved sources under the control of competent health authorities that can verify HACCP controls. Retail operators should develop and use written purchase specifications designed to ensure food safety and desired quality levels. Retail operators should be responsible to maintain quality and safety of products.

Proper storage temperature after receipt is critical to maintain product safety and essential quality. Chilled products should be stored in a hygienic manner at temperatures less than or equal to 4℃ (40°F), MAP products at 3℃ (38°F) or lower, while frozen products should be stored at temperatures less than or equal to -18℃ (0°F).

Preparation and packaging should be carried out in a manner consistent with the principles and recommendations found in Section 3, Prerequisite Programmes and Codex Labelling Standards. Product in an open full display should be protected from the environment such as use of display covers (sneeze guards). At all times, displayed seafood items should be held at temperatures and conditions that minimize the development of potential bacterial growth, toxins and other hazards in addition to loss of essential quality.

Consumer information at the point of purchase, for example placards or brochures, that inform consumers about storage, preparation procedures and potential risks of seafood products if mishandled or improperly prepared, is important to ensure that product safety and quality is maintained.

A system of tracking the origin and codes of fish, shellfish and their products should be established to facilitate product recall or public health investigations in the event of the failure of preventive health protection processes and measures. These systems exist for molluscan shellfish in some countries in the form of molluscan shellfish tagging requirements.

18.1 RECEPTION OF FISH, SHELLFISH AND THEIR PRODUCTS AT RETAIL – GENERAL CONSIDERATIONS

Potential Hazards: *see Reception* 7.1, 8.1

Potential Defects: *see Reception* 7.1, 8.1

Technical Guidance:

- The transport vehicle should be examined for overall hygienic condition. Products subject to filth, taint or contamination should be rejected.
- The transport vehicle should be examined for possible cross contamination of ready to eat fish and fishery products by raw fish and fishery products. Determine that cooked-ready-to-eat product has not been exposed to raw product or juices or live molluscan shellfish and that raw molluscan shellfish have not been exposed to other raw fish or shellfish.

· Seafood should be regularly examined for adherence to purchasing specifications.

· All products should be examined for decomposition and spoilage at receipt. Products exhibiting signs of decomposition should be refused When a log of the cargo hold temperature for the transport vehicle is kept, records should be examined to verify adherence to temperature requirements.

18.1.1 Reception of Chilled Products at Retail

Potential Hazards: Pathogen growth, microbiological contamination, chemical and physical contamination, Scombrotoxin formation, C. botulinum toxin formation

Potential Defects: Spoilage (decomposition), Contaminants, Filth

Technical Guidance:

· Product temperature should be taken from several locations in the shipment and recorded. Chilled fish, shellfish and their products should be maintained at or below 4℃ (40℉). MAP product, if not frozen, should be maintained at or below 3℃ (38℉).

18.1.2 Reception of Frozen Products at Retail

Potential Hazards: Unlikely

Potential Defects: Thawing, Contaminants, Filth

Technical Guidance:

· Incoming frozen seafood should be examined for signs of thawing and evidence of filth or contamination. Suspect shipments should be refused.

· Incoming frozen seafood should be checked for internal temperatures, taken and recorded from several locations in the shipment. Frozen fish, shellfish and their products should be maintained at or below −18℃ (0℉).

18.1.3 Chilled Storage of Products at Retail

Potential Hazards: Scombrotoxin formation, microbiological contamination, pathogen growth, chemical contamination, C. botulinum toxin formation

Potential Defects: Decomposition, Contaminants, Filth

Technical Guidance:

· Products in chilled storage should be held at 4℃ (40℉). MAP product should be held at 3℃ (38℉) or below.

· Seafood should be properly protected from filth and other contaminants through proper packaging and stored off the floor.

· A continuous temperature recording chart for seafood storage coolers is recommended.

· The cooler room should have proper drainage to prevent product contamination.

· Ready-to-eat items and molluscan shellfish should be kept separate from each other and other raw food products in chilled storage. Raw product should be stored on shelves below cooked product to avoid cross contamination from drip.

· A proper product rotation system should be established. This system could be based on first in, first out usage, production date or best before date on labels, sensory quality of the lot, etc, as appropriate.

18.1.4 Frozen Storage of Products at Retail

Potential Hazards: Unlikely

Potential Defects: Chemical decomposition (rancidity), Dehydration

Technical Guidance:

· Product should be maintained at -18℃ (0℉) or less. Regular temperature monitoring should be carried out. A recording thermometer is recommended.

· Seafood products should not be stored directly on the floor. Product should be stacked to allow proper air circulation.

18.1.5 Preparation and Packaging Chilled Product at Retail

Refer to Section 8.2.3, "Labelling".

Potential Hazards: Microbiological contamination, Scombrotoxin formation, pathogen growth, physical and chemical contamination, allergens

Potential Defects: Decomposition, Incorrect Labelling

Technical Guidance:

· Care should be taken to ensure that handling and packaging product is conducted in accordance to guidelines in Section 3, Pre-requisite Programmes.

· Care should be taken to ensure that labelling is in accordance to guidelines in Section 3, Prerequisite Programmes and Codex Labelling Standards especially for known allergens.

· Care should be taken to ensure that product is not subjected to temperature abuse during packaging and handling.

· Care should be taken to avoid cross contamination of ready-to-eat and raw shellfish, shellfish and their products at the work areas or by utensils or personnel.

18.1.6 Preparation and Packaging of Frozen Seafood at Retail

Refer to Section 8.2.3, "Labelling".

Potential Hazards: Microbiological contamination, chemical or physical contamination, allergens

Potential Defects: Thawing, Incorrect Labelling

Technical Guidance:

· Care should be taken to ensure that allergens are identified, in accordance to Section 3, Prerequisite Programmes and Codex Labelling Standards.

· Care should be taken to avoid cross contamination of ready-to-eat and raw product.

· Frozen seafood products should not be subjected to ambient room temperatures for a prolonged period of time.

18.1.7 Retail Display of Chilled Seafood

Potential Hazards: Scombrotoxin formation, microbiological growth, microbiological contamination, C. botulinum toxin formation.

Potential Defects: Decomposition, Dehydration

Technical Guidance:

· Products in chilled display should be kept at 4℃ (40℉) or below. Temperatures of product should be taken at regular intervals.

· Ready-to-eat items and molluscan shellfish should be separated from each other and from raw food products in a chilled full service display. A diagram of display is recommended to ensure that cross contamination does not occur.

· If ice is used, proper drainage of melt water should be in place. Retail displays should be selfdraining. Replace ice daily and ensure ready-to-eat products are not placed on ice upon which raw product was previously displayed.

· Each commodity in a full service display should have its own container and serving utensils to avoid cross contamination.

· Care should be taken to avoid arranging product in such a large mass/depth that proper chilling cannot be maintained and product quality is compromised.

· Care should be taken to avoid drying of unprotected products in full service displays. Use of an aerosol spray, under hygienic conditions is recommended

· Product should not be added above the "load line" where a chilled state cannot be maintained in self-service display cases of packaged product.

· Product should not be exposed to ambient room temperature for a prolonged period of time when filling/stocking display cases.

· Seafood in full service display cases should be properly labelled by signs or placards to indicate the commonly accepted name of the fish so the consumer is informed about the product.

18. 1. 8 Retail Display of Frozen Seafood

Potential Hazards: *Unlikely*

Potential Defects: *Thawing*, *Dehydration* (*Freezer Burn*)

Technical Guidance:

· Product should be maintained at −18℃ (0℉) or less. Regular temperature monitoring should be carried out. A recording thermometer is recommended.

· Product should not be added above the "load line" of cabinet self-service display cases. Upright freezer self-service display cases should have self-closing doors or air curtains to maintain a frozen state.

· Product should not be exposed to ambient room temperature for a prolonged period of time when filling/stocking display cases.

· A product rotation system to ensure first in, first out usage of frozen seafood should be established.

· Frozen seafood in retail displays should be examined periodically to assess packaging integrity and the level of dehydration or freezer burn.

ANNEX I

POTENTIAL HAZARDS ASSOCIATED WITH FRESH FISH, SHELLFISH AND OTHER AQUATIC INVERTEBRATES

1 EXAMPLES OF POSSIBLE BIOLOGICAL HAZARDS

1. 1. 1 Parasites

The parasites known to cause disease in humans and transmitted by fish or crustaceans are broadly classified as helminths or parasitic worms. These are commonly referred to as Nematodes, Cestodes and Trematodes. Fish can be parasitised by protozoans, but there are no records of fish protozoan disease being transmitted to man. Parasites have complex life cycles, involving one or more intermediate hosts and are generally passed to man through the consumption of raw, minimally processed or inadequately cooked products that contain the parasite infectious stage, causing foodborne disease. Freezing at – 20℃ or below for 7 days or –35℃ for about 20 hours for fish intended for raw consumption will kill parasites. Processes such as brining or pickling may reduce the parasite hazard if the products are kept in the brine for a sufficient time but may not eliminate it. Candling, trimming belly flaps and physically removing the parasite cysts will also reduce the hazards but may not eliminate it.

Nematodes

Many species of nematodes are known to occur worldwide and some species of marine fish act as secondary hosts. Among the nematodes of most concern are *Anisakis* spp. , *Capillaria* spp. , *Gnathostoma* spp. , and *Pseudoteranova* spp. , which can be found in the liver, belly cavity and flesh of marine fish. An example of a nematode causing disease in man is *Anisakis simplex*; as the infective stage of the parasite is killed by heating (60℃ for 1 minute) and by freezing (-20℃ for 24 hours) in the fish core.

Cestodes

Cestodes are tapeworms and the species of most concern associated with the consumption of fish is *Dibothriocephalus latus*. This parasite occurs worldwide and both fresh and marine fish are intermediate hosts. Similar to other parasitic infections, the foodborne disease occurs through the consumption of raw or underprocessed fish. Similar freezing and cooking temperatures as applied to nematodes will inactivate the infective stages of this parasite.

Trematodes

Fish-borne trematode (flatworm) infections are major public health problems that occur endemically in about 20 countries around the world. The most important species with respect to the numbers of people infected belong to the genera *Clonorchis* and *Ophisthorchis* (liver flukes), *Paragonimus* (lung flukes), and to a lesser extent *Heterophyes* and *Echinochasmus* (intestinal flukes) . The most important definitive host of these trematodes is man or other mammals. Freshwater fish are the second intermediate host in the life cycles of *Clonorchis* and *Ophistorchis*, and freshwater crustaceans in the case of *Paragonimius*. Foodborne infections take place through the consumption of raw, undercooked or otherwise under-processed products containing the infective stages of these parasites. Freezing fish at -20℃ for 7 days or at -35℃ for 24 hours will kill the infective stages of these parasites.

1.1.2 Bacteria

The level of contamination of fish at the time of capture will depend on the environment and the bacteriological quality of the water in which fish are harvested. Many factors will influence the microflora of finfish, the more important being water temperature, salt content, proximity of harvesting areas to human habitations, quantity and origin of food consumed by fish, and method of harvesting. The edible muscle tissue of finfish is normally sterile at the time of capture and bacteria are usually present on the skin, gills and in the intestinal tract.

There are two broad groups of bacteria of public health importance that may contaminate products at the time of capture - those that are normally or incidentally present in the aquatic environment, referred to as the indigenous microflora, and those introduced through environmental contamination by domestic and /or industrial wastes. Examples of indigenous bacteria, which may pose a health hazard, are *Aeromonas hydrophyla*, *Clostridium botulinum*, *Vibrio parahaemolyticus*, *Vibrio cholerae*, *Vibrio vulnificus*, and *Listeria monocytogenes*. Non-indigenous bacteria of public health significance include members of the Enterobacteriaceae, such as *Salmonella* spp., *Shigella* spp., and *Escherichia coli*. Other species that cause foodborne illness and which have been isolated occasionally from fish are *Edwardsiella tarda*, *Pleisomonas shigeloides* and *Yersinia enterocolitica*. *Staphyloccocus aureus* may also appear and may produce heat resistant toxins.

Indigenous pathogenic bacteria, when present on fresh fish, are usually found in fairly low numbers, and where products are adequately cooked prior to consumption, food safety hazards are insignificant. During storage, indigenous spoilage bacteria will outgrow indigenous pathogenic bacteria, thus fish will spoil before becoming toxic and will be rejected by consumers. Hazards from these pathogens can be controlled by heating seafood sufficiently to kill the bacteria, holding fish at chilled temperatures and avoiding postprocess cross-contamination.

Vibrio species are common in coastal and estuarine environments and populations can depend on water depth and tidal levels. They are particularly prevalent in warm tropical waters and can be found in temperate zones during summer months. *Vibrio* species are also natural contaminants of brackish water tropical environments and will be present on farmed fish from these zones. Hazards from *Vibrio* spp. associated with finfish can be controlled by thorough cooking and preventing cross-contamination of cooked products. Health risks can also be reduced by rapidly chilling products after harvest, thus reducing the possibility of proliferation of these organisms. Certain strains of *Vibrio parahaemolyticus* can be pathogenic.

1.1.3 Viral Contamination

Molluscan shellfish harvested from inshore waters that are contaminated by human or animal faeces may harbour viruses that are pathogenic to man. Enteric viruses that have been implicated in seafood-associated illness are the hepatitis A virus, caliciviruses, astroviruses and the Norwalk virus. The latter three are often referred to as small round structured viruses. All of the seafood-borne viruses causing illness are transmitted by the faecal-oral cycle and most viral gastro-enteritis outbreaks have been associated with eating contaminated shellfish, particularly raw oysters.

Generally viruses are species specific and will not grow or multiply in foods or anywhere outside the host cell. There is no reliable marker for indicating presence of the virus in shellfish harvesting waters. Sea-foodborne viruses are difficult to detect, requiring relatively sophisticated molecular methods to identify the virus.

Occurrence of viral gastro-enteritis can be minimized by controlling sewage contamination of shellfish farming areas and pre-harvest monitoring of shellfish and growing waters as well as controlling other sources of contamination during processing. Depuration or relaying are alternative strategies but longer periods are required for shellfish to purge themselves clean of viral contamination than for bacteria. Thermal processing (85 ~ 90℃ for 1.5 min.) will destroy viruses in shellfish.

1.1.4 Biotoxins

There are a number of important biotoxins to consider. Around 400 poisonous fish species exist and, by definition, the substances responsible for the toxicity of these species are biotoxins. The poison is usually limited to some organs, or is restricted to some periods during the year.

For some fish, the toxins are present in the blood; these are ichtyohaemotoxin. The involved species are eels from the Adriatic, the moray eels, and the lampreys. In other species, the toxins are spread all over the tissues (flesh, viscera, skin); these are ichtyosarcotoxins. The tetrodotoxic species responsible for several poisonings, often lethal, are in this category.

In general these toxins are known to be heat-stable and the only possible control measure is to check the identity of the used species.

Phycotoxins

Ciguatoxin

And the other important toxin to consider is ciguatoxin, which can be found in a wide variety of mainly carnivorous fish inhabiting shallow waters in or near tropical and subtropical coral reefs. The source of this toxin is dinoflagellates and over 400 species of tropical fish have been implicated in intoxication. The toxin is known to be heat stable. There is still much to be learnt about this toxin and the only control measure that can reasonably be taken is to avoid marketing fish that have a known consistent record of toxicity.

PSP/DSP/NSP/ASP

Paralytic Shellfish Poison (PSP), Diarrhetic Shellfish Poison (DSP), Neurotoxic Shellfish Poison (NSP), and Amnestic Shellfish Poison complex (ASP) are produced by phytoplankton. They concentrate in bivalve molluscan shellfish which filter the phytoplankton from the water, and also may concentrate in some fish and crustacea.

Generally, the toxins remain toxic through thermal processing so the knowledge of the species identity and/or origin of fish or shellfish intended for processing is important.

Tetrodotoxin

Fish mainly belonging to the family Tetradontidea (" puffer fishes") may accumulate this toxin which is responsible for several poisonings, often lethal. The toxin is generally found in the fish liver, roe and guts, and less frequent in the flesh. Differently from most other fish biotoxins that accumulate in the live fish or shellfish, algae do not produce this toxin. The mechanism of toxin production is still not clear, however, apparently there are often indications of the involvement of symbiotic bacteria.

1.1.5 Scombrotoxin

Scombroid intoxication, sometimes referred to as histamine poisoning, results from eating fish that have been incorrectly chilled after harvesting. Scombrotoxin is attributed mainly to *Enterobacteriaceae* which can produce high levels of histamine and other biogenic amines in the fish muscle when products are not immediately chilled after catching. The main susceptible fish are the scombroids such as tuna, mackerel, and bonito, although it can be found in other fish families such as *Clupeidae*. The intoxication is rarely fatal and symptoms are usually mild. Rapid refrigeration after catching and a high standard of handling during processing should prevent the development of the toxin. The toxin is not inactivated by normal heat processing. In addition, fish may contain toxic levels of histamine without exhibiting any of the usual sensory parameters characteristic of spoilage.

1.2 CHEMICAL HAZARDS

Fish may be harvested from coastal zones and inland habitats that are exposed to varying amounts of environmental contaminants. Of greatest concern are fish harvested from coastal and estuarine areas rather than fish harvested from the open seas. Chemicals, organochloric compounds and heavy metals may accumulate in products that can cause public health problems. Veterinary drug residues can occur in aquaculture products when correct withdrawal times are not followed or when the sale and use of these compounds are not controlled. Fish can also be contaminated with chemicals such as diesel oil, when incorrectly handled and detergents or disinfectants when not properly rinsed out.

1.3 PHYSICAL HAZARDS

These can include material such as metal or glass fragments, shell, bones, etc.

鱼贝类实验室感官评价指南

CODEX GUIDELINES FOR THE SENSORY EVALUATION OF FISH AND SHELLFISH IN LABORATORIES

(CAC/GL 31 -1999)

1 指南的范围及目的

本指南旨在指导检测人员进行感官检测。尽管指南原是依照食品法典的要求而制订的，但也包括了一些专门为这些标准中未涵盖但按要求在水产品检测中应进行感官评价的产品而制订的条款[1]。本指南适用于为测定产品加工过程中的缺陷而对样品进行的实验室中的感官检验程序（包括蒸煮），而这类感官检验程序在实验室外不能正常进行。另外，本指南还提供了用于此种检验或用于培训检验员的实验设施方面的技术资料。

本指南的目的是为在以检查为目的的感官检验所需的设备及步骤提出建议，以确保标准在应用中的一致性。

本指南中所述的“鱼”意指：鱼类、甲壳类及软体动物类水产品。

2 感官评价的设备

2.1 总论

感官评价应由受训合格的人员进行（见第四部分）操作，他们使用一套感官评价方法，对特定范围的产品进行评价。

2.2 用于感官评价的实验室

2.2.1 位置及布局

图1给出一个适于进行水产品检验的实验室规划图。该规划图表明原则上预处理区与鉴定区应当分离。

办公用品、贮藏室、人员设备及其他可能需用的检验设备应在此前提下放置在其他的地方。鉴定区内不能进行化学和微生物分析实验，但有些检验分析可在预处理区进行。

2.2.2 预处理区

此区域用于处理和存放水产品，以及进行感官评价用样品的准备工作。应按照渔业设施设计及构造的良好操作规范的要求建造。房间的设计应确保蒸煮的风味不会干扰感官分析。

2.2.3 鉴定区

除了在蒸煮前对样品的最后的修整，此区域不应进行任何样品准备工作。

---

[1] 如果食品法典委员会提出新的建议，可以添加附加标准。

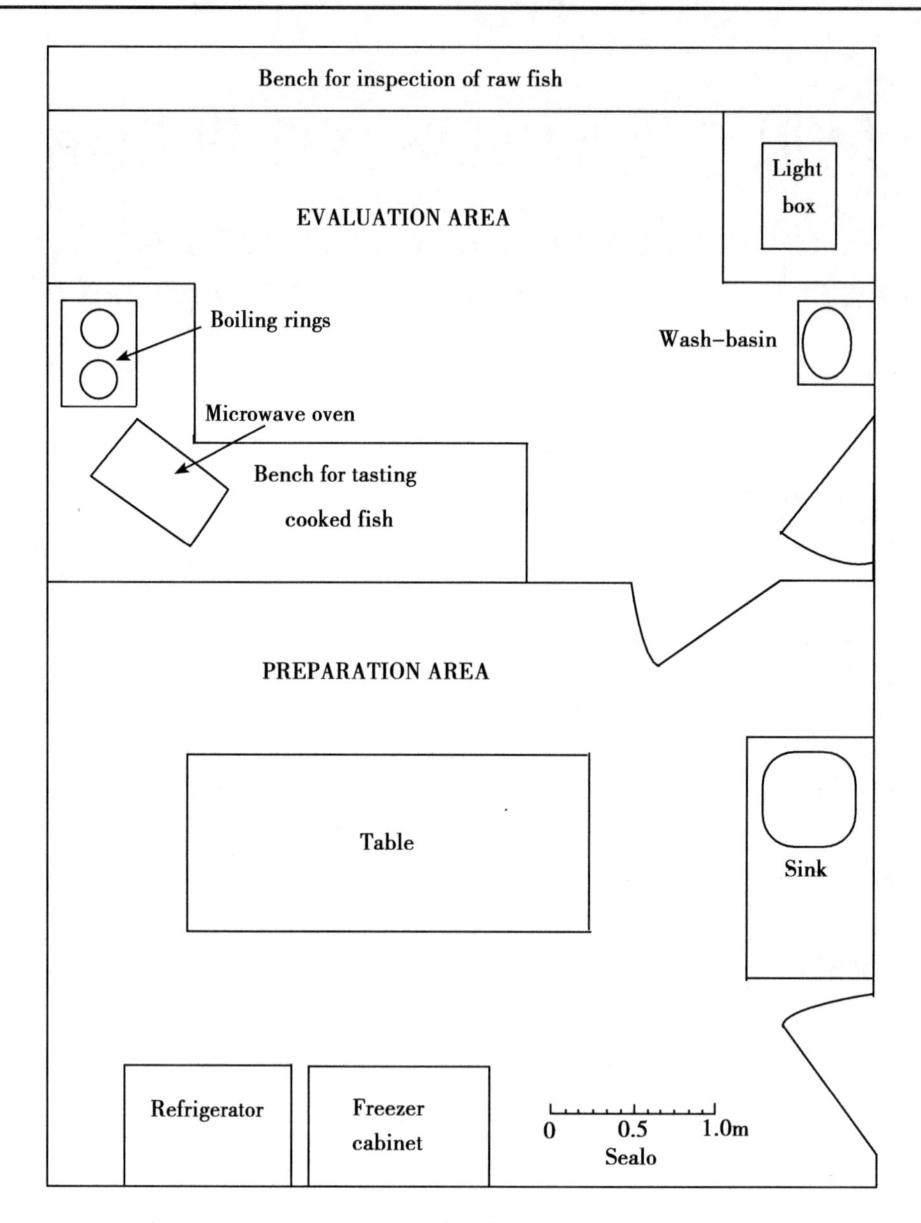

图1 水产品检验实验室规划图

面积、通风、鉴定步骤及样品检验顺序的设计应当使其对感官刺激的干扰程度减至最低，同样，也应将鉴定同伴及其他人员的影响和干扰降至最小，鉴定区的色彩应为中性色彩。

工作台面照明应使用日光或人造日光，应符合标准中所有特定条件的要求。

2.2.4 设备

所需设备的类型及数量取决于被测产品的性质以及检验的数量和频率。

3 感官评价的程序

3.1 样品的选择及运输

在大多数情况下，对水产品进行感官评价是为了对一批水产品做出处理决定，例如，接受或拒绝进口货物的交付、市场中水产品新鲜等级的划分等。按照指南对从批次产品中取样，并对样品进行检验，根据检验结果做出决定。指南通常会指定常规检验或商业检验如何进行抽样。

当选择样品进行检查时，检查人员应确保取样程序及随后对样品的处理不会对其感官特性造成重大影响。

检查人员应检查样品是否按需要包装完好，在将样品发送到实验室之前，对其进行温度控制。如果样品在运输过程中没有专业人员进行监督，检查员应确保样品在途中不会被干扰破坏。

鉴定实验室接受样品后，如果不立即进行鉴定，则应在适当条件下贮藏样品。但鲜品或冷藏产品应于运达当天进行检验。冷藏或冻藏产品应进行适当包裹以防止干耗或脱水。

3.2 样品检验的准备工作

附件1中的表1列出了在一些品种和产品在鉴定中可能用到的水产品特征，应当根据产品类型选择合适的样品准备方法。以下介绍了一些生鲜和冷冻的鱼的预处理步骤。

如果是完整的鱼，应去除内脏，并将内脏保存起来。去头，从一侧取肉，将此部分置于盘中用于检验。

速冻（QF）产品可以放在鉴定区的实验台上，但如果将样品放于盘中则便于进行性状评价和之后的清理工作。

冷冻产品应首先在冻结状态下进行检验，然后将全部或部分样品解冻，并进行感官评价。样品是否能够或需要进行再分割，取决于产品的性质，单体速冻（IQF）虾或鱼片的包装可以打开，进行二次取样。如果鱼或肉块较厚的话，取样就比较困难，在这种情况下可使用锯。

对冷冻产品的解冻应尽可能快速地进行，但应避免全部或部分样品的温度上升，这样会导致腐败。最简单的方法是将样品单位散放在预处理区的实验台面或桌面上，在环境温度下解冻。应遮盖起来以避免产品脱水或被污染。对解冻过程进行监控，当判定解冻结束时，立即对产品进行鉴定或将其转移至冰箱，放入冰箱贮存前须将产品用塑料膜覆盖。应限制贮藏的时间以保持样品的完整性。如果可能的话，样品应放在托盘中解冻，以便于解冻汁液的数量和性质都能得以评价。

如果产品采用适当的防水包装，或者与水的接触不会对产品的感官性质产生重大影响，可以将原料浸入水中加速解冻，必须小心防止其进一步腐败或细菌滋生。较小的样品单位，例如，单体速冻（IQF）鱼片、小包装的虾或贝肉，可以用微波炉的解冻设置进行解冻，但要注意不要采用过高的温度设定，否则产品的某些部分可能会被过度加热。

较大的冻鱼或大块冷冻产品室温下解冻需要数小时，甚至超过一个正常的工作日，因此不能完全监控整个解冻过程。解决方法之一是在第一个工作日结束时开始对产品进行解冻，次日早晨，解冻可完成或基本完成。或者在第一个工作日开始时尽早开始解冻，并在此工作日结束时将产品移至冷藏室，在低温下完成解冻过程。在不损坏原料的条件下，可以在大块产品解冻时将其分成几块，以加快解冻过程。

3.3 蒸煮

对在解冻状态下不能判定气味或凝胶状态的样品，则应从有争议的产品中截取一小部分鱼肉（约200g），使用下述的一种蒸煮方法，快速地确定其气味、口味和凝胶状态。在蒸煮过程中将产品内部温度加热到65～70℃。不能过度蒸煮，蒸煮时间随产品大小和采用的温度而不同。准确的蒸煮时间和条件应依据预先实验来确定。

烘焙：用铝箔包裹产品，并将其均匀放入扁平锅或浅平锅上。

蒸：用铝箔包裹产品，并将其置于带盖容器中沸水之上的金属架上。

袋煮：将产品放入可煮薄膜袋中加以密封，浸入沸水中煮。

微波：将产品放入适于微波加热的容器中，若用塑料袋，应检查确定塑料袋不会发出任何气味。根据设备说明加热。

3.4 产品评价步骤

水产品的标准和规格规定了产品可被评价的特征，以便接收或拒收产品以及对产品进行分级。附件1中的表1列出了可应用于标准和质量分级方案的水产品感官属性。在产品检测过程中为使质量标准的应用前后一致，需要以一致而系统的态度进行感官评价，应根据相关的品种特性进行评价。

评估员应特别注意标准中规定的产品特征，根据这些特征可以确定产品是否符合标准要求，他们也应适当评价和记录其他相关的样品特性。

3.4.1 生鲜产品的评估

生鲜产品，通常根据外观和气味进行评估。在冷藏期间发生变质，会以多种方式在外观上体现出来，仅根据外观将冰藏的水产品进行分级并不困难。附件1中的表1列出了相关的产品特征。

3.4.2 冻品的评估

冻品应于冻结状态进行检验，评估员应注意所有包装材料及冰衣的性质及状态，以及产品的脱色及脱水程度等情况。评估员应注意产品是否出现任何已解冻或重新结冰的迹象，冻块是否有塌陷或扭曲变形的迹象，包装材料中是否有聚集的冻结起来的汁液（注意要与那些可能在冻结前就已在鱼上的水相区别），以及冰衣是否有部分损失。

必要时，可以用相应的未冻产品代替解冻样品进行测试。很难根据解冻后的整鱼鉴定其新鲜度，因为冻结和解冻的过程会改变眼睛、皮肤以及鳃和血液的颜色，即使是在对产品质量没有什么影响的短期冻藏后，鱼鳃也会变得似皮革状或出现轻微的酸败味。

3.4.3 蒸煮样品的评估

除非立即评估，样品应保存在密封容器中，冷却至适宜的品尝温度，保温。已蒸煮过的产品（如熟虾）应略微重新加热。

评估员应注意产品外观变化，记录任何反常的特征。对气味进行评估，记录其特征和强度，特别是任何类似于化学污染的气味。应鼓励评估人员品尝蒸煮过的食品，因为一些化合物（如轻度的腐烂或燃料污染）只能用嘴才可以检测出来。

样品的口味应当能证实根据其气味作出的评估，并提供更多的信息。例如，大部分食品添加剂（如食盐、山梨酸盐、多磷酸盐等）不能通过气味检测出，但通过品尝就可以察觉。只通过感官分析是不能测定食品添加剂的成分和含量的，也不能测定产品中是否使用了禁用的食品添加剂或者食品添加剂的用量是否超标。这些，在必要时应采用适当的化学分析方法予以确认。

4 评估员的培训

4.1 客观感官评价能力的培训

4.1.1 客观感官评价能力培训的考虑因素

本部分提供了用于筛选和训练评估员的测试材料。

感官测试的目标是测试评估人员对于样品进行感官分析的固有能力。为了对水产品进行客观的感官分析，必须挑选有能力履行感官检验工作的评估员，对其进行检验方法的培训，并且监控其能力不断增强以承担感官评价任务。因此，感官培训包括以下内容：

（1）评估员的挑选，要根据其基本的感官敏锐性以及描述、分析感觉的能力（也就是不带个人偏见）。评估员候选人不能对海产品及食品添加剂有过敏性反应。

(2) 随着对检验程序的熟悉、分析能力的提高、在复杂的食品体系中辨别和鉴别样品感官属性的能力的提高以及敏锐度和记忆力的提高，因此他（她）可提供准确的、前后一致的、标准的感官评价，并且结果可重复。

(3) 通过对感官分析结果的频繁的、定期的评估，监督评估员的表现及其分析结果前后的一致性。

4.1.2 评估员候选人的挑选

评估员候选人的要求是：

(1) 没有嗅觉缺失症（无法感觉到气味），可感觉出腐败及其他不好的气味，并用前后一致的方式进行描述；

(2) 没有味觉缺失症（无法感觉到基本味道），可尝出与腐败及其他不好的味道，并能用前后一致的方式进行描述；

(3) 色视觉正常，能够检测出水产品外观的异常；

(4) 感官感觉可信，能够将其正确表达；

(5) 能够学习关于新的或不熟悉的关于感觉（气味、滋味、外观、组织）的术语，并能使用其进行报告；

(6) 能够对感官刺激做出解释，并据此找出产生这种刺激的根本原因。

前五项可以通过测试测定，第六项则需要通过专门的培训和练习来提高。进行试验时，重复基本的味觉及嗅觉试验很有用。必须确保候选人接受的是基本能力的测试，而且不会受陌生试验情况的影响。在每次测试中应选用不同的代码及不同的测试次序。

4.1.2.1 依据味觉的筛选

食品风味变化多样，这就要求候选测试者要去识别并描述食品口味的差异，尤其是识别并描述腐败缺陷带来的异味，这表明其具备了能去识别基本味觉的能力。在挑选和培训测试者时，其区别苦味和酸味的能力尤为重要，因为没有经验的测试者都容易将这两种味道混淆。这两种味道对于水产品的测试十分关键，因为它们常出现于腐败的开始阶段。

许多标准资料中都记述了一个相应的标准试验，采用正常品尝者都可感觉到的浓度。应表明测试中使用的混合物是可以被感觉到的。

表1 公布的筛选和培训测试者的测验用溶液

口味	使用的标准化合物（水中）	DFO 筛选测试（1986－96）	Meilgaard etal.（由弱到强）（1991）	Jellinek（1985）	ASTM（1981）	Vaisey Genser and Moskowitz（1977）
苦	咖啡因	0.06%	0.05%～0.2%	0.02%、0.03%	0.035%、0.07%、0.14%	0.150%
酸	柠檬酸	0.06%	0.05%～2.0%	0.02%、0.03%、0.04%	0.035%、0.07%、0.14%	0.01%
咸	氯化钠	0.02%	0.2%～0.7%	0.08%、0.15%	0.1%、0.2%、0.4%	0.1%
甜	蔗糖	2.0%		0.40%、0.60%	1.0%、2.0%、4.0%	1.0%
鲜*	谷氨酸钠	0.08%				

*尽管还存在争议，但一些专家已经证明鲜味是第五种基本味道。这可以在挑选测试员的过程中使用，但在培训中应明确使用，并阐明鱼类中核（糖核）苷酸是产生这一味道的原因

4.1.2.2 依据嗅觉的筛选

有多种试验方法可供选用以完成此项筛选工作。

因为人们能够感觉到大量的分开的气味，因此应挑选使用那些既能够代表候选者可能知道的普通气味，同时也能够代表水产品变质后产生的气味的样品。附件2介绍了在评价产品气味中适用的方法和样品。

4.1.2.3 正常色视觉的筛选

通过使用眼科标准测试（包括石原氏（Ishihara）色盲测试法、Farnsworth－Munsell 100色测试法）中的一种测定评估员候选人是否色盲。这些试验用品可以通过医疗供应途径购买，同时应当配有完整的使用说明。必须严格按照说明中指定的条件进行操作。

4.1.2.4 组织评估的筛选试验

这适用于鱼产品因组织原因而被拒收的情况。这项测试中，对原料的触摸是必不可少的。需要进行评估的特征包括：

（1）紧实性：对于新鲜的鱼、贝（虾）类；

（2）弹性：对于新鲜的鱼类。

Tilgner设计了一种实验方法（1977年），并在Jellinek作了报告（1985年）。此种测试法是使用一系列紧实程度逐渐增加的样品，允许候选人用其惯用的手的食指按压样品以进行检测，并按紧实程度从低到高的顺序排列样品。这需要评估中具有紧实程度的概念以及感官属性增长幅度的概念。虽然有一些样品也可以由适当的食品样品制得，但实验中所用的样品是由聚氯乙烯浇铸成的永久性的样品。

4.1.3 评估员的培训

推荐使用水产品感官评价评估员培训课程提纲。下面是一个训练提纲的示例：课程中基础感官科学培训的时间长度可由下面给出的10小时（1.5天），直至大学水平培训全部课程的时间长度。推荐采用参与实践练习的方法，每部分在讨论中对概念进行举例说明（例如，在味觉这部分，可准备基本味的溶液给学生品尝）。附件3给出了推荐使用的水产品感官评价评估员的培训课程提纲。

4.1.4 对评估员的考察

通过对评估员做出的感官结论进行考察，以确认感官培训以及感官评估一致性的效果。有多种途径进行考察，可单独使用也可结合起来使用。

（1）首先是使用性质已知的检验样品，在评估员每日的测试训练中进行检验。检验结果会被送回样品总协调处进行分析。这种方法的优点是样品会在真实的实验室条件下进行评估。使用的样品是根据4.2节“样品准备与处理”的步骤进行制备，也可以使用性质已知且数量充足的商品。

（2）另一种确认评估员表现的方法是通过采用真实的鉴定实验和校准程序。这需要在一个中心实验室中进行，实验室要足够大，可以安排下全部参加试验的评估员。样品准备根据4.2节“样品准备与处理”中的步骤进行，也可以使用性质已知且数量充足的商业产品。此方法应定期重复使用，以确保评估员鉴定产品的能力没有发生改变，而且评估员表现必须达到“及格/接受”和“不及格/拒绝”样品中事先规定的要求。

（3）鉴定评估员表现的另一补充方法就是积累这些按时间进行的检查结果，并同其他已知的样品信息相比较，如再次检验结果、消费者投诉、化学分析等。

4.1.5 参考文献

参见附录Ⅱ。

4.2 样品准备及处理

4.2.1 样品类型

需认真考虑的唯一一个重要因素就是用于与水产品相关的感官技能的个人培训的样品。在感官培训中使用正确的样品是绝对必要的。

在感官评估员或检查员的培训中可考虑使用两种样品。

（1）受控腐败样品：这些样品应能够体现并代表全部质量区间，以及与气味、风味、外观、组织有关产品性质的正常范围。

在这些样品的准备中提供质量完美的样品作参考点是必要的。

出现的质量缺陷应当是自然产生的，如果可能的话，应当展示出所用产品的典型的感官特性，如果样品是人工腐败或污染的，在培训中它们既不能表现合格个体的典型感官特性，也不能表现不合格的典型感官特性。

对准备样品的工作人员来说，具备从收获到冷冻的过程中哪些商业加工会对商品造成损害，以及清楚哪些方法和条件会产生损害等方面的知识非常重要。理解腐败的主要途径对准备受控腐败样品非常有用。

在可能的情况下，用于可控腐败的样品应在收获地进行制备，考虑其品种、细菌群落等因素，根据腐败时典型的气味及其他特性，模仿商品试样，重现正常的腐败条件对其进行加工。

（2）商品试样：在任何可能的条件下，应在检测人员的感官培训中使用商品试样。很多时候，质量缺陷（气味、风味、外观、组织等）以及腐败（发霉或腐烂的气味、风味，酸败、石油馏出物等）的商业生产的试样，可以将它们很好的展现出来。这些商业生产的样品可使我们在用“真正”的样品进行培训的过程中对感官评估人员进行考察。这些样品也可以在感官科学中用于衡量与做出正确判断相关的个人的记忆力。

很多时候，受控腐败样品无法表现出所有程度的质量缺陷或腐败，而商业生产的样品可以显示出轻度、中度和重度腐败。

4.2.2 样品包的准备

样品应当用充足的时间进行准备，以便能够使产品产生大量的质量缺陷，同时如果需要的话允许对产品进行加工。

如果可能，应使（整）鱼发生自然腐败，这样可以形成典型的腐败气味。

4.2.2.1 基准

获得所有品种、产品形式的有已知历史记录、没有商业上违规的优质原料是必要的，为讲习班的参与者提供一个始终一致的参考。只要可能，准备受控腐败样品时应包括鲜品和冻品两种产品形式。在一开始这批样品的品质就应是相同的。

准备腐败样品时保存正确的记录是必要的。样品应依据其批次采用一致的代码，其后每一批体现产品长期于环境条件或冰藏条件保存的样品中，所取的每个代码的样品应当前后一致。有必要在腐败的进程中监控温度，防止其波动。

要想得到真实的腐败效果，必须要在合适的温度和环境污染条件下进行。如果一开始时的原料尺寸、质量是一致的，并且在腐败过程中使个体间保持接触，可以将个体间腐败程度的差异减至最低。

鱼类以不同的速率变质，所以应当每隔一定时间对产品进行一次检查，并且在加工之前将性质相似的产品归在一起。在这个阶段常常需要专家对样品进行鉴定。

需要的数量是依据培训目的和检验品种而定的，最少 5 个，最多 8 个。一批中至少有 50% 是合格产品。

4.2.2.2 腐败

一般来说，高温和低温下的分解、腐败都应包括在内，但是，有关品种的知识、标准的加工方法以及在加工过程的哪一点最有可能发生腐败，这些都将决定主要采用哪种腐败发生方法。应当避免方便而走“捷径”，这点很重要。如果要研究预冷却过程中的腐败，就应避免使用冻鱼。必须仔细对温度进行监控。

4.2.2.3 包装和贮藏

确定保质期时，处于腐败过程的应当考虑其品种和产品的类型。

罐装产品在使用前应至少放置30天。应放在14~18℃的凉爽干燥的地方，否则其货架期会缩短。用于培训的罐装海产食品的保质期最长约为2年。超过这个时间，其逐步显示出来的性质可能会影响人的判断，或者会使培训用的样品变得没有意义。

除非准备示范冷藏的损害，生的及预煮的冻品应当适当的裹冰衣以防止脱水和出现冻斑。根据贮藏时间的长短，为确保质量，样品可能需要定期重新进行裹冰衣。如果可能的话，产品应真空包装以保证质量，这对某些鱼类以及预煮样品的贮藏都是非常必要的。

在用于讲习班之前，所有生的、预煮的以及罐装的可控腐败样品都应由具备资格的人进行鉴定。样品应具有化学分析和感官分析的结果，以确定增量样品的品质和一致性。

4.2.3 样品特性

4.2.3.1 感官特征

（1）样品必须具有正常的气味、风味、外观、组织等特性。

（2）如果产品通常呈现出的特性是由产地、饲料气味等原因造成的，可能的话，选择这些产品作受控腐败样品。

（3）呈现出腐败气味或有污染缺点的样品不能太强烈以致压倒参与者的感觉，或在培训中影响到对其他样品的判断。

（4）呈现出轻微至中等的腐败或污染气味的样品，分析难度更大，但能更好的表现出真实状况。

（5）每个增加的样品或每个代码的样品都必须显示出始终如一的或相似的特性，这些特性对于培训是有价值的。

4.2.3.2 化学特征

可靠试样的化学特征在培训中非常有用（见附件3第二部分 课程提纲模型的实践练习）。

（1）选择用于指示腐烂的化学指示剂（CID），这在新鲜产品是用不到的。

（2）选择能够对用于培训的特定产品其重要的腐败路径进行监控的CID。采用能够根据CID区分出合格、略不合格和腐败初期的方法。如果可能最好采用两个CID。

（3）应当在待检验的水产品的各加工形式中（洗过的、煮过的、装罐的或贮藏的）保留CID。

（4）CID的变化应能体现出水产品感官质量的变化。

（5）对于已准备的样品的每个增量，应有足够数量的二次抽样作分析，以衡量样品与增量样品之间的差异程度。这对于那些代表着从合格产品转向腐败初期的增量样品非常重要。

附件 1　感官评价中的水产品属性

性状	特征	标准及其描述
冷却的鱼类		
生鲜的整体，去内脏或未去内脏的	外表面	颜色：明亮、灰暗、变白 黏液：无色、变色
	皮	受损情况：无、穿孔、磨损
	眼睛	外形：凸起、平、凹陷 亮度：清晰、混浊 颜色：正常、变色
	腹腔	内脏（在完整的鱼中）：完整、消化 清洁度（在已去内脏的鱼中）：内脏被彻底去除并已清洗、未彻底去除内脏、未清洗 腹壁：明亮、干净、变色、消化 寄生物：无、有 血液：明亮、红色、棕色
	组织	皮：光滑、砂砾状；肉质：紧密、柔软
	鳃的外观和品质	颜色：亮红或淡红、变白、变色 黏液：光亮、不透明、变色
	鳃的气味	新鲜、独特、中性、微酸、轻微不新鲜、明显腐败、腐臭
生鱼片	外观	半透明、有光泽、自然色、不透明、暗淡、有血迹、变色
	组织	紧密、有弹性、柔软、可塑
	气味	海水、新鲜、中性、酸味、轻微不新鲜、腐败、腐臭
熟鱼片	气味	腐败程度：海水、新鲜、中性、霉味、酸味、腐败
		污染：没有、消毒剂、燃油、化学用品、硫化物
	口味	变质：甜、奶油味、清油味、中性、酸味、氧化味、腐臭、霉味、发酵味、酸败味、苦味
		污染：没有、消毒剂、燃油、很苦、碱性物、多磷酸盐、化学用品
	组织	多汁、紧密、柔软、糊状、凝胶状、干燥
冻结的鱼类		
冻结状态下	外观	冻斑程度：没有、轻微、表面的、大面积的、深度的 颜色：正常、多脂鱼变色（由黄色变成青铜色）
解冻生鱼片	组织	紧密、有弹性、可变形、非常紧密、坚硬、僵硬 液滴：少量、中等、大量 气味、腐败和污染：按照要求冷却的鱼类 冷藏气味：无冷藏气味、强烈、纸板气味、酸败气味
解冻鱼片	气味和口味	变质和污染：按照要求冷却的鱼类 冷藏气味：无冷藏气味、强烈、纸板气味、酸败气味
	组织	紧密、多汁的、坚韧、纤维状、干燥

续表

<table>
<tr><th>性状</th><th>特征</th><th>标准及其描述</th></tr>
<tr><td colspan="3">冷却的虾类</td></tr>
<tr><td rowspan="3">生鲜的</td><td>外观(带壳)</td><td>色泽鲜明、头部略黑、头和身体呈黑色</td></tr>
<tr><td>外观(去壳)</td><td>半透明、全白或浅灰、轻微变黑、严重变黑、非常透明、黏滑的、从带头的产品中取出的尾部肉的一端略变成淡黄色</td></tr>
<tr><td>气味</td><td>新鲜、海中的、霉味、氨气味、酸味、腐败、腐臭</td></tr>
<tr><td rowspan="4">蒸煮虾肉</td><td>外观</td><td>白色、不透明、有黑点、背部严重变黑、轻微半透明</td></tr>
<tr><td>气味</td><td>新鲜、煮沸牛奶味、霉味、氨气味、酸败味、酸味、腐败</td></tr>
<tr><td>口味</td><td>味美、乳酪味、中性、霉味、酸味、苦味、腐败味</td></tr>
<tr><td>组织</td><td>紧密的、有弹性的、柔软的、糊状的</td></tr>
<tr><td colspan="3">冻结的虾类</td></tr>
<tr><td colspan="3">对于冷冻虾类的分级标准及其详细说明与冷冻鱼类的分级是同样重要的</td></tr>
<tr><td colspan="3">生鲜或冷冻的头足类动物</td></tr>
<tr><td rowspan="4"></td><td>颜色</td><td>皮：明亮、灰暗、变白
肉：珍珠白、石灰色、粉色或浅黄色</td></tr>
<tr><td>黏着度</td><td>与肉黏着、易于与肉分离</td></tr>
<tr><td>组织</td><td>肉：非常紧密、紧密、轻微柔软
触须：较难撕开的、易于撕开的</td></tr>
<tr><td>气味</td><td>新鲜的、海藻味的、轻微或没有味、酸的</td></tr>
</table>

附件 2　依据嗅觉筛选评估人员的测试方法

一、以下是加拿大所用的一组样品

1. 罐装鲑鱼（鱼类）
2. 罐装沙丁鱼（鱼类/熏制）
3. 酵母（酵母的生长）
4. 咖啡（普通产品 - 为了阐明方法）
5. 橙子和菠萝（水果的气味）
6. 黄瓜和芦笋（蔬菜的气味）
7. 醋、桂皮、胡椒粉和丁香（易于区分的辛辣味）
8. 香草（甜香味）
9. 精制芥末（强醋成分，显示认知混合物中气味能力）
10. 丙酮，外用酒精（污染物，溶剂）
11. 石油产品（燃料油）
12. 老植物油（变质的油）

在本测试中，要求候选者在不通过视觉观察的情况下，只通过认知样品的气味鉴别样品。候选者需鉴别并论述样品。记录正确鉴别的数目。在此步骤中，准许候选者再次检测任何一个样品。此测试在 2 小时或 4 小时（在这期间可以进行其他测试或面试）后重新进行一次，并记录正确鉴别的数目。测试成绩的提高（除非在第一次测试中全部正确，否则不能只测一次）显示了候选者学习新术语并应用其描述感觉的能力。

二、宾夕法尼亚大学气味鉴别测试

嗅觉评估的标准测试已经被 Sensonics 联合公司（08033 美国，新泽西，HADDONFIELD，HADDON 大街 155 号）所采用。

附件 3　推荐使用的水产品感官评价评估员的培训课程提纲

一、课堂学习

第一部分：感官评价的理论原则和实验室实践（10 小时）

A　基础感官测试原则：

1　情感或主观测试（测试形式、获取信息、收集数据资料、回答形式和数字、从信息中判断可能性）；

2　分析或客观测试（测试形式、获取信息、收集数据资料、回答形式和数字，从信息中判断可能性）；

(1) 判断力测试：获取或未获取的信息的形式；

(2) 描述性测试：定性的和定量的。

3　水产品评估人员或产品专家的感官测试准则。

B　对水产品感官特性的感性认识和感觉的作用：

1　生理学感觉：视觉、嗅觉、味觉、触觉和听觉；

2　感官特性的认识：外表或颜色、气味、口味、组织；

3　感觉的相互作用。

C　样品评价技术：

1　气味评价技术；

2　风味评估技术；

3　组织评价（紧实度和弹性）；

4　海产品样品的特别技术。

D　感官评估中的心理物理学基础：

1　入门：察觉和认识；

2　强度：感性认识特性力度的对数性质；

3　饱和度：对现象的解释说明。

E　影响感官评价的因素：

1　生理学作用：混合，掩饰，遗留，增强和抑制；

2　心理学作用：期待、刺激、光环、秩序、接近、鼓舞、合乎逻辑的、建议、对比、集中和中央趋向；

3　对生理学和心理学作用的控制。

F　基础数据资料收集和分析：

1　判别方法：三选题（给定三个选项或平衡设计），二对三，五选二，作配对对比：

(1) 随机选取和设计形式；

(2) 信息分析。

2　描述方法，包括气味轮廓，组织轮廓，光谱，QDA：

(1) 规模：按种类、系列、数量评价；

(2) 可选信息和设计形式；

(3) 信息分析。

3　质量控制的感官方法（见总论）。

G　术语学和有关标准的应用。分析员应当懂得可以帮助提高长期感官记忆的和作为交流手段的感官描述的重要性（见附件 1）。

1　术语学的发展（包括国际已知术语认可过程的原始资料）；

2　定义的重要性；

3　相关标准的运用；

4　总览水产品质量方面的术语，特别注意与低分解水平相关的内容。

H　样品的处理和准备：

1　性状和编码；

2　样品的随机性，使用的目的和条件；
3　样品的同质性和使用温度；
4　样品的尺寸和数量。

第二部分：水产品品质下降（3 小时）

A　水产品的成分：
1　主要成分：蛋白质、脂肪、糖类、水；
2　少量成分：非蛋白质含氮物、矿物质、维生素。

B　质量下降的过程：
1　蛋白质、脂肪、非蛋白质含氮物质以及在某些“种”的水产品中糖类物质的含量下降；
2　微生物引起的腐败；
3　有关各类型腐败过程的术语。

C　水产品质量的化学指示物及其与感官信息的相互关系。

第三部分：污染和污渍（1 小时）

A　类型：
1　自然发生（泥土臭味）；
2　人为造成（石油，纸浆和制纸污水，其他生产污水）。

B　风味的机制和气味的改变。

C　污染和污渍的测试方法（特别注意）。

二、实践练习

第一部分：水产品术语的介绍、明确定义和涉及术语示例的参考资料（2 小时）

第二部分：腐败和分解（18 小时）

这一部分的课程提供了第一手的经验。建议每次只对一种鱼进行评估。

这一部分包括对整鱼、鱼片、罐装鱼、熏鱼以及其他类型的产品的评估。需要时（例如，用油装填的鱼罐头的包装可能掩盖了气味）受培训者应该如对气味评估一样对口味进行评估。

建议按照下列次序安排以下三部分的课程，总共需要大约 4 个小时。建议在测试受培训者进行下一项练习之前，测试其正确鉴定样品质量的能力，从而评估该项培训的成效：

（1）示范部分：由一位资深专家对已知质量的样品进行一组示范。依据感官结果、描述任何适用的化学迹象方面的资料，使用带有标签的样品按照从最高到最低的顺序表示出一个完整的质量等级。

（2）讨论部分：随机选择没有标签的样品，单独进行鉴定，并对其鉴定结果进行讨论。

（3）测试部分：单独对没有标签的样品进行评估，并与专家的结果进行比较。

资料收集及资料分析方面的内容、附带样品的详细讨论，要反馈给受培训者。

第三部分：冻藏水产品的品质下降（4 小时）

A　展示水产品外观、气味、风味和组织等方面由于冻藏而导致的不同程度的质量缺陷；

B　包括低脂肪和高脂肪水产品样品；

C　包括相关术语、定义以及氧化过程和组织变化方面的资料。

第四部分：罐装水产品的品质下降（4 小时）

除第二部分的内容外，还应包括加工前及加工后品质下降方面的信息。

第五部分：其他缺陷（2 小时）

A　由于样品使用了穿刺而导致的缺点的探测（只通过气味进行）；

B　展示视觉缺陷。

附录 I　水产品感官分析中部分相关术语的定义

术语	定义
外观 (Appearance)	物质/样品的所有可见的特征
检验员/评估员 (Analyst/Assessor)	参与感官测试的人员
舱底污水味 (Bilgy)	与厌氧细菌生长有关的气味，用舱底污水味表示这种恶臭气味。术语“舱底污水味”可用来描述被船甲板上的污水污染了的任何质量程度的水产品。舱底污水通常是盐、水、燃料以及废水的混合物
苦味 (Bitter)	四种基本味道之一，开始由舌根部感觉到，与咖啡因和奎宁味道相同，其感觉通常可持续一段时间（2～4 秒）
海水咸味 (Briny)	与清洁的海藻和海洋空气有关的气味
白垩的 (Chalky)	组织：由口感干燥的细小颗粒组成； 外观：干燥、不透明、白垩形
黄瓜味 (Cucumber)	某些品种的鱼类，在其非常新鲜时其气味与新鲜的黄瓜的气味类似
分解 (Decompose)	分解成许多小的组成成分
腐烂的 (Decomposed)	鱼表现出与变质有关的令人不快的或令人厌恶的气味、口味、颜色、组织或其他特征
明显的 (Distinct)	容易被察觉的
过度喂养 (Feedy)	这个术语用来描述鱼被过度喂养的情况。在鱼死后，胃分解酶首先侵蚀其内脏器官，然后是腹壁，再接下来是肌肉组织。如果胃分解酶渗透到鱼肉，则可能造成变质，生成二甲基化合物（DMS），可能由于食物链作用引起某种动物性浮游生物的污染。“过度喂养”水产品的气味被描述成类似于某些烹调后的含硫蔬菜（如椰菜、花椰菜、芜青及卷心菜）的气味
排泄物味 (Fecal)	与排泄物有关的气味
紧密 (Firm)	入口或触摸时有中等程度阻力的物质
鱼类 (Fish)	通常指所有水生冷血脊椎动物，包括鱼类、软骨鱼类及圆口纲脊椎动物。水生哺乳动物、无脊椎动物和两栖动物则不包括在其中
鱼腥味 (Fishy)	与年老的鱼有关的气味，类似于三甲胺（TMA）或鳕鱼肝油的气味。对于有些品种的鱼，这种味道的出现就可以作为其变质的迹象，有些则不可以
风味 (Flavor)	可以导致对味觉、嗅觉、视觉和触觉产生刺激的食品特性，通常有暖的、冷的以及微痛等
新鲜度 (Freshness)	由消费者、加工者、使用者或管理机构限定的与水产品的时间、加工过程以及特性相关的概念
水果味 (Fruity)	轻度发酵的水果的气味。此术语用来描述高温分解造成的气味。例：罐装菠萝
味重的 (Gamey)	指某些品种（如鲭鱼），其口味和（或）气味浓烈的特性。其类似于将鲜鸭肉与鲜鸡肉进行比较的情况
有光泽的 (Glossy)	有光泽的表面会使光线在其表面以 45°角反射

续表

术语	定义
多粒的 (Grainy)	某些分析员可以察觉的明显含有具有中等硬度的颗粒的产品。有时会在水产品罐头中发现
强度 (Intensity)	感官感觉的强烈程度
彩虹色的 (Iridescent)	像彩虹一样多种色彩排列在一起。类似于蛋白石或水上的油膜发出的光彩
掩饰 (Masking)	一种感官感觉遮蔽住其他一种或几中感官感觉，使之不易被察觉的现象
粗粉状 (Mealy)	用来描述口感像淀粉一样的产品
金属味 (Metallic)	指与硫酸亚铁或马口铁罐头有关的口味或气味
潮湿的 (Moist)	对产品释放出的水分的感觉。此感觉可来自于水或油
霉味 (Mouldy)	与霉变奶酪或面包有关的气味
糊口 (Mouth coating)	在口中覆以薄膜的感觉
满口 (Mouth filling)	味道分散到满口的感觉，如味精刺激的鲜味
软烂的 (Mushy)	柔软的、浓稠的、果肉状的硬度。对水产品进行触摸或用口加力可感觉到没有或只有很少的肌肉组织
有霉味的 (Musty)	发霉的潮湿地窖的气味。产品也可能有霉味
嗅觉 (Odour)	嗅觉神经通过鼻腔对易挥发物质的接受而产生刺激所得到的感觉
变味 (Off odour)	与有一定风味的产品变质有关的一种典型特征
不透明 (Opaque)	用来描述不透光的产品。在生鲜水产品的肌肉组织中，这常是由于蛋白质因其酸度下降而失去其反射光线的性质而造成的
糊状、膏状 (Pasty)	食品在口中和唾液混合后形成黏性的糊状或膏状物，会黏附在手指或口腔软组织的表面上
保持 (Persistent)	无重大变化地存在着，而不是短时间就消失的
辛辣的 (Pungent)	一种刺激的、强烈的或穿透性的感官感觉
腐臭 (Putrid)	与腐败的肉类有关的气味
质量 (Quality)	品质优秀的程度。赋予产品满足规定的和暗指的需求能力的特征的集合
酸败味的 (Rancid)	与油脂酸败有关的气味或风味。可能带来糊口感或使舌根部产生刺痛感。有时用“强烈的”或“过度的”来形容
参考 (Reference)	既可以是用来与其他产品进行比较的一个样品，又可以是与产品类型不同、原料相同，用于展示原料特征和属性的样品
腐烂蔬菜味 (Rotting vegetable)	腐烂蔬菜的气味，特别是含硫蔬菜，如熟椰菜，花椰菜，卷心菜

续表

术语	定义
有弹力的 (Rubbery)	在压力作用下产生变形，但去除压力时则恢复其原状的有回弹力的物质
咸味 (Salty)	可用舌品尝出的，与钠盐有关的口味
感觉的 (Sensory)	与感官的作用有关的
黏糊的 (Slimy)	黏性的、光滑的、有弹性的、胶黏的或果冻状的一种流体
酸味 (Sour)	通常缘于含有有机酸而产生的一种气味或口味
不新鲜 (Stale)	与湿纸板或冻藏有关的气味，产品的口味也可能有不新鲜味
STP (STP)	三聚磷酸钠，会产生一种似肥皂的、碱性的口感
甜味 (Sweet)	可用舌品尝出的，与糖有关的口味
味觉 (Taste)	感觉之一，其感受器在口中，并易被溶解成分激活。味觉包括甜、咸、酸、苦，有时还包括鲜味
术语学 (Terminology)	用来描述产品感觉属性的术语
半透明的 (Translucent)	用来描述可允许部分光线穿透，但通过其不能产生清晰影像的物体
透明的 (Transparent)	用来描述可允许光线穿透，并且通过其可以产生清晰影像的清洁的物体
鲜味 (Umami)	如谷氨酸一钠（味精 MSG）溶液等物质产生的口味。一种肉味的、味美的、满口的感觉
西瓜味 (Watermelon)	新鲜西瓜被切开时发出的气味。有时可在某些品种的非常新鲜的鱼类中发现相似的气味
发酵酵母 (Yeasty fermented)	与酵母及其他发酵产品（如面包或啤酒）有关的气味

附录 II

参考文献

ASTM Atlas of odour character profiles, publication DS 61, PCN 05 – 061000 – 36. Complied by Andrew Dravnieks.

ASTM Committee E – 18, 235, draft of terminology document.

ASTM Aroma and Flavor Lexicon for Sensory Evaluation DS 66. G. V Civille and B. G. Lyon, eds.

ASTM Committee E – 18 on Sensory Evaluation of Materials and Products, 1981. STP 758 – Guidelines for the Selection and Training of Sensory Panel Members. ASTM Committee E – 18 on Sensory Evaluation of Materials and Products, Terminology Committee, (date?). Draft definition for《Expert》and《Expert Assessor》.

Cardello, A. 1993. Sensory methodology for the classification of fish according to edibility characteristics. Lebensmittel – Wissenschaft – und – Technologie 16. 190 – 194.

Department of Fisheries and Oceans, Canada. Code of practice for fishery products.

Department of Fisheries and Oceans, Canada. Regulations respecting the inspection of processed fish and processing establishments.

Department of Fisheries and Oceans, Canada, Inspection Branch. 1986 to 1995. Notes from Sensory Methods in Fish Inspection – Sensory Training course given by the National Centre for Sensory Science, Inspection Branch, Department of Fisheries and Oceans, Canada.

Howgate, Peter 1992. Codex review on inspection procedures for the sensory evaluation of fish and shellfish. CX/FFP 92/14.

IFST – International Institute of Food Science and Technology. Sensory Quality Control: Practical Approaches in Food and Drink Production. Proceedings of a joint symposium at the U. of Aston, 6 – 7 – January, 1977. Session II, Measurement of Fish Freshness by an Objective Sensory Method. P. Howgate, p. 41.

ISO 5492 (1983) Sensory analysis – vocabulary.

ISO 8586 – 2 Sensory Analysis – General guidance for the selection, training and monitoring of assessors – Part 2. Experts.

Jellined, G. 1985. Sensory Evaluation of Food – Theory and Practice. Ellis Harwood, Ltd., Chichester, England.

Johnsen, et al., 1987. A lexicon of pond – raised catfish flavor descriptors. J. Sensory Studies 4, 189 – 199.

Laverty, 1991. Torry Taste Panels. In Nutrition and Food Science, Vol 129 No. 2 – 4. Includes terminology based on odour of gills in raw, iced cod.

Learson, Robert 1994, personal correspondence. NOAA/NMFS Research Laboratory, Gloucester, MA.

Multilingual guide to EC freshness grades for fishery products. Torry research station, Aberdeen, Scotland and the West European Fish Technologists Association (WEFTA). Compiled and edited by P. Howgate, A. Johnston, and K. J. White.

NOAA Handbook 25, part 1, Inspection.

NOAA/NMFS, Technical Services Unit.

Kramer and Liston, (eds) Seafood Quality Determination. Proceedings of the International Symposium on Seafood Quality Determination, Coordinated by the University of Alaska Sea Grant College Program, Anchorage. Alaska, 10 – 14 November 1986.

Learson and Ronsivalli, (1969), A new approach for evaluating the quality of fishery products.

Meilgaard, M., Civille, G. V., and. Larmond, E. 1991. Sensory Evaluation Techniques. CRC Press, Inc., Boca Raton, FL.

Poste, L., Mackie, D., Butler, G. and. Larmond, E. 1991. Laboratory Methods for Sensory Analysis of Food. Agriculture Canada Research Branch.

Prell and Sawyer, (1988). Flavor Profiles of 17 Species of North Atlantic Fish J. Food Science, 53, 1036 – 1042.

Prell and Sawyer (1988). Consumer evaluation of the Sensory Properties of Fish J. of Food Science 53. 12 – 28, 24.

Sawyer et al. , (1988) Consumer evaluation of the sensory properties of fish. J. OF food Science, Vol, 53. No. 1.

Sawyer, F. M. et al. 1981. A comparison of flavor and texture characteristics of selected underutilized species of North Atlantic fish and certain treatment of fish. International Institute of Refrigeration. Paris, France. P. 505.

Shewan et al. , (1953), The development of a numerical scoring system for the sensory assessment of the spoilage of wet white fish stored in ice. F. Sci. Food Agric. , 4 June.

Soldberg, et al. (1986), Sensory profiling of cooked, peeled and individually frozen shrimp. In Seafood Quality Determination, Elsevier Science Publishers.

Vaisey Genser, M. and Moskowitz, H. R. 1977. Sensory Response to Food. Forster Publishing Ltd. , Zurich, Switzerland.

Wilhelm, Kurt, 1994, personal correspondence. NOAA/NMFS Research Laboratory, GLOUCESTER, MA.

CODEX GUIDELINES FOR THE SENSORY EVALUATION OF FISH AND SHELLFISH IN LABORATORIES

(CAC/GL 31 - 1999)

1 SCOPE AND PURPOSE OF THE GUIDELINES

The guidelines are intended to be used by analysts who need to apply sensory methods when using criteria based on sensory attributes of the products. Although the guidelines have been written with the Codex requirements in mind they include some provisions for products not covered by these standards but where sensory evaluation is used in the testing of fishery products for conformity requirements. [1] These guidelines are to be used for sensory examination of samples in a laboratory to determine defects by procedures, including cooking, which are not normally done by analysts in the field. Technical information is provided on the laboratory facilities used for such analyses and training of analysts.

The objective of guidelines is to ensure uniformity of application of standards by making recommendations for inspection purposes concerning the facilities required in sensory testing and the procedures for carrying out sensory tests.

For the purpose of this document the use of fish means finfish, crustaceans, and molluscs.

2 FACILITIES FOR SENSORY EVALUATION

2.1 GENERAL OBSERVATIONS

Sensory evaluation should be carried out by adequately trained personnel (see Section 4). They evaluate a specialized range of products, and use one sensory methodology.

2.2 LABORATORIES FOR SENSORY EVALUATION

2.2.1 Location and Layout

Figure 1 illustrates a plan of a laboratory that would be suitable for use for examining fishery products. The plan illustrates the principle that the preparation area should be separate from the evaluation area.

Office accommodation, storage rooms, staff facilities, and possibly other test facilities should be provided elsewhere in the premises. The evaluation area must not be used for chemical and microbiological analyses. however, some types of analyses could be done in the preparation area.

2.2.2 Preparation Area

This area is to used for the handling and storage of fishery products, and for the preparation of samples

---

[1] Additional criteria may be included if new recommendations are made by the Committee.

for sensory evaluation. It should be constructed so as to comply with the requirements of good manufacture practices for the design and construction of fishery establishments. The rooms should be designed to ensure cooking odours do not interfere with sensory analysis.

2. 2. 3 Evaluation Area

There should be no preparation of products in this area other than final trimming of samples prior to cooking.

The area, ventilation, procedures and sample sequence should be organized to minimize disturbing sensory stimuli. Also influence and disturbances from fellow evaluators and other personnel should be minimized. The colour of evaluation area should be neutral.

The working surfaces should be illuminated by daylight or artificial daylight. Any specific conditions in standards should be met.

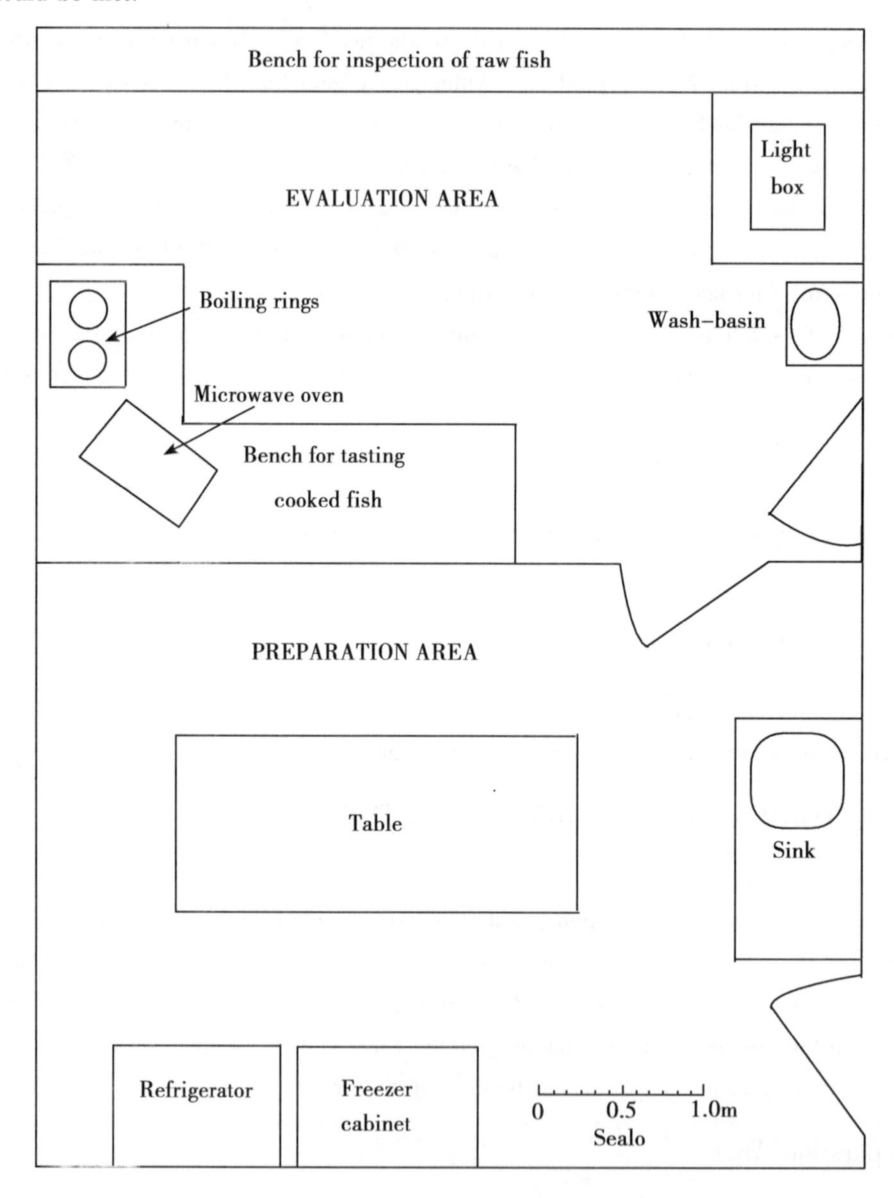

Figure 1 Illustrative Plan of a Laboratory for Sensory Enaluation of fishery Products

2.2.4 Equipment

The exact type and amount of equipment required will depend to some extent on the nature of products to be inspected and the volume and frequency of the examinations.

3 PROCEDURES FOR SENSORY EVALUATION

3.1 COLLECTING AND TRANSPORTING SAMPLES

In most circumstances where fishery products are subjected to sensory evaluation a decision is made about a batch of fish, for example, acceptance or rejection of a consignment of imported products, classification of batches of fish on a market into freshness grades. The decision is made on the basis of an examination of a sample drawn from the batch according to guidelines which will usually specify how a sample is to be taken for the intended regulatory or commercial purpose of the examination.

When collecting a sample for inspection the inspector should ensure that the procedures used for taking the sample, and the subsequent handling of the sample, do not materially affect its sensory properties.

The inspector should check that the sample is properly packed and where necessary, under temperature control before dispatching it to the inspection laboratory. If the sample is not under the supervision of officials during transport the inspector should ensure that sample can not be tampered with during the journey.

On receipt at the inspection laboratory, samples, if not evaluated immediately, should be stored under appropriate conditions. However, fresh and chilled products should be examined on the day they are received. Products in either chill or frozen storage should be appropriately wrapped to prevent drying out or desiccation.

3.2 PREPARATION OF SAMPLES FOR EXAMINATION

Table 1 in Annex 1presents attributes useful in evaluating some species and products. Procedures for preparation of samples should be appropriate for the product types. Some procedures relative to fresh or frozen finfish are described in the following paragraphs.

The fish, if entire, should be gutted and the guts retained. The head should be removed, and the fillet from one side to taken off. The portions should be assembled on tray for analysis.

QF Products can be laid out on the examination bench in the evaluation area, but it is often more convenient for presentation and for clearing up after if sample units are presented on trays.

Frozen products should be first examined in the frozen state. The complete sample unit or portions of the unit should then be thawed for sensory evaluation. Whether the sample can, or should be subdivided, depends on the nature of the products. Packs of IQF shrimps or fillets can be opened and subsamples taken. Portions could be sawn off large fish or off blocks, but this might be difficult in the case of thick material unless a bandsaw is available.

Frozen material should be thawed out as quickly as possible, but without raising the temperature of all or part of the product so that it might spoil. The simplest procedure is to spread out the sample units on the benches and tables in the preparation area and leave them to thaw at ambient temperature. They should be covered to prevent drying and contamination. The progress of thawing should be monitored and when it is judged that thawing is complete the products should be evaluated, or transferred to a refrigerator. Products should be covered with plastic film before storing in the refrigerator. Storage should be limited in order to maintain sample integrity. If possible sample units should be thawed out on trays so that the amount and nature of the thaw drip can be assessed.

Thawing can be accelerated by immersion of the material in water. This is acceptable if product is protected from the contact with water by suitable wrappings, or if contact with water does not materially affect the sensory properties of the product. Care must be taken to prevent further spoilage or bacterial growth. Small sample units such as IQF fillets or small packs of shrimps or shellfish meats could be thawed in a microwave cooker on the defrost setting, but care must be taken not to use too high power settings otherwise parts of the material will be overheated.

Large frozen fish or large blocks of frozen products will take many hours to thaw out at ambient temperature, longer than a normal working day, and they can not be properly monitored throughout the whole process of thawing. One solution is to lay the products out for thawing at the end of a working day when they will just be completely, or almost completely, thawed by the following morning. Alternatively the material can be put out to thaw as early as possible in the day and transferred to a chill room at the end of the day to complete the process at low temperature. It is helpful to break apart blocks of product when they are partially thawed to accelerate thawing if this can be done without damaging the material.

3.3 COOKING

In cases where a final decision on odour or gelatinous state cannot be made in the thawed uncooked state, a small portion of the disputed material (approximately 200g) is sectioned from the sample unit and the odour and flavour or gelatinous condition confirmed by cooking without delay by one of the following cooking methods. The following procedures are based on heating the product to an internal temperature of 65 ~ 70℃. The product must not be overcooked. Cooking times vary according to the size of the product and the temperatures used. The exact times and conditions of cooking for the products should be determined by prior experimentation.

Baking Procedure: Wrap the product in aluminum foil and place it evenly on a flat cookie sheet or shallow flat pan.

Steaming Procedure: Wrap the product in aluminum foil and place it on a wire rack suspended over boiling water in a covered container.

Boil-in-Bag Procedure: Place the product in a boilable film-type pouch and seal. Immerse the pouch in boiling water and cook.

Microwave Procedure: Enclose the product in a container suitable for microwave cooking. If plastic bags are used, check to ensure that no odour is imparted from the plastic bags. Cook according to equipment instructions. [1]

3.4 PROCEDURES FOR THE ASSESSMENT OF PRODUCTS

Standards and specifications for fishery products will specify the features of the product that are able to be evaluated, and the criteria for accepting or rejecting products or for allocating them to grades. Table 1 presented in Annex I lists sensory attributes and criteria which may apply to standards and quality grading schemes. In order to apply quality criteria consistently in the inspection of products it is necessary to con-

[1] General Standard for Quick Frozen Fish Fillets, Annex A "Sensory and Physical Examination".

duct the sensory assessments in a consistent and systematic manner. Samples should be assessed relative to the characteristics of the species concerned.

Assessors must pay particular attention to those features of the product which are referred to in any standards and which determinate conformance to the standard, but in addition they should assess and record other relevant attributes of the samples, as appropriate.

3.4.1 Assessment of Raw Products

Fresh fish will normally be assessed by appearance and odour. Fish change in appearance in a number of ways during spoilage in ice and it is not usually difficult to accurately grade iced fish by appearance alone. Characteristics to look for are listed in Table 1 of Annex I.

3.4.2 Assessment of Frozen Products

Frozen fish should be examined in the frozen state. The assessor should note the nature and state of any wrappings and glazes and the product should be examined for any discolourations and for the extent and depth of any dehydration. The assessor should note if there are any signs that product might have been thawed and refrozen. Signs of slumping or distortion of blocks, the collection of frozen drip in pockets in the wrappings, (not to be confused with water that might have been present on the fish at the time of freezing), and the partial loss of glaze.

Thawed samples should be presented and examined as for the corresponding unfrozen product where appropriate. It is not easy to evaluate the freshness of thawed whole fish by appearance because the freezing and thawing processes alter characteristics like the eyes, skin and colour of gills and blood. The gills have a leathery or slightly rancid odour even after short periods of frozen storage which have no significance for the quality of the product.

3.4.3 Assessment of Cooked Samples

Cooked samples should be held in a closed container, allowed to cool to a comfortable tasting temperature, and kept warm unless they are assessed immediately. Products which have already been cooked, for example cooked shrimps, should be warmed up slightly.

The assessor should note the appearance of the product and record any unusual features. The odour should be assessed and its character and strength recorded, particularly any unusual odours like chemical taints. Assessors should be encouraged to taste cooked samples as some compounds can only be detected by mouth (e.g. low levels of decomposition or fuel contamination).

The flavor of a sample in the mouth should confirm the assessment based on odour, but can give additional information. For example most additives such as salt, sorbates, polyphosphates, are not detectable by odour, but are detectable by taste. Sensory analysis alone should not be used to determine the presence of additives and any suspicion that non permitted additives have been used, or that excess amounts of permitted additives are present, should be confirmed by chemical analysis where appropriate.

4 TRAINING OF ASSESSORS

4.1 OBJECTIVE SENSORY TRAINING

4.1.1 Considerations for Objective Sensory Training

In the sections below examples are provided of test materials which have been used for screening and training analysts.

Objective sensory testing measures the intrinsic sensory attributes of a sample through the analytic sensory perceptions of human assessors. In order to conduct objective sensory analyses of fish and fish products, assessors must be selected for their ability to perform the sensory tasks required, must be trained in the application of the required test methods, and must be monitored for their ongoing ability to perform the sensory tasks. Thus, sensory training includes:

(a) The selection of assessors for basic sensory acuity and for the ability to describe perceptions analytically i. e. without the effect of personal bias. Allergies to seafood or to some food additives could eliminate an analyst candidate.

(b) The development of the analytical capability by familiarization with test procedures, improvement of ability to recognize and identify sensory attributes in complex food systems, and improvement of sensitivity and memory so that he/she can provide precise, consistent, and standardized sensory measurements which can be reproduced.

(c) The monitoring of the assessor's performance and the consistency of their analytic decisions by the frequent periodic assessment of the sensory decisions.

4.1.2 Selection of Candidate Assessors

A candidate for assessor training should demonstrate that he/she:

1. is not anosmic (unable to perceive odours) – so that odours of decomposition and other defects will be perceived and described in a consistent manner;

2. is not ageusic (unable to perceive basic tastes) – so that tastes associated with decomposition and other defects will be perceived and described in a consistent manner;

3. has normal colour vision and is able to detect anomalies in the appearance of fish and fish products in a consistent manner;

4. is able to rely on sensory perceptions and to report them appropriately;

5. is able to learn terminology for new or unfamiliar perceptions (odours, tastes, appearances, textures) and to report them subsequently; and

6. is able to define sensory stimuli and relate them to an underlying cause in the product.

The first five points can be measured in testing, the last ability is developed during specific product training.

In conducting the tests, it is useful to allow for repetition of the tests for basic taste and for odour perception. This is necessary to ensure that the candidate is being tested for basic ability and not responding to an unfamiliar testing situation. New code numbers and presentation sequences are used in each test method.

4.1.2.1 Screening for Perception of Basic Tastes

The diversity of flavours, especially of defects from decomposition, which the inspector will be required to perceive and describe make it essential that some indication of the general ability to perceive basic tastes be established. One area of particular importance in selection and training is the ability to discriminate bitter and sour tastes/flavours as this is a common area of confusion in inexperienced assessors. These tastes/flavours are critical in the examination of fish and fish products as they are evident in the early stages of decomposition.

A matching standards test using concentrations which should be perceived by a normal taster has been described by several standard sources. The concentrations used have been shown in testing to be perceptible.

Table 1 A selection of published test solutions used for screening an. d training analysts

Basic Tastes	Standard Compounds Used (in water)	DFO Screening Tests (1986 - 96)	Meilgaard et (slight to very strong) (1991)	Jellinek (1985)	ASTM (1981)	Vaisey Genset and Moskowitzb (1977)
Bitter	caffeine	0.06%	0.05 to 0.2%	0.02 & 0.03%	0.035, 0.07& 0.14%	0.150%
Sour	citric acid	0.06%	0.05 to 0.2%	0.02, 0.03 & 0.04%	0.035, 0.07 & 0.14%	0.01%
Salt	sodium chloride	0.02%	0.2 to 0.7%	0.08 & 0.15%	0.1, 0.2%, 0.4	0.1%
Sweet	sucrose	2.0%	2.0 to	0.40 & 0.60%	1.0, 2.0 & 4.0%	1.0%
umami *	monosodium glutamate	0.08%				

* This has been identified by some analysts as being a fifth basic taste, however, this remains controversial. This may be used as part of the selection procedure, but should definitely be used as part of the training sessions to illustrate the contribution to the flavours of fish contributed by the ribonucleotides.

4.1.2.2 Screening for Perception of Odours

In this case, several types of tests are available which will accomplish the selection procedure.

Because people are able to perceive a very large number of separate odour qualities, the samples used should be chosen to be both representative of common odours with which the candidate would likely have had experience, and also be representative of odour qualities which occur as defects in fish and fish products. Two examples two test methods which would be appropriate for use in assessing odour perception are presented in Annex II.

4.1.2.3 Screening of Normal Colour Perception

Colour blindness is measured by the use of one of several standard ophthalmologic tests including the Ishihara Colour Blindness Test and the Farnsworth - Munsell 100 - Hue Test. These tests may be purchased through medical supply sources and should come with complete instructions as to their use. They must be administered under the exact conditions specified in the instructions.

4.1.2.4 Screening Test for the Assessment of Texture

There can be cases when fish is rejected for texture. These are tests which are essentially done by touch on raw product. Characteristics which may be assessed include:

(a) firmness: in fresh fish and shellfish (shrimp); and

(b) springiness: in fresh fish.

One such test is the procedure designed by Tilgner (1977) and reported in Jellinek (1985). This test used a series of samples which increase slightly in firmness and uses pressure with the forefinger of the dominant hand to assess firmness and allow the candidate to rank the samples from least to most firm. This allows the assessment of the concept of firmness and the concept of increasing intensity in a sensory attribute. The samples used in the test described are permanent samples cast from polyvinyl chloride although a series of samples can also be generated from appropriate food samples.

4.1.3 Training of Assessors

A *Suggested Syllabus for a Training Course for Assessors in the Sensory Assessment of Fish and Fishery*

Products The following is a model training syllabus. The length of the basic sensory science training which is included in the course can vary from the 10 hours (1.5 days) shown below to full length courses of university level training. It is suggested that hands – on exercises accompany each section to demonstrate the concept under discussion (e. g. prepare basic taste solutions and have the students taste them during the lecture on taste). A Suggested Syllabus for a Training Course for Assessors in the Sensory Assessment of Fish and Fishery Products is presented in Annex Ⅲ.

4.1.4 Monitoring of Assessors

The validation of the effectiveness of sensory training and of the consistency of sensory assessments is achieved through ongoing monitoring of the sensory decisions made by the assessor. This may be accomplished in a variety of ways, either singly or in combination.

(a) The first is the use of check samples which are samples of known quality which are distributed to inspectors for examination in their day – to – day testing facility. The results are sent back to the central coordinator of the samples for analysis. The advantage of this method is that samples are being assessed under the actual laboratory conditions. Samples used for this are prepared using the procedures described in Section 4.2, Preparation and Handling of Samples. Also commercial product of known quality and which is available in sufficient quantity may be used.

(b) Another procedure which is used to validate the performance of an inspector is through actual accreditation testing and calibration procedures. These are conducted in a central location laboratory which is large enough to accommodate all of the inspectors participating in the test. Samples are prepared using the procedures described in Section 4.2 Preparation and Handling of Samples. Also commercial product of known quality and which is available in sufficient quantity may be used. This procedure must be repeated at regular intervals to ensure that no change has occurred in the inspectors' ability to evaluate products and the inspector must reach a pre-defined level of performance on both . pass/accept. samples and . fail/reject. samples.

(c) A supplementary method of evaluation of an inspector's performance is the accumulation over time of the on – going inspection results vs. any other known information on samples, e. g. reinspection results, consumer complaints, chemical analyses, etc.

4.1.5 Reference Documents

Reference documents are presented in Appendix Ⅱ.

4.2 PREPARATION AND HANDLING OF SAMPLES

4.2.1 Type of Samples

Samples used for the purpose of training individuals in sensory techniques concerning fishery products are the single most important factor to be considered. It is imperative that proper samples be provided in reference to sensory training.

There are two types of samples to be considered in the training of sensory analysts or inspectors.

1. Controlled spoilage samples: These samples should display or represent a full range of quality, as well as the normal range of product characteristics related to odour, flavor, appearance, and texture.

It is essential that samples of excellent quality be provided as a reference point during the preparation of such packs.

Quality defects should be naturally occurring, if possible, to exhibit sensory characteristics which are typical of the product to be used. If the samples are spoiled or contaminated artificially, they may not exhibit typical sensory properties for both the acceptable or unacceptable units to be used for training.

It is important for the individual preparing the samples to have knowledge of the normal commercial processing of the product to be spoiled from harvesting to freezing and be aware of processing methods and conditions under which spoilage usually occurs. Understanding the general pathways of decomposition would be useful in the preparation of controlled spoilage samples.

When possible, controlled spoilage, samples should be prepared where the product is harvested and processed to allow for the species, flora, etc. to duplicate normal spoilage conditions that allows for typical odours of decomposition as well as other characteristics that mimic commercial samples.

2. Commercial samples: Whenever possible, the use of commercial samples should be incorporated into the sensory training of individuals. Many times, quality defects (odour, flavor, appearance, texture, etc.), as well as taints (musty/mouldy odours, flavours, rancidity, petroleum distillates, etc.) can be best shown with commercially produced samples that have these defects. These commercially manufactured samples allow one to assess sensory personnel during training by providing . real life. samples. They can also be used to measure an individual's retention abilities as it relates to making correct decisions in sensory science.

Many times, quality defects and taints are not found in all intensities in controlled spoilage samples but can be shown in slight, medium, and strong intensity from commercially produced samples.

4.2.2 Preparation of Sample Packs

Sample preparation should be started in plenty of time to allow one to obtain the majority of defects as well as allowing product to go through a curing process if necessary.

If possible, the spoilage run should be conducted with fish . in the round to allow for natural spoilage to occur. This allows for typical spoilage odours to form.

(1) Baseline

It is essential that excellent quality material of all species and product forms of known history, without commercial abuse, be obtained to provide a constant reference to the workshop participants. Whenever possible, both fresh and frozen product forms should be included in the preparation of controlled spoilage samples. The lot should be uniform with respect to its quality at the start of the run.

Proper record keeping is essential in the preparation of spoilage samples. Samples of each code taken should be consistent within a set, each succeeding set representing a longer period of time that the product has been held under ambient or iced conditions. Temperature monitoring is essential to prevent fluctuations during each spoilage run.

Spoilage must be accomplished under appropriate conditions of temperature and environmental contamination if authentic spoilage effects are to be obtained. Variations in spoilage rates between individual units can be minimized if the starting material is of uniform size and quality and contact between individual units is maintained during spoilage.

Fish tend to spoil at different rates so one should examine product at regular intervals and group the product together that have similar characteristics prior to processing. Expert evaluation of the samples is constantly needed at this stage.

The number of increments needed will depend on the purpose of the training and the species to be examined but a minimum of 5 increments and as many as 8 may be needed. At least 50% of the pack should be of acceptable product.

(2) Spoilage

Generally, both high and low temperature decomposition spoilage should be included, but knowledge of the species and the standard processing method and at what point of the process is spoilage most likely to occur should determine the general spoilage method. It is important to avoid . shortcuts. for the sake of convenience. If pre-chilling spoilage is the issue, the use of frozen fish must be avoided. Careful temperature control is a necessity.

(3) Packaging and Storage

The species and type of product from a spoilage run should be taken into account to determine the amount of shelf life one can expect.

Canned products should be allowed to cure in the can for at least 30 days prior to use. They should be stored in a cool and dry location with a temperature range of 14 ~ 18℃, otherwise one can expect a much shorter storage life. Maximum shelf life of canned seafood products for training purposes is approximately 2 years. After this amount of time, characteristics develop that may affect one's judgement or render the samples of little value for training purposes.

Unless freezer storage damage is intended to be demonstrated, raw and pre – cooked frozen products should be properly glazed to prevent dehydration and freezer burn. Depending on the length of storage, the samples may require periodic reglazing to ensure the quality. If possible, product should be vacuum packed to ensure quality and is essential in the storage of some fish species as well as pre-cooked samples.

Both raw, precooked and canned controlled spoilage samples should be evaluated by a qualified individual prior to use in a workshop. The samples should have both chemical analysis and sensory results to determine the quality of the increment and the homogeneity of the increment.

4.2.3 Characteristics of Samples

4.2.3.1 Sensory Attributes

A. Must show normal odour, flavor, appearance, texture, etc. characteristics of the species to be used for samples;

B. If product forms normally show characteristics attributed to harvest location, feed odours, etc. include with the controlled spoilage samples if possible;

C. Samples which exhibit odours of spoilage or contamination defects must not be too intense to the point of overpowering the participant's senses and affecting judgement of other samples during a training session;

D. Samples showing slight to moderate odours of spoilage or contamination provide more of a challenge and better represent . real world. conditions;

E. Each increment or code must show consistent or similar characteristics to have value when used for training.

4.2.3.2 Chemical Attributes

Inclusion of chemical attributes of authentic pack samples can be useful in training (see Annex Ⅲ Section Ⅱ Practical Exercises from the model Syllabus).

A. Chemical indicators of decomposition (CID) are selected that are essentially absent in the fresh product;

B. A CID is selected that will monitor the decomposition pathway of interest in the particular products to be used for training. Methods are used which are capable of differentiating between the CID levels found in passable, slightly abused – passable and the first definite stage of decomposition. When possible it is preferable to use two CID's;

C. The CID should be retained in the processed forms (washed/cooked/canned/stored) of the fishery product to be examined;

D. The changes in a CID should track the changes in sensory quality in the fishery product;

E. A sufficient number of subsamples should be analyzed for each increment of prepared sample to measure the degree of variation within sample increments. This is especially important for those increments representing the transition from a passable product to the first definite stage of decomposition.

ANNEX I

Table 1 Examples of Attributes of Fishery Products Used in Sensory Evaluation[1]

	Vertebrate fish, iced	
Presentation	Feature	Criteria and description
Raw whole, gutted or ungutted	outer surface, skin	colour: bright, dull, bleached slime: colourless, discoloured damage: none, punctures, abrasions
	Eyes	shape: convex, flat, concave brightness: clear, cloudy colour: normal, discoloured
	belly cavity	guts (in intact fish): intact, digested cleanliness (in gutted fish): completely gutted and cleaned, incompletely gutted, not washed belly walls: bright, clean, discoloured, digested parasites: absent, present blood: bright, red, brown
	texture, appearance of gills	skin: smooth, gritty, flesh, firm, soft colour: bright red or pink, beached, discoloured mucus: clear, opaque, discoloured
	odour of gills	fresh, characteristic, neutral, slightly sour, slightly stale, definite spoilage, putrid
Raw fillets	Appearance	translucent, glossy, natural colour, opaque, dull, blood-stained, discoloured
	Texture	firm, elastic, soft, plastic
	Odour	marine, fresh, neutral, sour, stale, spoiled, putrid
Cooked fillets	Odour	spoilage: marine, fresh, neutral, musty, sour, spoiled taints: absent, disinfectant, fuel oil, chemicals, sulphides
	Flavour	spoilage: sweet, creamy, fresh oil, neutral, sour, oxidised, putrid, musty, fermented, rancid, bitter, taints: absent, disinfectant, fuel oil, very bitter, alkaline, polyphosphates, chemicals
	Texture	succulent, firm, soft, pasty, gelatinous, dry
	Vertebrate fish, frozen	
Frozen	Appearance	freezer burn: absent, slight, superficial, extensive, deep colour: normal, yellow to bronze discolouration in fatty fish

[1] References to be included for the clarification of sensory properties, as established by ISO.

Next table

	Vertebrate fish, iced	
Thawed fillets, raw	Texture	firm, elastic, flexible, very firm, hard, stiff drip: slight, moderate, abundant odour spoilage and taints: as for chilled fish cold storage: absence of cold storage odours, sharp, cardboardy, rancid
Thawed fillets	odour and flavour	spoilage and taints: as per chilled fish cold storage: absence of cold storage odours, sharp, cardboardy, rancid
	Texture	firm, succulent, tough, fibrous, dry
	Crustacean shellfish, chilled	
Raw	Appearance, shell on	bright colours, slight blackening on the head, blackening on head and body
	Appearance, peeled meats	translucent, overall white or light grey, slight black discolouration, extensive black discolouration, very translucent, slimy, yellowish discolouration on butt end of tail meat taken from head-on products
	Odour	fresh, marine, musty, ammoniacal, sour, spoiled, putrid
Cooked meats	Appearance	white, opaque, blacks spots, extensive back discolouration, slightly translucent
	Odour	fresh, boiled milk, musty, ammoniacal, rancid, sour, spoiled
	Flavour	sweet, creamy, neutral, musty, sour, bitter, spoiled
	Texture	firm, elastic, soft, mushy
	Crustaceanshellfish, frozen	
Criteria specific to the grading of frozen shellfish, and their descriptions, are essentially the same as those applied to the grading of frozen vertebrate fish.		
	Cephalopods, fresh or refrigerated	
	Colour	skin: bright, dull, bleached meat: pearly white, lime coloured, pinkish or light yellow
	Adherence	adherent to the meat, easily separating from the meat
	Texture	meat: very firm, firm, slightly soft tentacles: resistant to tearing off, can be torn off easily
	Odour	fresh, seaweed, slight or no odour, sour

ANNEX Ⅱ

EXAMPLES OF TEST METHODS WHICH WOULD BE APPROPRIATE FOR USE IN SCREENING ASSESSORS FOR ODOUR PERCEPTION

1. The following is a list of samples as used in Canada:

(a) canned salmon (fish)

(b) canned sardines (fish/smoke)

(c) yeast (growth of yeasts)

(d) coffee (common product – to illustrate the method)

(e) orange & pineapple (fruity odours)

(f) cucumber & asparagus (vegetable odours)

(g) vinegar, cinnamon, pepper & cloves (pungent odours which can be differentiated)

(h) vanilla (sweet odour)

(i) prepared mustard (strong vinegar component, illustrates ability to perceive in mixtures)

(j) acetone, rubbing alcohol (contaminants, solvents)

(k) petroleum product (fuel oils)

(l) old vegetable oil (rancid oil)

In this test, the candidate is asked to identify the samples only by the odour as all visual information is masked. The sample are then identified and discussed with the candidate and the number of correct identifications recorded. During this step the candidate is given the opportunity re – examine any of the samples. The test is repeated after a time period such as 2 or 4 hours (during which other selection tests or interviews may be given), and number of correct responses recorded. The improvement in test scores which should occur (unless all were correct on the first round) gives an indication of the ability of the candidate to learn new terms to describe sensory perceptions.

2. The University of Pennsylvania Smell Identification Test, a standardized test for assessment of odour perception, is available from Sensonics, Incorporated, 155 Haddon Avenue, Haddonfield, New Jersey, 08033 USA.

ANNEX Ⅲ

SUGGESTED SYLLABUS FOR A TRAINING COURSE FOR ASSESSORS IN THE SENSORY ASSESSMENT OF FISH AND FISH PRODUCTS

Ⅰ. LECTURES

Part Ⅰ: Theoretical Principals and Laboratory Practices of Sensory Assessment (10 Hours)

A. Basic Sensory Testing Principles:

1. Affective or subjective testing (test types, information gained, data collection, respondent type and numbers, decision-making possible from this information).

2. Analytical or objective testing (test types, information gained, data collection, respondent type and numbers, decision – making possible from this information).

(i) Discriminative testing: types of information that is gained and that is not.

(ii) Descriptive testing: qualitative and quantitative.

3. The role of the fish and seafood assessor or product expert in sensory testing.

B. Action of the Senses and the Perception of Sensory Properties of Fishery Products:

1. The physiology of the senses – sight, smell, taste, touch and hearing;

2. The perception of sensory properties – appearance/colour, odour, flavor, texture; and

3. Sensory interactions.

C. Sample Evaluation techniques:

1. Odour evaluation techniques.

2. Flavor evaluation techniques.

3. Texture evaluation (firmness and springiness).

4. Special techniques for seafood samples.

D. Basic Psychophysics of Sensory Assessment:

1. Thresholds; detection and recognition.

2. Intensity; the logarithmic nature of character strength perception.

3. Saturation; explanation of the phenomenon.

E. Factors Influencing Sensory Judgements:

1. Physiological effects; blending; masking, carry – over, enhancement and suppression.

2. Psychological effects; expectation, stimulus, halo, order, proximity, stimulus, logical, suggestion, contrast and convergence, and central tendency.

3. Control of physiological and psychological effects.

F. Basic Data Collection and Analysis:

1. Discriminative methods: triangle (3-alternative forced choice or balanced design), duo-trio, two-out-of-five, paired comparison):

(i) Ballot information and design types

(ii) Analysis of data

2. Descriptive methods: Flavor Profile, Texture Profile, Spectrum, QDA:

(i) Scales; category, line, magnitude estimation

(ii) Ballot information and design types

(iii) Analysis of data

3. Sensory methods for quality control – general discussion.

G. Terminology and the use of reference standards. the analyst should. understand the role of sensory descriptors as an aid to developing long term sensory memory and as a means of communicating results.. (see Appendix 1):

1. Terminology development (including internationally recognised sources for known terms).

2. The importance of definitions

3. The use of reference Standards

4. Overview of terms relevant to seafood quality, with specific attention to those associated with low levels of decomposition.

H. Sample Handling and Preparation:

1. Presentation and coding.

2. Randomization of samples; purpose and occasion for use.

3. Homogeneity of samples and serving temperature.

4. Sample size and quantity.

Part Ⅱ: Deterioration of Fish and Fish Products (3 Hours)

A. Composition of Fish and Shellfish:

1. Major components: protein, fat, carbohydrate, water;

2. Minor components; non – protein nitrogenous compounds, minerals, vitamins.

B. Pathways of Quality Deterioration:

1. Breakdown of protein, fat, non – protein nitrogenous compounds, and, for some species, carbohydrates;

2. Microbial spoilage;

3. Terminology associated with each type of spoilage pathway.

C. Chemical Indicators of Fish Quality and the Correlation of these with Sensory Data.

Part Ⅲ: Contamination and Taint (1 Hour)

A. Types:

1. Naturally-occurring (muddy-earthy off – flavours);

2. Man-made (petroleum, pulp and paper effluent, other processing effluents).

B. Mechanism of flavor and odour changes.

C. Testing methods for contamination and/or taint (special considerations).

Ⅱ. PRACTICAL EXERCISES

Part Ⅰ: Presentation of Seafood Related Terminology, Clear Definitions, and References Which Demonstrate the Terms (2 hours)

Part Ⅱ: Spoilage and Decomposition (18 hours)

This portion of the course provides hands – on experience. It is suggested that only one species at a time be evaluated.

This section may include whole fish, fillets, canned fish and/or smoked fish and other specialty products. Whenever possible, trainees should evaluate flavor as well as odour, e. g. especially in products such as canned fish packed in oil as the packing medium can mask odours.

The following sequence of three session formats are suggested for each species and will require approximately 4 hours in total. It is suggested that the effectiveness of the training be evaluated by testing the trainee's ability to assess sample quality correctly before moving on to another species:

(a) Demonstration session: Group demonstrations of samples of known quality by an experienced product expert. The labelled samples should represent a full range of quality, in order from highest to lowest quality, with discussion of sensory results, descriptors, as well as any data from chemical indicators of quality which are appropriate for that species;

(b) Discussion session: Random presentation of blind - coded samples for individual evaluation an group discussion of the results;

(c) Testing session: individual evaluation of blind-coded test samples and comparison of results with product expert.

The collection and analysis of data with detailed discussions of the samples will provide feed-back to the trainees.

Part Ⅲ: Deterioration in Frozen Stored Fish and Shellfish (4 hours)

A. Demonstration of varying degrees of defects in appearance, odour, flavor, and texture caused by frozen storage of seafood products.

B. Include both low-fat and high - fat fish and seafood samples.

C. Have available terminology, definitions, and references for the oxidation process and for textural changes.

Part Ⅳ: Deterioration in Canned Fish and Shellfish (4 hours)

A. As for section II, and also to include information on pre - and post - processing deterioration.

Part Ⅴ: Other Defects (2 hours)

A. Detection of taints using spiked samples (assess by odour only).

B. Demonstration of visual defects.

APPENDIX Ⅰ

DEFINITIONS OF SOME OF THE TERMS USED IN SENSORY ANALYSIS OF SEAFOOD

Definition	
Appearance	All the visible characteristics of a substance/sample
Analyst/Assessor	Any person taking part in a sensory test
Bilgy	The aromatic associated with anaerobic bacterial growth, which is illustrated by the rank odour of bilge water. The term . bilgy. can be used to describe fish of any quality which has been contaminated by bilge water on board a vessel. Bilge water is usually a combination of salt water fuel, and waste water
Bitter	One of the four basic tastes, primarily perceived at the back of the tongue, common to caffeine and quinine. There is generally a delay in perception (2 – 4 seconds)
Briny	The aroma associated with the smell of clean seaweed and ocean air; Chalky In reference to texture, a product which is composed of small particles which imparts a drying sensation in the mouth. In reference to appearance, a product which has a dry, opaque, chalk like appearance
Cucumber	The aroma associated fresh cucumber, similar aromas can be associated with certain species of very fresh raw fish
Decompose	To break down into component parts
Decomposed	Fish that has an offensive or objectionable odour, flavor, colour, texture, or substance associated with spoilage
Distinct	Capable of being readily perceived
Feedy	Feedy. is used to describe the condition of fish that have been feeding heavily. After death, the gastric enzymes first attack the internal organs, then the belly wall, then the muscle tissue. If the enzymes have penetrated into the flesh, they are capable of causing quality changes dimethyl (DMS), and may be attributed to certain zooplankton as it passes through the food chain. The odour of . feed. fish has been described as similar to certain sulfur containing cooked vegetables, such as broccoli, cauliflower, turnip, or cabbage
Fecal	Aroma associated with feces
Firm	A substance which exhibits moderate resistance when force is applied in the mouth or by touch
Fish	Means any of the cold-blooded aquatic vertebrate animals commonly known as such. This includes Pisces, Elasmobranchs and Cyclostomes. Aquatic mammals, invertebrate animals and amphibians are not include
Fishy	Aroma associated with aged fish, as demonstrated by trimethylamine (TMA) or cod liver oil. May or may not indicate decomposition, depending on species
Flavor	An attribute of foods resulting from the stimulation of taste, smell, sight, pressure, and often warmth, cold or mild pain
Freshness	Concept relating to time, process, or characteristics of seafood as defined by a buyer, processor, user, or regulatory agency
Fruity	Aroma associated with slightly fermented fruit. Term is used to describe odours resulting from high temperature decomposition. Example = canned pineapple
Gamey	The aroma and/or flavor associated with the heavy, gamey characteristics of some species such as mackerel. Similar to the relationship of fresh duck meat as compared to fresh chicken meat

Nent table

Definition	
Glossy	A shiny appearance resulting from the tendency of a surface to reflect light at 45 degree angle
Grainy	A product in which the assessor is able to perceive moderately hard, distinct particles. Sometimes found in canned seafood products
Intensity	The perceived magnitude of a sensation
Iridescent	An array of rainbow like colours, similar to an opal or an oil sheen on water
Intensity	The perceived magnitude of a sensation
Masking	The phenomenon where one sensation obscures one or several other sensations present
Mealy	Describes a product that imparts a starch – like sensation in the mouth
Metallic	Aroma and/or taste associated with ferrous sulphate or tin cans
Moist	The perception of moisture being released from a product. The perception can be from water or oil
Mouldy	Aroma associated with mouldy cheese or bread
Mouth coating	The perception of a film in the mouth
Mouth filling	The sensation of a fullness dispersing throughout the mouth. A umami sensation, as stimulated by MSG
Mushy	Soft, thick, pulpy consistency. In seafood little or no muscle structure discernible when force is applied by touch or by mouth
Musty	The aroma associated with a mouldy, dank cellar. Product can also have a musty flavor
Odour	Sensation due to stimulation of the olfactory receptors in the nasal cavity by volatile material. Same as aroma
Off odour/	Atypical characteristics often associated with deterioration or transformation of a flavor product
Opaque	Describes product which does not allow the passage of light. In raw muscle tissue of fishery products, this is usually due to the proteins loosing their light reflecting properties due to falling pH
Pasty	A product which sticks together like paste in the mouth when mixed with saliva. Forms a cohesive mass which may adhere to the soft tissue surfaces of the mouth or fingers
Persistent	Existing without significant change; not fleeting
Pungent	An irritating, sharp, or piercing sensation
Putrid	Aroma associated with decayed meat
Quality	A degree of excellence. A collection of characteristics of a product that confers its ability to satisfy stated or implied needs
Rancid	Odour or flavor associated with rancid oil. Gives a mouth – coating sensation and/or a tingling perceived on the back of the tongue. Sometimes described as . sharp. or . painty
Reference	Either a sample designated as the one to which others are compared, or another type of material used to illustrate a characteristic or attribute
Rotting vegetable	Aroma associated with decayed vegetables, in particular the sulfur containing vegetables, such as cooked broccoli, cabbage, or cauliflower
Rubbery	A resilient material which may be deformed under pressure, but returns to its original form once the pressure is released
Salty	The taste on the tongue associated with salt or sodium
Sensory	Relating to the use of the sense organs
Slimy	A fluid substance which is viscous, slick, elastic, gummy, or jelly-like
Sour	An odour and/or taste sensation, generally due to the presence of organic acids
Stale	Odour associated with wet cardboard or frozen storage. Product can have a stale flavor as well

Nent table

Definition	
STP	Sodium tripolyphosphate. Can produce a soapy, alkaline feel and taste in the mouth
Sweet	The taste on the tongue associated with sugar
Taste	One of the senses, the receptors for which are located in the mouth and activated by compounds in solution. Taste is limited to sweet, salty, sour, bitter and sometimes umami
Terminology	Terms used to describe the sensory attributes of a product
Translucent	Describes an object which allows some light to pass, but through which clear images can not be distinguished
Transparent	Describes a clear object, which allows light to pass and through which distinct images appear
Umami	Taste produced by substances such as monosodium glutamate (MSG) in solution. A meaty, savory, or mouth filling sensation
Watermelon	Aroma characteristic of fresh cut watermelon rind. Similar odours are sometimes found in certain species of very fresh raw fish
Yeasty/fermented	Aroma associated with yeast and fermented products such as bread or beer

APPENDIX Ⅱ

REFERENCE DOCUMENTS

ASTM Atlas of odour character profiles, publication DS 61, PCN 05 – 061000 – 36. Complied by Andrew Dravnieks

ASTM Committee E – 18, 235, draft of terminology document

ASTM Aroma and Flavor Lexicon for Sensory Evaluation DS 66. G. V. Civille and B. G. Lyon, eds

ASTM Committee E – 18 on Sensory Evaluation of Materials and Products, 1981. STP 758 – Guidelines for the Selection and Training of Sensory Panel Members

ASTM Committee E – 18 on Sensory Evaluation of Materials and Products, Terminology Committee, (date?). Draft definition for. Expert. and. Expert Assessor

Cardello, A. 1993. Sensory methodology for the classification of fish according to edibility characteristics. *Lebensmittel-Wissenschaft-und-Technologie* 16, 190 ~ 194

Department of Fisheries and Oceans, Canada. Code of practice for fishery products

Department of Fisheries and Oceans, Canada. Regulations respecting the inspection of processed fish and processing establishments

Department of Fisheries and Oceans, Canada, Inspection Branch. 1986 to 1995. Notes from. *Sensory Methods in Fish Inspection.* – Sensory Training course given by the National Centre for Sensory Science, Inspection Branch, Department of Fisheries and Oceans, Canada

Howgate, Peter 1992. Codex review on inspection procedures for the sensoric evaluation of fish and shellfish. CX/FFP 92/14

IFST – International Institute of Food Science and Technology.. Sensory Quality Control: Practical Approaches in Food and Drink Production.. Proceedings of a joint symposium at the U. of Aston, 6 ~ 7January, 1977. Session II, . Measurement of Fish Freshness by an Objective Sensory Method.. P. Howgate, p. 41

ISO 5492 (1983) Sensory analysis – vocabulary

ISO 8586 – 2 Sensory Analysis – General guidance for the selection, training and monitoring of assessors – Part 2. Experts

Jellined, G. 1985. *Sensory Evaluation of Food-Theory and Practice.* Ellis Horwood, Ltd., Chichester, England

Johnsen, et al., 1987. A lexicon of pond-raised catfish flavor descriptors. J. Sensory Studies 4, 189 ~ 199

Laverty, 1991. Torry Taste Panels. In Nutrition and Food Science, Vol 129 No. 2 ~ 4. Includes terminology based on odour of gills in raw, iced cod

Learson, Robert 1994, personal correspondence. NOAA/NMFS Research Laboratory, Gloucester, MA

Multilingual guide to EC freshness grades for fishery products. Torry research station, Aberdeen, Scotland and the West European Fish Technologists Association (WEFTA). Compiled and edited by P. Howgate, A. Johnston, and K. J. White

NOAA Handbook 25, part 1, Inspection

NOAA/NMFS, Technical Services Unit

Kramer and Liston, (eds) Seafood Quality Determination. Proceedings of the International Symposium on Seafood Quality Determination, Coordinated by the University of Alaska Sea Grant College Program, Anchorage. Alaska, 10 – 14 November, 1986

Learson and Ronsivalli, (1969), A new approach for evaluating the quality of fishery products

Meilgaard, M., Civille, G. V., and Carr, B. T. 1991. *Sensory Evaluation Techniques.* CRC Press, Inc., Boca Raton, FL

Poste, L., Mackie, D., Butler, G. and. Larmond, E. 1991. Laboratory Methods for Sensory Analysis of Food. Agriculture Canada Research Branch

Prell and Sawyer, 1988. Flavor Profiles of 17 Species of North Atlantic Fish. J. Food Science, 53, 1036 ~ 1042

Prell and Sawyer (1988). Consumer evaluation of the Sensory Properties of Fish. J. of Food Science 53, 12 ~ 28, 24

Reilly, T. I. and York, R. K. 1993. Sensory analysis application to harmonize expert assessors of fish products. Proceedings of. Quality Control and Quality Assurance of Seafood., May 16 ~ 18, 1993, Newport, Oregon (Eds. Sylvia, G., Shriver, A. L. and Morrisey, M. T.)

Sawyer et al., (1988). Consumer evaluation of the sensory properties of fish.. J. of Food Science, Vol. 53. No. 1

Sawyer, F. M. et al. 1981. A comparison of flavor and texture characteristics of selected underutilized species of North Atlantic fish and certain treatment of fish. International Institute of Refrigeration. Paris, France. p. 505

Shewan et al., (1953), The development of a numerical scoring system for the sensory assessment of the spoilage of wet white fish stored in ice. J. Sci. Food Agric., 4 June

Soldberg, et al. (1986), Sensory profiling of cooked, peeled and individually frozen shrimp. In Seafood Quality Determination, Elsevier Science Publishers

Vaisey Genser, M. and Moskowitz, H. R. 1977. Sensory Response to Food. Forster Publishing Ltd., Zurich, Switzerland

Wilhelm, Kurt, 1994, personal correspondence. NOAA/NMFS Research Laboratory, Gloucester, MA

水产及水产加工品产品认证证书模式

MODEL CERTIFICATE FOR FISH AND FISHERY PRODUCTS

(*CAC/GL48 - 2004*)

1 引言

认证是一种被进出口国家管理机构用来证明其水产及水产加工品检查体系控制的方法。为帮助促进国际贸易，认证证书的数量和类型应该得到限制，并通过国际（法典）认证证书模式得到促进。在可能的地方，尤其当一个出口国的检验系统和要求经评价与进口国等效时，也应考虑官方的或官方认可的认证[1]以外的认证。双边或多边协定的建立，如双边认可协定可能为终止认证证书的发布提供合理的基础。

2 范围

认证证书模式适用于国际贸易中符合食品安全卫生标准以及进口国生产要求的水产及水产加工品，不包括动、植物健康问题。假如进口国和出口国的相关权威机构满意认证体系的安全性，则在管理上和经济上可行的地方，证书可以以电子格式发布。

证书应当根据检验机构定期检测的结果，充分描述一批或几批产品与相关法规要求的符合性。其他检测、分析结果、质量保证程序的评价或产品说明等也需要进行证明。

3 定义

认证[2]：是一个官方认证机构或官方认可认证机构，以书面或其他等效方式担保水产及水产加工品或它们的控制体系满足要求的程序。水产及水产加工品的认证，恰当地说，是建立在一系列检查活动基础上的，这些活动包括在线检查、质量保证体系审核和最终产品的检验等。

认证机构：是指官方认证机构和被主管当局官方认可的认证机构。

认证官员[3]：被授权完成和发布认证的认证机构雇员。

检查：是指水产及水产加工品或水产及水产加工品控制体系、原料、加工、和销售的检查，包括加工过程中的及已完成产品的检测，以验证其是否符合要求。

检查体系[4]：是指官方的或官方认可的检查体系。

官方检查体系或官方认证体系：是由拥有管辖权的政府机构实施管理的体系，该机构被授权行

---

[1] 认证是指官方认证和官方认可的认证。

[2] 食品进出口检查和认证原则（CAC/GL20 - 1995）。

[3] 通用官方认证形式和证书制成及发布指南（CAC/GL 38 - 2001）。

[4] 关于食品进出口检查和认证体系等同发展指南（CAC/GL 34 - 1999）。

使制定规章制度或执行或两者兼有的职能。

官方认可的检查体系或官方认可的认证体系：是指已由拥有管辖权的政府机构正式批准或认可的体系。

官方认证：是由出口国官方认证机构根据进口国或出口国要求发布的认证。

官方认可的认证：是由出口国官方认可的认证机构根据认可条件和进口或出口国要求发布的认证证书。

要求：是水产及水产加工品贸易相关的主管部门设立的标准，其内容包括公共健康保护、消费者和公平贸易环境的保护。

4 关于认证证书的制作和发布的一般考虑

4.1 建议水产及水产加工品认证证书的制作和发布均需按照以下原则和适用的内容进行：

（1）《通用官方证书格式和证书的制作与发布指南》（CAC/GL 38 - 2001）；

（2）《食品进出口检查和认证原则》（CAC/GL 20 - 1995）；

（3）《食品进出口检查和认证体系的规划、运作、评价和资格鉴定指南》（CAC/GL26 - 1997）；

（4）《食品进出口检查和认证体系对等协议编制指南》（CAC/GL 34 - 1999）；

（5）《国际食品贸易行为规范修订草案》（CCGP 修订本）。

4.2 证书语言的选择应充分考虑进口国的需要、认证官员的理解，并尽量减少对于出口国不必要的障碍

5 认证模式的格式和使用

5.1 格式

5.1.1 卫生认证证书模式（附件 I）：当制作证书以证明一批货物中水产及水产加工品是在一个受控的企业中生产的，并满足出口国的法律和要求，或其生产条件满足等效或 COMPLIANCE 协议的规定时，应考虑使用卫生认证证书模式。

5.2 使用

卫生认证证书模式的每一个空格均需填写，或者进行标记以防止证书更改。认证证书模式需要包含或按照下列内容完成：

5.2.1 证书编号对于每一种证书来说都是唯一的，并经过出口国主管机构的批准。如果临时要求附加信息，可以附录或证明的形式体现。假如有附录，它必须具有与原始证书相一致的证书编号以及与卫生认证证书相同的认证官员的签名。

5.2.2 发货国就认证证书模式而言，指定具有验证和认证生产企业是否符合要求的职能的主管机构所在国家的名称。

5.2.3 主管机构是授权行使各种职能的官方组织。其职责可能包括整个地区或当地的官方检查或认证系统的管理。

5.2.4　认证机构是指官方认证机构和主管机构官方认可的机构。

5.2.5　加工状态或方式描述了水产及水产加工品目前的表现形式（例如，新鲜、冷冻、罐头等）及/或所用的加工方法（例如，熏制、沾面包屑等）。

5.2.6　包装类型可以是纸箱、盒子、袋子、箱、鼓形圆桶、桶、托盘等。

5.2.7　批次检验人/日期代码是加工者统计他们的水产及水产加工品生产而建立的批次识别系统，从而在公众健康调查和召回事件中，便于产品的可追溯性。

5.2.8　运输方式应当描述出飞机、火车、卡车或其他运输工具的编号，以及飞机、船的名称等。

5.2.9　证明是确认产品或数批产品来源于受到该国主管机构良好管理的企业，并且该产品是在有效的 HACCP 和卫生程序下进行加工或处理的。

5.2.10　证书原件应可识别，并应以“原件”标记恰当地标明，或者当需要证书复印件时，可以在证书复印件上加上“复印件”标记或具有同等效果的其他术语。术语“替代件”被指定用于由于任何合理和充分的理由（例如，证书在运输过程中发生损坏）而由认证官员颁发的替代证书上。

5.2.11　页码应在证书超过一页时使用。

5.2.12　印章和签名其应用方式应尽可能降低欺诈风险。

附件 I

水产及水产加工品认证证书模式草案

（抬头或专用标识） 证号编号：______

发货国：
主管机构：
认证机构：

Ⅰ 水产品详细内容

产品名称	品种（学名）	加工状态或方式	包装类型	批次检验人/日期代码	包装数量	净重
				总计：		

贮存和运输温度要求：______℃

Ⅱ 水产品原产地

主管机构批准出口的生产单位的地址和（或）注册号：______

______

发货人姓名和地址：______

______

Ⅲ 水产品目的地

水产品发自于：______
（派发地）
到达：______
（目的国家和地点）
通过下列运输方式：______
收货人姓名及目的地地址：______

______

Ⅳ 证明

在下面签名的认证官员特此证明：
（1）上述产品来源于批准的企业，该企业被证实或确认受到出口国主管机构的良好管理；
（2）在执行有效的HACCP和卫生程序条件下，根据《水产及水产加工品操作法典规范》（CAC/RCP 52－2003）的要求进行操作、准备或加工、鉴定、贮存和运输等。

认证在 ______________________ 于 ______________ 20 ______________________

（地点） （日期）

（盖章）

______________________ ______________________

（认证官员签名） （名称和官方职位）

电话：

传真：

电子邮件（可选项）：

MODEL CERTIFICATE FOR FISH AND FISHERY PRODUCTS

(*CAC/GL 48 – 2004*)

1 INTRODUCTION

Certification is one method that can be utilized by regulatory agencies of importing and exporting countries to compliment the control of their inspection system for fish and fishery products. To help facilitate international trade, the numbers and types of certificates should be limited and could be promoted through international (Codex) model certificates. Notwithstanding, alternatives to the use of official and officially recognized certificates❶ should be considered wherever possible, in particular where the inspection system and requirements of an exporting country are assessed as being equivalent to those of the importing country.

The establishment of bilateral or multilateral agreements, such as mutual recognition agreements may provide the logical basis for discontinuing with the issuance of certificates.

2 SCOPE

The model certificates apply to fish and fishery products presented for international trade that meet food safety, wholesomeness and conformity to food production requirements of the importing country. Animal and plant health matters are not covered. Where administratively and economically feasible, certificates may be issued in an electronic format provided that the relevant authorities of both the importing and exporting country are satisfied with the security of the certification system.

Certificates should adequately describe one or several lots or batches of product's compliance with regulatory requirements based on regular inspections by the inspection service. Additional examinations, analytical results, evaluation of quality assurance procedures or product specifications may also be attested to.

3 DEFINITIONS

Certification❷ is the procedure by which official certification bodies or officially recognized certification bodies provide written or equivalent assurance that fish and fishery products or their control systems conform to requirements. Certification of fish and fishery products may be, as appropriate, based on a range of inspection activities which may include continuous on – line inspection, auditing of quality assurance systems, and examination of finished products.

Certifying Bodies are official certification bodies and officially recognized bodies by the competent authority.

Certifying officers❸: employees of certifying bodies authorized to complete and issue certificates.

---

❶ For the purpose of this document, "certificates" shall mean "official certificates" and "officially recognized certificates".

❷ Principles for Food Import and Export Inspection and Certification (CAC/GL 20 – 1995).

❸ Guidelines for Generic Official Certificates Formats and the Production and Issuance of Certificates (CAC/GL 38 – 2001).

Inspection[1] is the examination of fish and fishery products or systems for control of fish and fishery products, raw materials, processing, and distribution including in - process and finished product testing, in order to verify that they conform to requirements.

Inspection system[2] means official and officially recognized inspection systems.

Official inspection systems and official certification systems are systems administered by a government agency having jurisdiction empowered to perform a regulatory or enforcement function or both.

Officially recognized inspection systems and officially recognized certification systems are systems which have been formally approved or recognized by an government agency having jurisdiction.

Official Certificates are certificates issued by an official certification body of an exporting country, in accordance with the requirements of the importing or exporting country.

Officially Recognized Certificatesf are certificates issued by an officially recognized certification body of an exporting country, in accordance with the conditions of that recognition and in accordance with the requirements of the importing or exporting country.

Requirements are the criteria set down by the competent authorities relating to trade in fish and fishery products covering the protection of public health, the protection of consumers and conditions of fair trading.

4 GENERAL CONSIDERATIONS CONCERNING THE PRODUCTION AND ISSUANCE OF CERTIFICATES

4.1 It is recommended that the production and issuance of the certificates for fish and fishery products should be carried out in accordance with the principles and appropriate sections of the:

(1) Guidelines for Generic Official Certificate Formats and the Production and Issuance of Certificates (CAC/GL 38 - 2001);

(2) Principles for Food Import and Export Inspection And Certification (CAC/GL 20 - 1995);

(3) Guidelines for the Design, Operation, Assessment and Accreditation of Food Import and Export Inspection and Certification Systems (CAC/GL 26 - 1997);

(4) Guidelines for the Development of Equivalence Agreements Regarding Food Import and Export Inspection and Certification Systems (CAC/GL 34 - 1999);

(5) Proposed Draft Revised Code of Ethics for International Trade in Foods (under revision by the CCGP)

4.2 The selection of the appropriate language (s) of certificates should be based on adequacy for the importing country's purpose, comprehension by the certifying officer and minimizing unnecessary burden on the exporting country

---

[1] Principles for Food Import and Export Inspection and Certification (CAC/GL 20 - 1995).

[2] Guidelines for the Development of Equivalence Agreements Regarding Food Import and Export Inspection and Certification Systems (CAC/GL 34 - 1999).

5 THE FORMAT AND USE OF MODEL CERTIFICATES

5.1 FORMAT

5.1.1 Model Sanitary Certificate (ANNEX I) – The format of the model sanitary certificate should be considered when developing a certificate to attest that fish and fishery products contained in a consignment were produced in establishments that are under the control of and produced to the laws and requirements of the exporting country, or under conditions cited in equivalence or compliance agreements.

5.2 USE

Each field of the Model Sanitary Certificate must be filled in or else, marked in a manner that would prevent alteration of the certificate. The Model Certificate should contain and be completed as follows:

5.2.1 Identification Number should be unique for each certificate and should be authorized by the competent authority of the exporting country. Should additional information be required on temporary basis, this may be incorporated as an addendum or an attestation. If there is an addendum, it must have the same identification number as the primary certificate and the signature of the same certifying officer signing the sanitary certificate.

5.2.2 Country of Dispatch for the purposes of the model certificate, designates the name of the country of the competent authority which has the competence to verify and certify the conformity of the production establishments.

5.2.3 Competent authority is the competent official organisation empowered to execute various functions. Its responsibility may include the management of official systems of inspection or certification at the regional or local level.

5.2.4 Certifying Bodies are official certification bodies and bodies officially recognized by the competent authority.

5.2.5 State or type of processing describes the state in which the fish and fishery product is presented (i. e. fresh, frozen, canned, etc.) and/or the processing methods used (i. e. smoked, breaded, etc.).

5.2.6 Type of packaging could be cartons, boxes, bags, cases, drums, barrels, pallets, etc.

5.2.7 Lot identifier / Date code is the lot identification system developed by a processor to account for their production of fish and fishery product thereby facilitating traceability/product tracing of the product in the event of public health investigations and recalls.

5.2.8 Means of transport should describe the flight/train/truck/container number, as appropriate and the name of the air carrier, vessel, etc.

5.2.9 Attestation is a statement confirming the product or batches of products originate from an establishment that is essentially in good regulatory standing with the Competent Authority in that country and that the products were processed and otherwise handled under a competent HACCP and sanitary programme.

5.2.10 Original Certificate should be identifiable and this status should be displayed appropriately with the mark "ORIGINAL" or if a copy is necessary, this certificate should be marked as "COPY" or terms of this effect. The term "REPLACEMENT" is reserved for use on certificates where, for any good and sufficient reason (such as damage to the certificate in transit), a replacement certificate is issued by the certifying officer.

5.2.11 Page numbering should be used where the certificate occupies more than one sheet of paper.

5.2.12 Seal and signature should be applied in a manner that minimizes the risk of fraud.

ANNEX I

MODEL SANITARY CERTIFICATE
COVERING FISH AND FISHERY PRODUCTS

(*LETTERHEAD or LOGO*) Identification number: ____________

Country of Dispatch:

Competent Authority:

Certifying Body:

Ⅰ Details identifying the fishery products

Description of product	Species (scientific name)	State or type of processing	Type of packaging	Lot Identifier/ date code	Number of packages	Net weight
				Sum :		

Temperature required during storage and transport: ______________________________℃

Ⅱ Provenance of the fishery products

Address (es) and/or the Registration number (s) of production establishment (s) authorized for exports by competent authority:

____

Name and address of consignor: ____

____

Ⅲ Destination of the fishery products

The fishery products are to be dispatched from: ______________________________

(Place of dispatch)

to: ____

(Country and place of destination)

by the following means of transport: ______________________________

Name of consignee and address at place of destination: ______________________________

____

Ⅳ Attestation

The undersigned certifying officer hereby certifies that:

(1) The products described above originate from (an) approved establishment (s) that has been approved by, or otherwise determined to be in good regulatory standing with the competent authority in the exporting country; and

(2) have been handled, prepared or processed, identified, stored and transported under a competent HACCP and sanitary programme consistently implemented and in accordance with the requirements laid down in Codex Code of Practice for Fish and Fishery Products (CAC/RCP 522003) .

Done at ______________________ on ______________ 20 ______________

(Place) (Date)

(SEAL)

______________________ ______________________

(Signature of certifying officer) (Name and official position)

Tel:

Fax:

E－mail (optional) :